Nucleic Acids and Molecular Biology

22

Series Editor

H. J. Gross
Institut für Biochemie
Biozentrum
Am Hubland
97074 Würzburg
Germany

Caroline Köhrer, Uttam L. RajBhandary (Eds.)

Protein Engineering

With 118 Figures and 9 Tables

Springer

Editors
Caroline Köhrer
Department of Biology
Massachusetts Institute of Technology
77 Massachusetts Avenue
Cambridge, MA 02139
USA

Uttam L. RajBhandary
Department of Biology
Massachusetts Institute of Technology
77 Massachusetts Avenue
Cambridge, MA 02139
USA

ISBN: 978-3-540-70937-4 e-ISBN: 978-3-540-70941-1

Nucleic Acids and Molecular Biology ISSN 0933-1891

Library of Congress Control Number: 2008931055

Cover design: WMXDesign GmbH, Heidelberg, Germany

Printed on acid-free paper

9 8 7 6 5 4 3 2 1

springer.com

Preface

Site-specific mutagenesis of DNA, developed some 30 years ago, has proved to be one of the most important advances in biology. By allowing the site-specific replacement of any amino acid in a protein with one of the other nineteen amino acids, it ushered in the new era of "Protein Engineering". The field of protein engineering has, however, evolved rapidly since then and the last fifteen years have witnessed remarkable advances through the use of new chemical, biochemical and molecular biological tools towards the synthesis and manipulation of proteins. The chapters included in this book reflect the rapid evolution of protein engineering and its many applications in basic research, biotechnology, material sciences and therapy. It is our hope that this book will provide the reader with an introduction to state-of-the-art concepts and methods and will be of use to anyone interested in the study of proteins, in academia as well as in industry.

Beginning with studies of enzyme mechanisms involving hybrid proteins generated by domain swapping (chapter by Goodey and Benkovic), the next two chapters (by Merkel et al. and Imperiali and Vogel Taylor) describe the ligation of chemically or biologically synthesized peptides to generate proteins carrying a variety of post-translational modifications and their use as biological probes. The following chapter (by van Kasteren et al.) reviews novel methods for chemical modification of amino acid side chains of proteins for the generation of mimics of post-translational modifications. The next several chapters cover different strategies towards expansion of the genetic code and the synthesis, in vivo and in vitro, of proteins carrying a variety of unnatural amino acids. These strategies involve mis-aminoacylated transfer RNAs generated either through chemical synthesis or enzymatic mis-aminoacylation. The unnatural amino acids include those that are photoactivatable or fluorescent, those that carry heavy atoms such as iodine, chemically reactive groups such as keto- or azido- groups, spectroscopic probes, or those that mimic phosphoamino acids. The potential applications of this newly emerging technology towards studies of proteins and the generation of proteins carrying novel chemical, physical and biological properties are virtually unlimited. One of these chapters (the one by Dougherty) focuses on applications of unnatural amino acid mutagenesis to the study of receptors and channel proteins of the central nervous system. Another chapter (the one by Hecht) includes modification of ribosomal RNA for the synthesis of proteins carrying D-amino acids in vitro. Two other chapters (by Hirao et al. and Leconte and Romesberg) discuss the use of new DNA and RNA base pair systems for expansion of the genetic code through expansion of the genetic alphabet. One of these chapters (the one by Leconte and Romesberg) also includes a discussion of directed evolution for the identification of DNA polymerases proficient in the use of unnatural base pairs in DNA synthesis. The final chapter (by Slusky et al.) deals with the important topic of membrane proteins, including a discussion of computational concepts for the design of peptide inhibitors to probe protein-protein interactions in membrane proteins.

The chapters are written by experts who have contributed much to the areas covered. It was a pleasure working with these colleagues and we thank them for their suggestions, their infinite patience and most importantly for their contributions to this book. We also thank Prof. Hans J. Gross and Ursula Gramm, Life Science Editor at Springer-Verlag, for their continuous support throughout the development of this project.

Caroline Köhrer
Uttam L. RajBhandary

Contents

Contributors

Songon An
Department of Chemistry, University of Minnesota, Minneapolis, MN 55455, USA, and Department of Chemistry, The Pennsylvania State University, 104 Chemistry Building, University Park, PA 16802, USA

Kimberly E. Beatty
Division of Chemistry and Chemical Engineering, California Institute of Technology, Pasadena, CA 91125, USA

Stephen J. Benkovic
Department of Chemistry, The Pennsylvania State University, 414 Wartik Laboratory, University Park, PA 16802, USA

Nediljko Budisa
Max Planck Institute of Biochemistry, Am Klopferspitz 18, 82152 Martinsried, Germany

Benjamin G. Davis
Department of Chemistry, University of Oxford, Chemistry Research Laboratory, 12 Mansfield Road, Oxford OX1 3TA, UK

William F. DeGrado
Department of Biochemistry and Biophysics, School of Medicine, University of Pennsylvania, Philadelphia, PA 19104, USA, and Department of Chemistry, University of Pennsylvania, Philadelphia, PA 19104, USA

Dennis A. Dougherty
Division of Chemistry & Chemical Engineering, California Institute of Technology, Pasadena, CA 91125, USA

Philippe Garnier
GlycoForm Ltd, Unit 44c, Milton Park, Abingdon OX14 4RU, UK

Nina M. Goodey
Department of Chemistry, The Pennsylvania State University, 414 Wartik Laboratory, University Park, PA 16802, USA, and Department of Chemistry and Biochemistry, Montclair State University, Upper Montclair, NJ 07043, USA

Sidney M. Hecht
Departments of Chemistry and Biology, University of Virginia, Charlottesville, VA 22904, USA

Ichiro Hirao
Protein Research Group, RIKEN Genomic Sciences Center, 1-7-22 Suehiro-cho, Tsurumi-ku, Yokohama, Kanagawa 230-0045, Japan

Barbara Imperiali
Department of Chemistry and Department of Biology, Massachusetts Institute of Technology, 77 Massachusetts Avenue, Cambridge, MA 02139, USA

Takashi Kanamori
Department of Medical Genome Sciences, Graduate School of Frontier Sciences, University of Tokyo, FSB401, 5-1-5 Kashiwanoha, Kashiwa, Chiba 277-8562, Japan

Caroline Köhrer
Department of Biology, Massachusetts Institute of Technology, 77 Massachusetts Avenue, Cambridge, MA 02139, USA

Aaron M. Leconte
Department of Chemistry, The Scripps Research Institute, 10550 North Torrey Pines Road, La Jolla, CA 92037, USA

Susan A. Martinis
Department of Biochemistry, University of Illinois at Urbana-Champaign, Urbana, IL 61801, USA

Anjali Mascarenhas
Department of Biochemistry, University of Illinois at Urbana-Champaign, Urbana, IL 61801, USA

Lars Merkel
Max Planck Institute of Biochemistry, Am Klopferspitz 18, 82152 Martinsried, Germany

Luis Moroder
Max Planck Institute of Biochemistry, Am Klopferspitz 18, 82152 Martinsried, Germany

Karin Musier-Forsyth
Departments of Chemistry and Biochemistry, The Ohio State University, Columbus, OH 43210, USA

Uttam L. RajBhandary
Department of Biology, Massachusetts Institute of Technology, 77 Massachusetts Avenue, Cambridge, MA 02139, USA

Floyd E. Romesberg
Department of Chemistry, The Scripps Research Institute, 10550 North Torrey Pines Road, La Jolla, CA 92037, USA

Abbey E. Rosen
Department of Chemistry, University of Minnesota, Minneapolis, MN 55455, USA

Joanna S. Slusky
Department of Biochemistry and Biophysics, School of Medicine, University of Pennsylvania, Philadelphia, PA 19104, USA

David A. Tirrell
Division of Chemistry and Chemical Engineering, California Institute of Technology, Pasadena, CA 91125, USA

Takuya Ueda
Department of Medical Genome Sciences, Graduate School of Frontier Sciences, University of Tokyo, FSB401, 5-1-5 Kashiwanoha, Kashiwa, Chiba 277-8562, Japan

Sander I. van Kasteren
Department of Chemistry, University of Oxford, Chemistry Research Laboratory, 12 Mansfield Road, Oxford OX1 3TA, UK

Elizabeth Vogel Taylor
Department of Chemistry and Department of Biology, Massachusetts Institute of Technology, 77 Massachusetts Avenue, Cambridge, MA 02139, USA

Hang Yin
Department of Chemistry and Biochemistry, University of Colorado at Boulder, Boulder, CO 80309, USA

Understanding Enzyme Mechanism through Protein Chimeragenesis

N.M. Goodey and S.J. Benkovic(✉)

Contents

Abstract The preparation of chimeras, proteins that contain segments from two or more different parent proteins, is a valuable tool in protein engineering yielding structures with novel properties. In addition to the obvious practical value of hybrid proteins as catalysts and biopharmaceuticals, their careful analysis can be used to understand the role of specific domains in enzymatic catalysis and protein evolution in a unique way that complements other structure-function studies. The study of hybrid enzymes can reveal, for example, the role specific subunits and/or domains play in dictating substrate specificity, catalytic activity, processivity, and stability. Popular chimeragenesis methods, including noncombinatorial and combinatorial methods, that can be used to generate hybrid proteins, are discussed here and four case studies are presented that beautifully demonstrate how hybrids can be studied to gain detailed understanding about substrate selectivity, enzymatic activity,

---

S.J. Benkovic
Department of Chemistry, The Pennsylvania State University, 414 Wartik Laboratory, University Park, PA 16802, USA, e-mail: sjb1@psu.edu

C. Köhrer and U.L. RajBhandary (eds.) *Protein Engineering.*
Nucleic Acids and Molecular Biology 22,

enzyme chemistry, and protein evolution. The examples highlight the power of chimeragenesis as a tool for gaining insights into enzyme mechanism as well as the need to combine this technology with other methods such as random mutagenesis and DNA shuffling, especially because the replacement of a domain can yield unpredicted perturbations to structural and functional parameters.

1 Introduction

Understanding how enzymes form stable structures, catalyze chemical reactions at greatly enhanced rates, show remarkable substrate specificity, and evolve is a long-standing quest in biochemical research. Proteins influence practically all biological processes and only by investigating protein structure and function can we begin to fully appreciate how they are integrated into living organisms. Elucidation of structure–function relationships in enzymatic catalysis was traditionally accomplished by observing the effect of reaction conditions, substrate structure, and/or natural or unnatural amino acid replacements on the rate, specificity, or three-dimensional structure of an enzyme. Recently, such investigations of enzymes have been enhanced by advances in structural, computational, biophysical, and protein engineering methods (Eisenmesser et al. 2002; Garcia-Viloca et al. 2004; Mittermaier and Kay 2006). This review focuses on the use of hybrid approaches, which are providing increasingly important avenues for obtaining novel insights into structure–function relationships in enzymatic catalysis (Armstrong 1990). Discussed herein is the preparation and use of hybrid enzymes with altered function or mechanism to gain detailed insights into enzyme structure, specificity, catalytic efficiency, and molecular evolution. The importance of combining novel hybrid generation technologies with other methods, such as rational and random mutagenesis, novel screening and selection approaches, X-ray crystallography, gene alignments, and pre-steady-state and steady-state kinetics, is highlighted.

1.1 Terminology

Throughout this review, the terms "hybrid" and "chimera" are used interchangeably. Hybrid proteins contain segments (domains or subdomains) from at least two different natural or man-made parent proteins (Armstrong 1990). Domains and subdomains are loosely defined terms that refer to structural motifs of various sizes and complexity, including small units of approximately ten to 30 amino acids, folded functional units, and large domains of several hundred amino acids that may have enzymatic activity (Ostermeier and Benkovic 2001). There are several types of hybrids. Single crossover hybrids consist of the N-terminal section of one protein and the C-terminal section of another (Fig. 1). In multiple crossover hybrids, one or more internal stretches of amino acid sequence have been replaced by the corre-

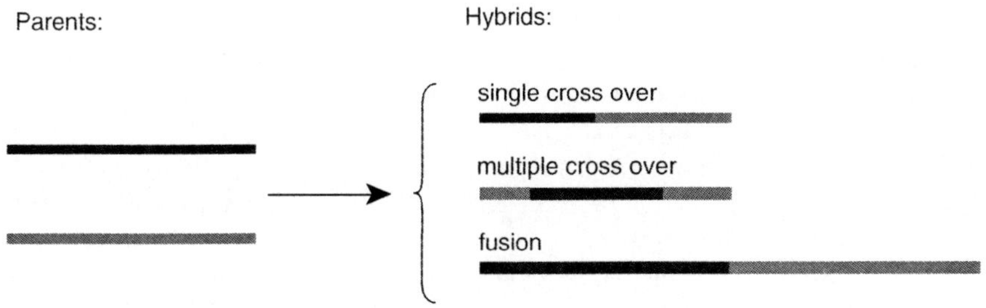

Fig. 1 Common types of hybrids created from two parent enzymes

sponding segment(s) from another enzyme(s). In fusion hybrids, a functional domain of one protein is linked with a domain(s) from another protein(s), creating a fused product that is larger than any of the parents alone. Figure 1 illustrates some of the types of hybrids; for the sake of simplicity, only hybrids derived from two parents are shown. Hybrids can of course contain amino acid mutations, deletions, and/or insertions.

1.2 What Can Hybrid Proteins Contribute to Understanding Enzyme Catalysis?

Section 2 of this review briefly outlines the common experimental methods for hybrid generation. In Sect. 3, recent examples from the literature are provided to demonstrate how chimeragenesis, combined with thorough biochemical studies, yields novel insights into enzyme mechanism. The case studies in Sect. 3 are not meant to be comprehensive, but rather are presented to illustrate the types of hybrid approaches used and the kind of information that can be obtained about protein function. We discuss how careful study of hybrid enzymes can lead to the identification of structural/functional domains and help determine which domains between two structurally similar proteins are interchangeable (Mas et al. 1986; Gurvitz et al. 2001). Hybrid studies can answer the question of whether a structural module can contribute a defined functional characteristic to the hybrid enzyme. The study of hybrid enzymes can reveal the role subunits/domains play in dictating substrate specificity, catalytic activity, processivity, and stability in a protein (Brock and Waterman 2000; Du et al. 2001; Stevenson et al. 2001; Lee et al. 2003).

Rational design of hybrids requires the preliminary identification of functionally important modules (Hopfner et al. 1998; Schneider et al. 1998) by inspection of the three-dimensional structure of either parent enzyme. In the absence of structural information, amino acid sequence alignments can be used to identify potentially important segments. Moreover, DNA sequences can be inspected to locate exon–intron interfaces, which may define the boundaries of structural or functional units. Recent advances in hybrid methods have made it possible to generate chimeric libraries in a random fashion, removing the need for structure or sequence alignments (Ostermeier et al. 1999b; Stevenson and Benkovic 2002). Thus, random methods make it possible to examine the contribution of protein segments to function without preconceived bias. These methods also have the advantage that the fusion points between structural or functional domains or subunits can be located precisely. Finally, inspection of hybrids can provide valuable understanding about the elusive but important residue–residue contacts and long-distance residue networks which are crucial for protein activity (Agarwal et al. 2002; Rajagopalan et al. 2002; Benkovic and Hammes-Schiffer 2003). In Sect. 2, we outline some of the popular chimeragenesis methods used to prepare hybrid proteins for the purpose of further understanding structure–function relationships.

2 Chimeragenesis Methods

Several excellent reviews in the literature comprehensively cover current chimeragenesis methods (Nixon et al. 1998; Lutz and Benkovic 2000, 2002; Stevenson and Benkovic 2002; Horswill et al. 2004). This section presents a short overview, focusing on techniques that have been used to generate chimeric proteins for structure–function relationship studies. These techniques can be divided into noncombinatorial and combinatorial approaches. Noncombinatorial methods are considered "rational" because it is necessary to choose both the domain targeted for swapping and the crossover points that define the domain. Combinatorial or random methods, on the other hand, produce large libraries of chimeric genes with fragments of random size inserted into or deleted from random positions of the target scaffold. A disadvantage of rational methods is that choosing the module for swapping and the precise end points of the fragment is difficult, and success depends on the level of understanding of the structure, folding, and catalytic mechanism of the enzyme under study. Much work is required to create a single or a few hybrid proteins, while combinatorial methods can yield more than 10^8 different hybrids in one experiment. However, rational approaches often lead to more definite information regarding the role of a specific domain in protein function, while combinatorial methods often do not yield instantly recognizable answers and rely heavily on development and execution of genetic selection protocols.

2.1 Noncombinatorial Domain Swapping Methods

The actual construction of rationally designed domain-swapped genes can be accomplished by several techniques. In the first method, *cassette mutagenesis*, restriction sites flanking the DNA particular sequence to be replaced are digested with the cognate restriction enzymes and a replacement sequence is inserted between these sites (Wells et al. 1985).

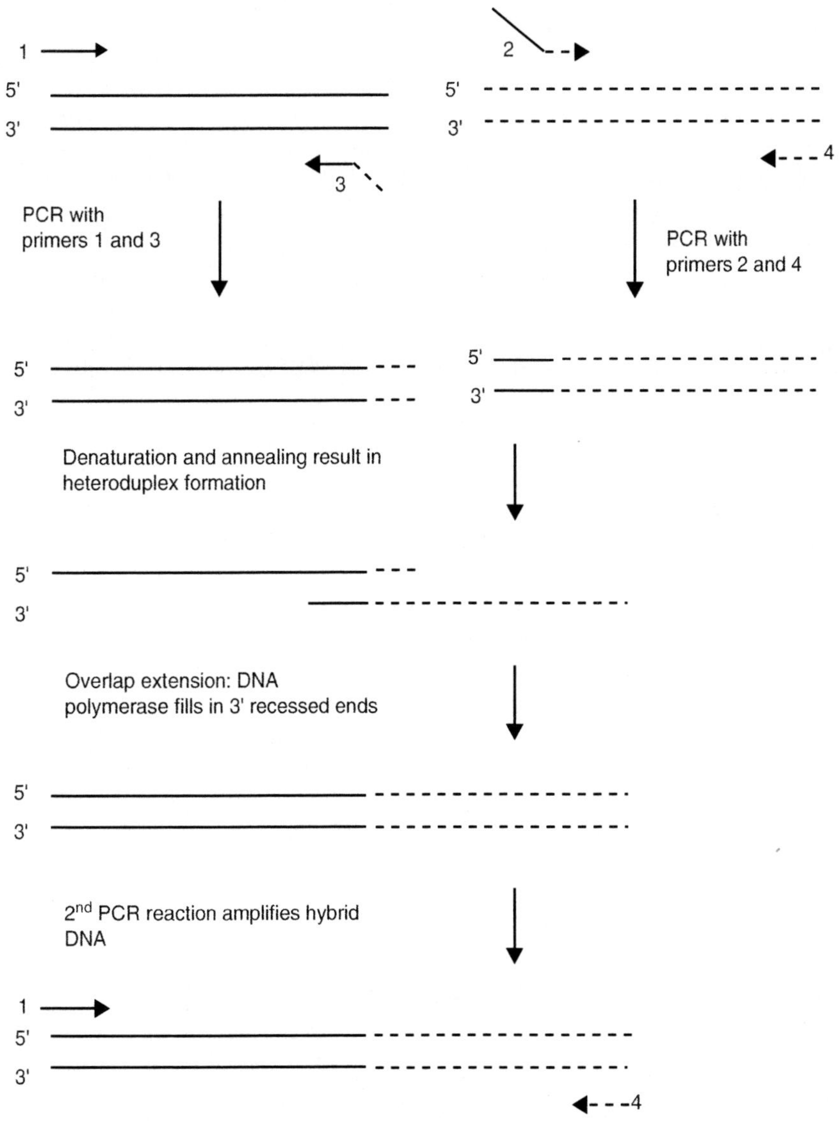

Fig. 2 Gene splicing by overlap extension polymerase chain reaction (PCR). *Arrows* represent primers (*1–4*), the *solid line* and *dotted lines* represent DNA sequences from two different parent genes

The restriction sites can be either naturally occurring or artificially introduced to the desired location in the target gene by oligonucleotide-directed mutagenesis. The choice of restriction sites is based on their uniqueness to the plasmid and conservation of the final amino acid coding sequence. In the second method, *gene splicing by overlap extension*, DNA molecules are recombined without the use of restriction endonucleases (Fig. 2) (Horton et al. 1989). Fragments from the genes that are to be recombined (parent genes) are generated in separate polymerase chain reactions (PCRs). The primers for these reactions are designed such that the ends of the products contain complementary sequences. When these PCR products are mixed, the strands having matching sequences at their 3′ ends overlap and act as primers for each other, and extension by DNA polymerase produces a sequence in which the original sequences are linked together. This method makes it possible to join two fragments in one cloning step; however, multiple cloning steps are required to perform an internal domain swap. In a variation of this method, *three-step PCR protocol for construction of chimeric proteins*, only one cloning step is required for the swapping of an internal domain (Grandori et al. 1997). Such methods can be used to replace a particular domain in protein A by the analogous domain in protein B. The study of the resulting domain-swapped hybrid can help determine the contribution of that particular domain to protein function.

2.2 *Combinatorial Methods*

Using in vitro random approaches, predicting which protein fragments should be fused and where the crossover positions should be located is not necessary. One random chimeragenesis approach is *random insertion*. As the name suggests, this method involves insertion of DNA sequences into random locations of a target gene (Heffron et al. 1978; Luckow et al. 1987; Hallet et al. 1997; Manoil and Bailey 1997). Investigation of hybrids generated by random insertion can be used to identify sites that are amenable to functional insertion of a protein domain into a scaffold (Betton et al. 1997). Another approach, *circular permutation of a protein*, results in the relocation of the N- and C-termini (Graf and Schachman 1996). This method can be used to understand evolutionary pathways and to systematically identify permissive sites for circular permutation, revealing information about the modularity of the scaffold (Baird et al. 1999; Hennecke et al. 1999). The remaining combinatorial approaches can be divided into two groups: homologous and nonhomologous recombination.

2.2.1 Homologous Recombination

Homologous recombination of gene fragments signifies the reassembly of DNA fragments from multiple parent genes in a way that incorporates DNA from multiple parents into a final gene product. The reassembly mechanism is based on DNA sequence homology and genes with low (less than 70%) homology cannot be combined (Sieber et al. 2001). As a result, there is an inherent bias present in homologous recombination libraries: the crossover positions fall in regions of high, rather than low, DNA similarity. Even so, these methods can be very useful in domain swapping, since often the parent genes share a high level of homology. Furthermore, homologous recombination can be used to combine fragments from more than two parent genes, making

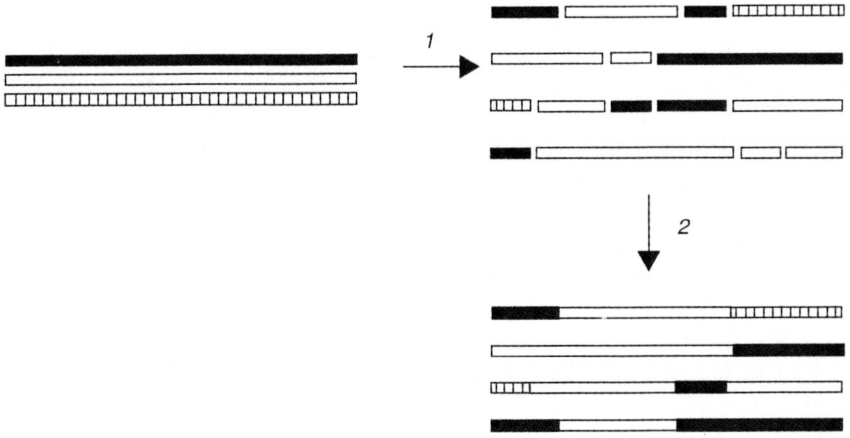

Fig. 3 DNA shuffling. Parent genes (only three are shown) are randomly fragmented with DNaseI (*1*). The fragments are reassembled in a primerless PCR reaction in which the fragments serve as primers as well as the template (*2*). Replication yields hybrid DNA strands that have components from different parent genes

it possible to explore the sequences of an entire family of enzymes in one experiment. The basic protocol for in vitro homologous recombination of parent genes is referred to as *DNA shuffling* (Fig. 3) (Stemmer 1994a, b). A group of homologous genes is randomly cleaved into small fragments by digestion with the restriction enzyme DNaseI. The fragments are subsequently reassembled by PCR in which they serve as both the templates and the primers. The fragments align and cross-prime each other for replication to give a hybrid DNA strand with components from several original parent genes.

DNA shuffling led the way to the development of related techniques which offer advantages in certain situations. For example, in a simple modification of the original protocol, *restriction enzyme based shuffling*, Kikuchi et al. (1999) employed a mixture of restriction endonucleases instead of DNaseI to cleave the parent genes. Compared with the original DNA shuffling protocol, restriction-enzyme-based shuffling has the advantage of yielding a higher frequency of chimeras because the use of DNaseI is avoided. DNaseI hydrolyzes double-stranded DNA preferentially at sites adjacent to pyrimidine nucleotides and can consequently introduce sequence bias into the recombination. Restriction-enzyme-based shuffling, however, suffers from the disadvantage that the crossover sites are biased to coincide with existing restriction sites. *Staggered extension process* (StEP) is another variation of DNA shuffling (Zhao et al. 1998; Volkov et al. 2000). In this procedure, terminal primers are employed to replicate the target DNA using PCRs with very short extension times to produce short strands of replicated DNA. The growing DNA strand acts as the primer in successive cycles of replication and changes templates multiple times, thus accumulating components from different parent genes. Other variations of DNA shuffling include *random priming recombination* (Shao et al. 1998), *DNA reassem-*

bly by interrupting synthesis (Short 1997), and *random chimeragenesis on transient templates* (RACHITT) (Coco et al. 2001). DNA shuffling and related methods have many practical applications in industrial protein engineering, but the outcomes are often difficult to rationalize because the resulting chimeras have multiple parents and multiple crossover positions. This is not significant when the goal is the end product, i.e., the engineered enzyme with desired properties. On the other hand, when the goal is to understand structure–function relationships, rational methods or methods described later, such as incremental truncation for the creation of hybrid enzymes (ITCHY), are often more suitable. The homologous recombination methods described above, however, have been successfully used in combination with other methods to offer insights into the determinants of substrate specificity and enzymatic activity of various families of proteins (Griswold et al. 2005; Park et al. 2006).

2.2.2 Nonhomologous Recombination

As discussed above, the main limitation of homologous recombination methods is that only genes with high homology can be combined. Often a group of proteins shares a similar three-dimensional structure, yet their DNA sequence identities are low. Recombination of such a group of proteins by homologous recombination methods would result in a library where the crossover positions would lie exclusively in the small regions of high DNA sequence homology. Conversely, the nonhomologous recombination methods described below do not rely on sequence homology of parental genes because instead of homologous fragment hybridization, a blunt-ended ligation step is used to bring the gene fragments together. Consequently, no bias toward the composition of the gene fragments is encountered. With nonhomologous recombination methods, it is possible to make a library of chimeras of two completely unrelated genes.

Most nonhomologous recombination methods are based on *incremental truncation*, which in its simplest form denotes the creation of libraries of proteins that have one or more amino acids deleted from either the C- or the N-terminus (Fig. 4) (Ostermeier et al. 1999a). Incremental truncation and related methods are dependent on the exonuclease III (*Exo*III) protein and its properties. *Exo*III catalyzes the digestion of gene fragments from the 3′ to the 5′ end at a controlled, uniform, and synchronous rate (Wu et al. 1976). Small aliquots of the reaction mixture are removed during the digestion step to create a library of genes with different numbers of DNA base pairs deleted. Incremental truncation technology has inspired the invention of more advanced chimeragenesis strategies. The first of these strategies, *incremental truncation for the creation of hybrid enzymes* (ITCHY), generates single crossover hybrid protein libraries between two parent genes (Fig. 4). Application of ITCHY results in incremental gene truncation libraries with randomly distributed crossover positions. These libraries contain hybrids with DNA insertions and deletions, and products of different sizes are produced. *Thio-ITCHY*, a variation of the ITCHY method, uses nucleotide triphosphate analogs (Lutz et al. 2001a). In a PCR, α-phosphothioate dNTPs are incorporated randomly and at low frequencies into the region of DNA targeted for truncation. The resulting phosphothioate internucle-

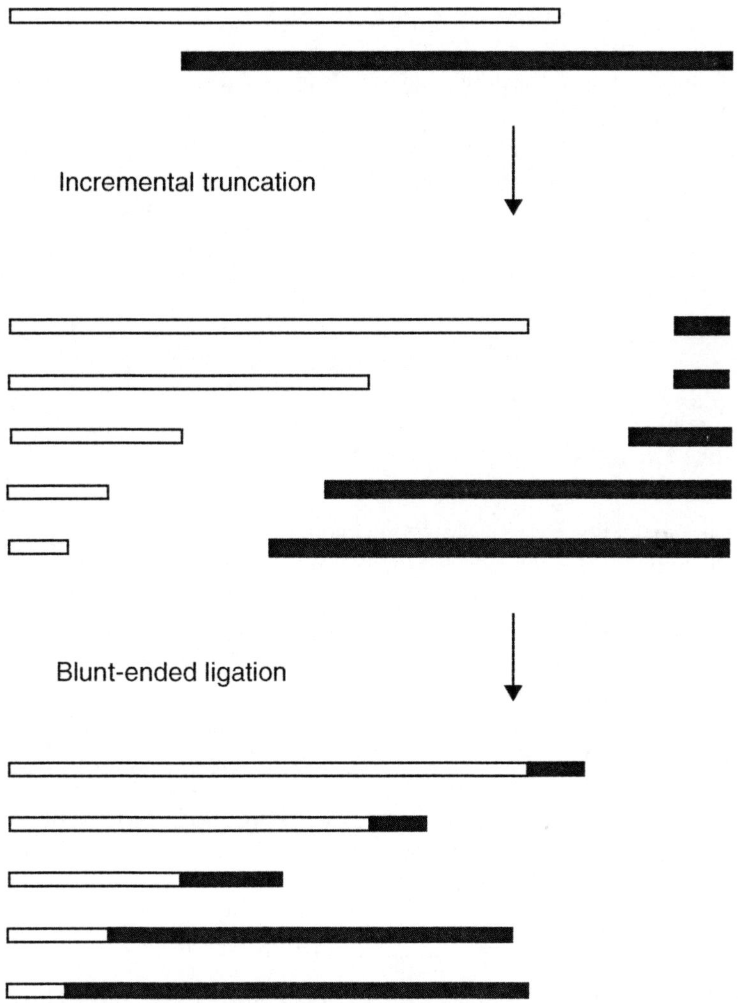

Fig. 4 Incremental truncation (*first arrow*) and incremental truncation of hybrid enzymes (ITCHY) (entire diagram). In ITCHY, incremental truncation libraries are generated by DNA digestion with exonuclease III. The resulting fragments are then ligated together to yield the final library

otide linkages are resistant to 3′–5′ exonuclease hydrolysis, rendering the target DNA resistant to degradation in a subsequent *Exo*III treatment. Thio-ITCHY has been successfully used to recombine genes with only 36% sequence identity to yield active chimeras (Saraf et al. 2006). The main limitation of ITCHY and thio-ITCHY is that only single crossover hybrids from only two parents can be generated, limiting the potential sequence diversity. To create multiple crossover libraries, Benkovic and coworkers combined ITCHY and DNA shuffling in a sequential manner in a method termed "SCRATCHY" (Lutz et al. 2001b). Sieber and coworkers developed a different technique for creating single crossover chimeric libraries between two parental genes without bias for the crossover positions (Sieber et al.

2001). This recombination method, *sequence homology independent protein recombination* (SHIPREC), begins with the production of a gene dimer which is subsequently fragmented by DNaseI. The parental-sized fragments are recovered and inverted by circular ligation and restriction digest to yield a library of hybrid enzymes with an even distribution of crossover positions. These methods have been used to explore structure–function relationships and the mechanism of evolution in proteins (Griswold et al. 2005; Peisajovich et al. 2006). When combined with genetic selections or screening techniques, the methods described in Sect. 2 can be used to gain novel insights into enzymes. Next we presented four interesting case studies that take advantage of the methods described so far to understand enzyme specificity, activity, and evolution.

3 Case Studies

The four outstanding examples presented here reflect on how hybrids can be prepared and studied to gain understanding about substrate selectivity, enzymatic activity, enzyme chemistry, and protein evolution. The first example specifically discusses how chimeragenesis can lead to novel insights into substrate selectivity in a family of enzymes (Griswold et al. 2005); the second focuses on understanding the determinants of metallo β-lactamase (MBL) activity in a new scaffold (Park et al. 2006); the third illustrates how specific mechanistic details, such as the structural origin of a pK_a value, can be investigated when hybrid methodology is combined with thorough kinetic analysis (Horswill, Guibao, Gerth, Lutz, and Benkovic, unpublished data); finally, the fourth case study demonstrates the role of hybrids in evaluating gene rearrangements that may be responsible for protein evolution (Peisajovich et al. 2006).

3.1 Hybrids in Understanding Protein Substrate Selectivity

Hybrid approaches are useful in studying the determinants of substrate selectivity (Chandrasegaran and Smith 1999; Smith et al. 1999; Cheon et al. 2004; Mani et al. 2005). Specifically, domains of two or more enzymes with similar structures can be interchanged and the effect on selectivity determined. Alternatively, hybrids can be generated randomly and the resulting library screened for activity on a certain substrate. The selectivities of the active clones are then measured and sequence alignments can reveal, for example, which domains are responsible for activity toward each substrate.

The first example highlights the use of chimeragenesis to alter and understand substrate specificity determinants in the glutathione transferase (GST) class of enzymes (Griswold et al. 2005). GSTs serve an important function in cellular detoxification by conjugating reactive, electrophilic compounds to the tripeptide glutathione (GSH)

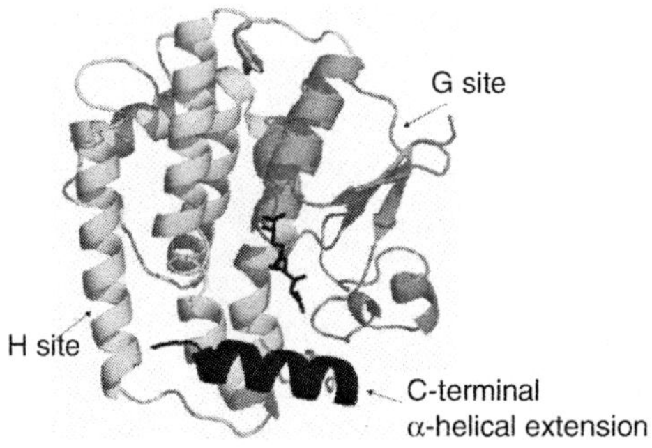

Fig. 5 Theta-class human GSTT2-2 (Protein Data Bank code 1LJR). The glutathione (GSH)-binding domain (G site) is shown in *dark gray*; the electrophilic substrate binding domain (H site) is shown in *light gray*, and the C-terminal α-helical extension is shown in *black*. GSH is shown in *black*

(Hayes et al. 2005). Benkovic, Iverson, Georgiou, and coworkers uncovered substrate selectivity determinants in the mammalian theta-class GST enzymes, human GSTT1-1 and rat GSTT2-2, by generating hybrids and determining their selectivities toward various electrophilic substrates (Griswold et al. 2005). Theta-class GST enzymes consist of an N-terminal GSH binding site (G site) and a C-terminal electrophilic substrate binding domain (H site), followed by a C-terminal α-helical extension, which covers the H and G sites (Fig. 5) (Ketterer 2001). The authors particularly focused on the roles of the H site and the C-terminal α-helical extension in electrophile selectivity (Griswold et al. 2005). The N-terminal G sites of the two enzymes share 79.2% amino acid identity, while the C-terminal H sites exhibit only 41.4% amino acid identity. The authors suspected that the low amino acid identity in the H sites was responsible for the divergent selectivities for electrophilic compounds: the specific activities of the human and rat GSTs were measured with the electrophiles 7-amino-4-chloromethyl coumarin (CMAC), 1-chloro-2,4-dinitrobenzene (CDNB), phenethyl isothiocyanate (PEITC), and ethacrynic acid (Fig. 6). Of the four electrophiles examined, only PEITC is conjugated to GSH by the human enzyme. The rat enzyme, on the other hand, has a more promiscuous nature, and catalyzes the conjugation of CMAC, CDNB, and PEITC.

The homology-independent techniques ITCHY and SCRATCHY (Ostermeier et al. 1999b; Lutz and Benkovic 2000) (Sect. 2) and the low-homology technique of recombination-dependent exponential amplification were used to create libraries of single and multiple crossover chimeras of human and rat GSTs in both orientations [rat–human (r–h) and human–rat (h–r)]. Since CMAC produces a fluorescent cytoplasmically retained product upon GSH conjugation, the library was screened by a high-throughput flow cytometric screening protocol for variants that conjugate CMAC to GSH (Kawarasaki et al. 2003). The active chimeras from the single crossover r–h library consisted entirely of the rat parental sequence except for the C-terminal α-helical extensions, which originated from the human enzyme (Fig. 7). The specificity constants (k_{cat}/K_M) toward CMAC of two active hybrids from this library were measured to be 20 and 60 mM^{-1} min^{-1}. These values lie between the k_{cat}/K_M values measured for the rat (650 mM^{-1} min^{-1}) and human (1.6 mM^{-1} min^{-1}) enzymes, showing that

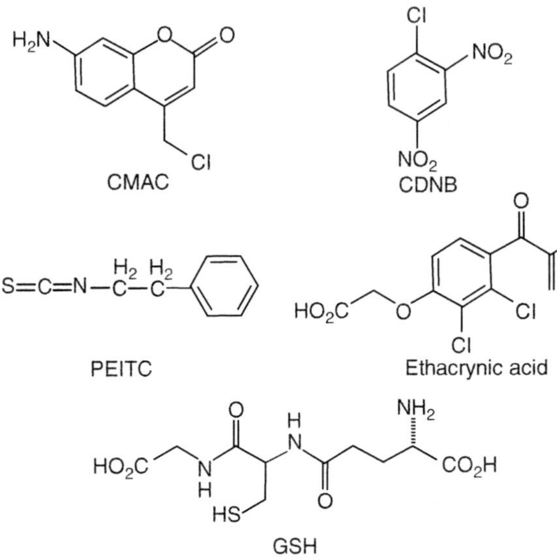

Fig. 6 Structures of glutathione transferase (GST) substrate GSH and four electrophiles; the electrophiles are 7-amino-4-chloromethyl coumarin (*CMAC*), 1-chloro-2,4-dinitrobenzene (*CDNB*), phenethyl isothiocyanate (*PEITC*), and ethacrynic acid

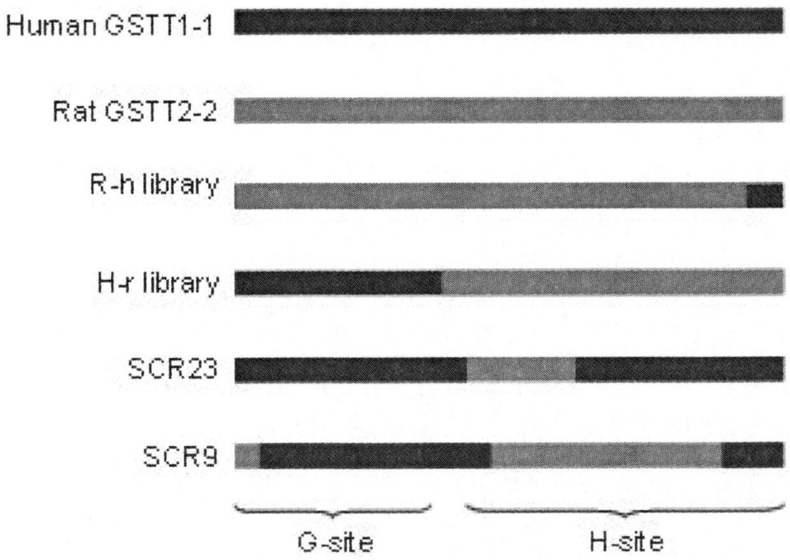

Fig. 7 Hybrids between human GSTT1-1 and rat GSTT2-2. Amino acid sequences from human GSTT1-1 and rat GSTT2-2 are shown in *black* and *gray*, respectively. Segments corresponding to the G site and the H site are indicated. *R-h* rat–human, *H-r* human–rat

replacement of the C-terminal α-helical extension by the human sequence in the rat enzyme lowers CMAC conjugation activity but not to the level of the human enzyme. These results demonstrate that CMAC specificity is modulated, but not exclusively determined, by the structure of the C-terminal α-helical extension.

When the complementary h–r ITCHY library was screened for CMAC conjugation activity, all active chimeras were found to have a full-length human N-terminal G site fused to a C-terminal rat H site with the crossover positions in the flexible loop that connects the G site to the H site. Clone HR-216 was studied in detail. Compared with the rat enzyme, HR-216 was found to have a threefold higher k_{cat} for CMAC (k_{cat} = 7.2 and 23 min^{-1} for rat GSTT2-2 and HR-216, respectively) as well as elevated specific activities for both CDNB and PEITC. On the basis of these results, the authors proposed that the human N-terminal domain (the G site) is indirectly enhancing rat-like catalytic efficiency, perhaps because it results in optimal interactions between the two domains.

As described above, single recombination events within the H site did not produce chimeras with high levels of CMAC activity. On the other hand, when the generation of multiple crossovers was engineered, enzymes that efficiently conjugate CMAC to GSH containing crossovers within the H site were discovered. For example, chimeras SCR23 and SCR9 consist of a human–rat–human sandwich structure (Fig. 8). The authors used homology modeling to show that the SCR23 structure has two α-helices within the H site that are derived from the rat parent sandwiched between the C-terminal and N-terminal human sequence. The SCR9 human–rat–human sandwich structure consists of a human N-terminal G site, followed by a mostly rat sequence derived H site, and then by a human C-terminal α-helix. Both SCR23 and SCR9 are efficient catalysts of GSH to CMAC conjugation: the specificity constants of SCR23 and SCR9 are 600 and 510 mM^{-1} min^{-1}, respectively, compared with 650 mM^{-1} min^{-1}

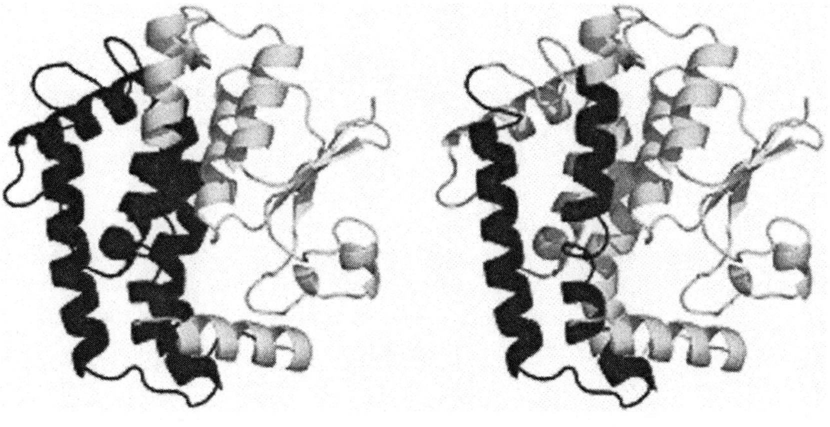

Fig. 8 SCR chimeras. SCR9 (*left*) and SCR23 (*right*) amino acid sequences mapped onto the crystal structure of human GSTT2-2 (Protein Data Bank code 1LJR). Human- and mouse-derived sequences are shown in *gray* and *black*, respectively

for rat GSTT2-2. The improved activity of human GSTT1-1 upon replacement of two helices in the H site by rat sequence, as in SCR23, demonstrates that these two helices contain residues that play important roles in electrophile selectivity. The authors note that the presence of human sequence in the G site restores the rat-like activity with CMAC for SCR9, canceling the detrimental effect of the human C-terminal α-helix.

Although both SCR23 and SCR9 are efficient catalysts of GSH to CMAC conjugation, their reactivities with the other substrates studied differ. This work determined that SCR9 has a similar selectivity profile toward the electrophile substrates studied as rat GSTT2-2, with the exception that SCR9 has higher activity toward CDNB. The rat parent and SCR9 both efficiently conjugate GSH to CDNB, whereas the human parent and SCR23 enzymes do not. Furthermore, unlike both parent enzymes and SCR9, SCR23 conjugates ethacrynic acid to GSH. The different selectivities of SCR9 and SCR23 show that the promiscuity of the rat enzyme is not due to a "loose" binding site but rather is due to specific determinants for selectivity toward different electrophiles. Furthermore, sequence alignment of SCR9 and SCR23 with the two parents revealed that SCR9 shares a 12 amino acid sequence with the rat enzyme, which is not found in SCR23 or the human enzyme. This result suggests that the 12 amino acid stretch might be critical for CDNB selectivity but not for CMAC selectivity. This example demonstrates how the careful study of hybrid enzymes can lead to the identification of very specific determinants of enzyme specificity.

3.2 *Hybrids in Understanding Determinants of Enzyme Activity*

One of the more challenging goals of protein engineering for some time has been the introduction of a new enzymatic activity into a protein scaffold to create a novel enzyme. Chimeragenesis can be used to replace segments of an enzyme by those from another enzyme with a similar structure but a different activity. If a new activity is obtained, it is clear that the segment that was replaced plays an important role in enzyme chemistry. Replacing larger segments rather than individual residues is often more useful in obtaining a new enzymatic activity because the short-range residue–residue interactions within a given domain are interrupted to a lesser extent and more sequence space is sampled.

In the second example, protein engineering of glyoxalase II (GlyII) by Kim and coworkers illustrates how innovative chimeragenesis techniques combined with rational mutagenesis and directed evolution can uncover catalytic activity determinants (Park et al. 2006). The authors introduced MBL activity into the human GlyII αβ/αβ metallohydrolase scaffold. The resulting variant was devoid of the function performed by GlyII, hydrolysis of the thioester bond of *S*-D-lactoylglutathione, but was newly endowed with the unrelated MBL function, hydrolysis of the β-lactam amide bond of cefotaxime, that leads to bacterial resistance against this β-lactam antibiotic (Mei et al. 1998). To plan the strategy for mutagenesis, the authors com-

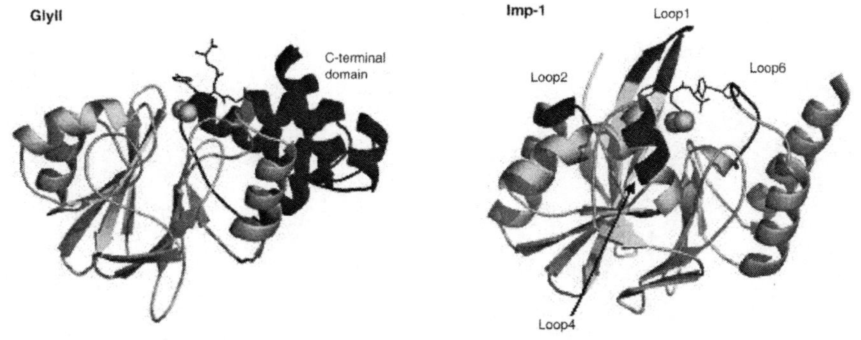

Fig. 9 Structures of glyoxalase II (*GlyII*) (Protein Data Bank code 1QH5) and IMP-1 (Protein Data Bank code 1DD6). The C-terminal domain of GlyII that was deleted is shown in *black*. The zinc atoms are represented by *gray spheres*. In IMP-1, functional loops 1, 2, 4, and 6 are shown in *black*. Bound ligands of GlyII and IMP-1, *S*-(*N*-hydroxy-*N*-bromophenylcarbamoyl) glutathione and mercaptocarboxylate, respectively, are represented by *black sticks* to show the location of the active site

pared the structure of human GlyII with that of IMP-1, a MBP family member from *Pseudomonas aeruginosa* (Fig. 9). GlyII and IMP-1 share only 13% amino acid sequence identity, but their αβ/αβ sandwich structures are similar.

While the three-dimensional structures of GlyII and IMP-1 are similar, their active sites differ in metal coordination and substrate binding. Previously published work showed that the IMP-1 substrate binding site consists of loops 1, 2, and 6 (Fig. 10) (Wang et al. 1999). Loops 1 and 2 make up a hydrophobic binding site for the β-side-chain substituents of antibiotics, whereas loop 6 contains residues that are necessary for the binding and activation of the β-lactam substrate during catalysis (Yanchak et al. 2000). Loop 4 may play a role in β-lactam binding or catalysis from a distance. The GlyII substrate binding site shares no similarities with the IMP-1 binding site described above; in GlyII, loop 4 and the C-terminal helical domain are indicated in substrate binding (Cameron et al. 1999). Both the IMP-1 and the GlyII active sites contain binuclear metal ions that are essential for catalysis, but the coordination sites have disparities. IMP-1 contains two zinc ions, whereas GlyII can have various metal ions in its binuclear site. Zn1 in IMP-1 and metal 1 in GlyII are both coordinated with three histidine residues. On the other hand, Zn2 in IMP-1 is coordinated to an aspartic acid, a cysteine, and a histidine residue, whereas metal 2 in GlyII is coordinated to two aspartic acid and two histidine residues.

To replace GlyII activity by MBP activity in the GlyII scaffold, the authors first eliminated GlyII catalysis by deleting the C-terminal helical domain suspected to be involved in substrate binding. The fact that deletion of this helical domain from GlyII led to the loss of GlyII activity confirmed that this domain contains substrate binding determinants required for GlyII activity. Secondly, a new active-site binding pocket in the modified GlyII scaffold was constructed on the basis of sequence

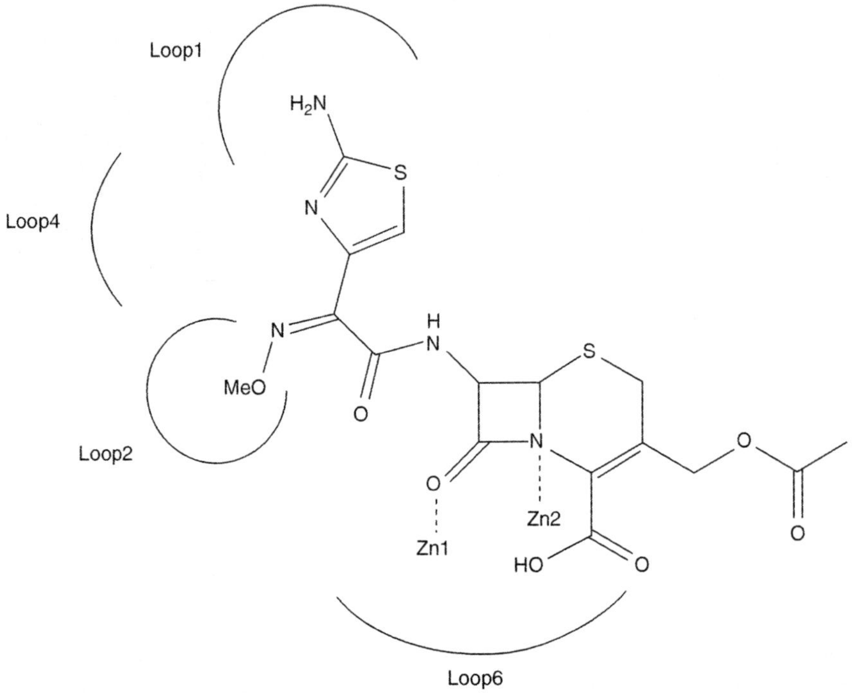

Fig. 10 Active site of IMP-1. Interactions of loops 1, 2, 4, and 6 with β-lactam antibiotic cefotaxime and coordinated zinc atoms (Zn1, Zn2) are shown

alignments of loops 1, 2, 4, and 6 in GlyII, IMP-1 and two other MBL family enzymes. Loops 1, 2, and 4 of the modified GlyII were designed to contain completely or partially conserved residues, which are found in IMP-1 and the two other MBL family enzymes, as well as several random residues for fine-tuning. Since there is significant variation in loop 6 among the MBL enzymes that were aligned, three different possible sequences were designed for this loop, each containing some residues that are found in MBL family enzymes and some random residues. The designed loops 1, 2, 4, and 6 were incorporated into the modified GlyII scaffold by PCR under error-prone conditions, allowing for point mutations. The resulting chimeric library was subjected to cefotaxime selection and the 13 surviving clones that exhibited antibiotic resistance were isolated and sequenced. Interestingly, all 13 clones contained segment FIKAXXXGNXXDA (X indicates a random residue) in loop 6, and none contained the other two designed possibilities, indicating that one or many of the fixed residues in segment FIKAXXXGNXXDA are critical for MBL activity. In the 13 cefotaxime resistant clones, two to nine additional amino acid replacements were found to be distributed in regions away from the active site. If the loops were integrated without the additional point mutations, no active clones were found, thus revealing that certain amino acid replacements distal from the active site are critical for MBL activity in the GlyII scaffold.

This finding is in accord with numerous studies that emphasize the importance of residues far away from the active site in protein function.

To further increase their catalytic activities, cefotaxime-resistant clones were subjected to seven rounds of DNA shuffling (Stemmer 1994a, b). The cefotaxime concentration was increased during each round, and after the seventh round, the clone showing the most improved growth on selective plates was chosen. This final clone provided *Escherichia coli* cells with 100-fold improved antibiotic resistance against cefotaxime compared with wild type cells/parent cells. Sequencing of the final clone revealed that it shared only 59% of amino acid identity with the original GlyII scaffold and 25% with the target IMP-1. More than 60% of the mutations were concentrated in the catalytic and substrate binding regions. The authors found that replacement of the residues designed for metal coordination by alanine led to loss of catalytic activity. This result suggests that the final selectant retained the designed metal coordination of IMP-1 and that this coordination is critical for cefotaxime hydrolysis. Molecular modeling of the final clone revealed an active-site architecture similar to that of the target IMP-1 with a well-organized binuclear metal binding site; loops 2 and 6 constitute two walls of the substrate binding pocket and loop 1 forms the ceiling. Thus, the substitution of loops 1, 2, 4, and 6 in GlyII by analogous loops in the MBL family replaced the catalytic function and active-site architecture of GlyII by those of IMP-1. Not only did the authors accomplish the impressive feat of introducing a new catalytic activity into the GlyII scaffold, they also uncovered important information about the determinants of activity in the GlyII and IMP-1 proteins in the process. This work showcases the potential of protein segment replacement combined with directed evolution in converting an enzyme into a new family member to create variants that catalyze diverse reactions and yield increased understanding of the determinants of enzyme chemistry.

3.3 Hybrids in Understanding Enzyme Mechanism

Combining structural elements belonging to different enzymes is a powerful method for generating new kinetic properties. Subsequent analysis can provide detailed insights into the subtle contribution of protein domains to different features of enzyme chemistry. In the third example, work on *E. coli/B. subtilis* dihydrofolate reductase (DHFR) hybrids by Benkovic and coworkers illustrates the power of studying enzyme chimeras by thorough kinetic analysis (Horswill, Guibao, Gerth, Lutz, and Benkovic, unpublished data). The authors chose four chimeras of well-characterized *E. coli* and *B. subtilis* DHFR enzymes that were originally identified from an activity-based selection (Saraf et al. 2004). The *E. coli* and *B. subtilis* enzymes have different kinetic properties and share 44% sequence identity on the protein level. Chimeras EB32 and EB122 have the *E. coli* sequence in their N-terminus and the *B. subtilis* sequence in the C-terminus (EB); and the number (32 or 122) reflects the residue at which crossover from the *E. coli* to the *B. subtilis* sequence occurs (Fig. 11). Complementary chimeras BE32 and BE123 have the

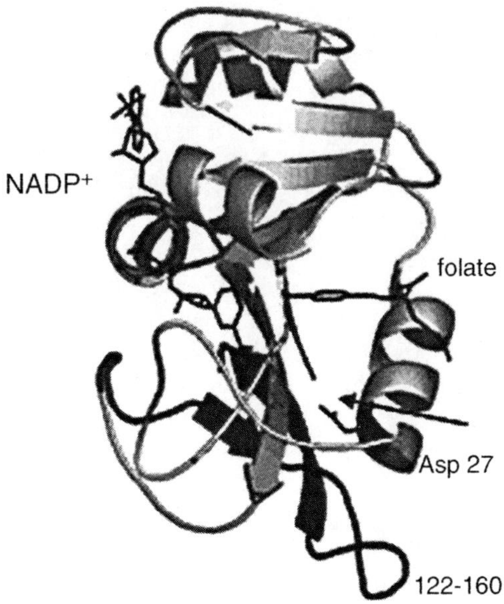

Fig. 11 Dihydrofolate reductase (DHFR) hybrid EB122. Residues 122–160 are shown in *black*. The structure of *Escherichia coli* DHFR bound to ligands NADP+ and folate (Protein Data Bank code 1rx2) was used to generate the figure. The catalytic residue Asp27 is shown

B. subtilis sequence in the N-terminus. The ligand dissociation rates, steady-state, and pre-steady-state kinetics of these hybrids were studied, and the rates determined for the chimeras were compared with those of the wild-type *E. coli* and *B. subtilis* enzymes.

DHFR catalyzes the stereospecific reduction of dihydrofolate to form tetrahydrofolate (THF) using the cofactor NADPH. The kinetic scheme of this reaction in the *E. coli* and *B. subtilis* enzymes has been thoroughly analyzed (Fig. 12 for *E. coli*) (Fierke et al. 1987). The following ligand dissociation rates were measured for the four selected hybrids: release of product NADP+ from E-NADP+ and the release of THF from the E-THF, E-THF-NADP+, and E-THF-NADPH complexes. The chimera EB122 consists mostly of the *E. coli* enzyme and accordingly can be considered a perturbation in/of *E. coli* DHFR. Interestingly, the four ligand dissociation rates mentioned above for this hybrid are more similar to those of *B. subtilis* DHFR than to those of the *E. coli* enzyme. This observation suggests that the determinants for the abovementioned ligand dissociation rates are encoded in the C-terminal domain consisting of residues 122–160 and that this domain is modular, allowing the ligand dissociation properties to be transferred from one DHFR scaffold to another. The results with BE123, however, did not support this hypothesis, suggesting that the modularity of the C-terminal domain as a determinant of ligand dissociation properties is more complex than initially suspected, illustrating the idea that when a domain is found to be "modular" in one set of enzymes, it does not necessarily implicate that this modularity is reciprocated in a different set. Clearly, there are other interactions across the protein that determine whether or not a particular domain swap will result in the desired or expected properties.

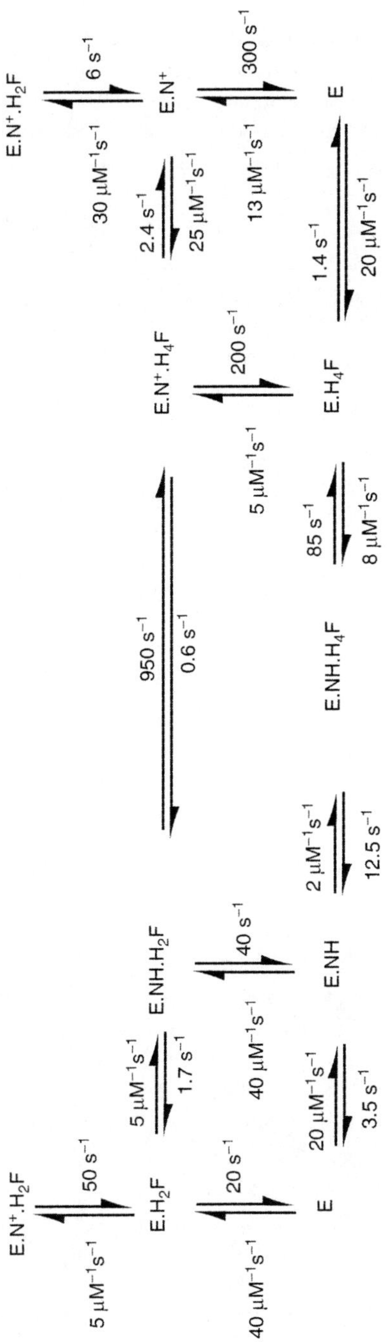

Fig. 12 pH-independent kinetic scheme of *E. coli* DHFR

This work also exemplifies how meticulous kinetic analysis of chimeras can give insight into complex aspects of catalysis, such as the environmental modulation of the pK_a of a catalytic residue. There has been an ongoing debate about how the local environment in DHFR raises the pK_a of the catalytic acid Asp27 in the E-DHF-NADPH complex to modulate the rate of its conversion to E-THF-$NADP^+$ from approximately 3.9 to approximately 6.5. This complex issue has been examined via kinetics, X-ray crystallography, spectroscopy, and computational studies (Adams et al. 1989; Appleman et al. 1990; Karginov et al. 1997; Rod and Brooks 2003). The work on chimeras by Benkovic and coworkers adds to this story. They discovered that the pK_a values for the hydride transfer step in chimeras EB32 and BE123 are similar to that of the *B. subtilis* enzyme. In both chimeras, *B. subtilis* DHFR is the dominant parent. Despite that in EB32 the residues adjacent to Asp27 originate from the *E. coli* parent, this chimera still has a pK_a for the hydride transfer step similar to that of *B. subtilis*. This shows that the active-site pK_a is not determined by the residues adjacent to Asp27 but rather by residues further remote and represents an excellent example of how chimeragenesis can be used to probe subtle mechanistic questions about enzyme chemistry.

3.4 Hybrids in Understanding Protein Evolution

Hybrid approaches are ideally suited for studying natural evolution of protein function because nature often evolves proteins by the redesign of existing frameworks, a process akin to domain swapping. The fourth example demonstrates how hybrid technology can also be used to explore this mechanism of protein evolution (Peisajovich et al. 2006). Many evolutionarily related proteins are *circular permutants*, which could technically originate from the ligation of the N- and C-termini of a protein, followed by the opening of the chain at another site to yield a new topology (Ponting and Russell 1995; Aravind et al. 2002; Koonin et al. 2002). In nature, however, such genomic rearrangements are unlikely and the steps leading to naturally occurring circular permutations remain a subject of study. *Permutation by duplication* is a widely accepted model explaining the origin of circular permutants in nature (Jeltsch 1999). It postulates that gene duplication is first followed by in-frame fusion and then by partial degeneration of the 5′ and 3′ coding regions in the first and second copies of the duplicated gene, respectively, leading to a new stop codon and a new topology (Fig. 13). These events are unlikely to happen simultaneously; rather a series of evolutionary intermediates must be formed along the way. These intermediates must retain the original protein activity to some extent for the organism to be sufficiently fit to avoid elimination by evolution. The permutation by duplication model has been questioned, especially since the intermediates would likely have exposed hydrophobic surfaces and been unstable (Bujnicki 2002).

Tawfik and coworkers used hybrid technology to test whether new protein topologies in DNA methyltransferases might have evolved gradually by the *permutation by duplication* mechanism (Peisajovich et al. 2006). DNA methyltransferases are composed of a target-recognition domain and a catalytic domain with a

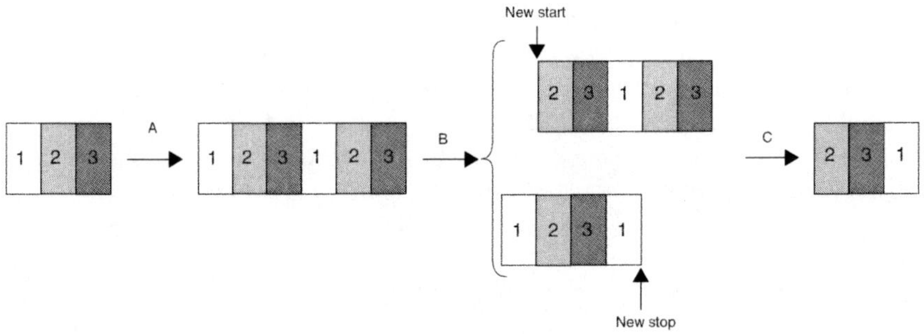

Fig. 13 The permutation by duplication model. Gene duplication and in-frame fusion lead to a fused dimer (*A*). Partial degeneration of the 5′ coding region in the first copy creates a new start codon in the N-terminally truncated intermediate. Partial degeneration of the 3′ coding region in the second copy introduces a new stop codon in the C-terminally truncated intermediate (*B*). This process leads to a circular permutant with new start and stop codons and a new topology (*C*)

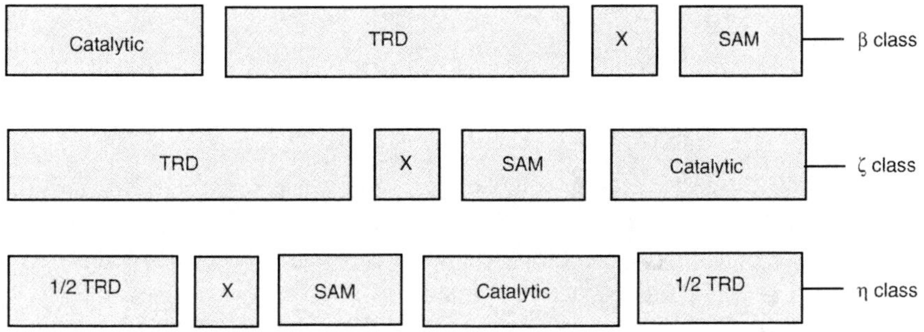

Fig. 14 DNA methyltransferases are divided into at least seven classes according to the linear order of sequence motifs. A schematic of three classes is shown, with the following domains: SAM-binding subdomain (*SAM*), the catalytic subdomain (*catalytic*), the target-recognition domain (*TRD*), and the C-terminal helix (*X*). Class η with a divided TRD was predicted by laboratory evolution experiments and was subsequently identified in natural genomes

Rossman fold structure (Martin and McMillan 2002). The order of nine conserved sequence motifs in the catalytic domain and the location of the target-recognition domain define at least seven families that might stem from the permutation by duplication mechanism (Fig. 14). The authors used gene duplication followed by the ITCHY method (Ostermeier et al. 1999c; Lutz et al. 2001a) to generate a library of *Hae*III methyltransferase hybrids consisting of duplicated fused genes and duplicates partially truncated at the 5′ or 3′ coding regions. The library was subjected to a selection and genes with duplications and truncations that retained in vivo activity were identified (Szomolanyi et al. 1980).

The selectants resembled natural methyltransferases from three known families in that their N- and C-termini coincided with those of natural methyltransferases; therefore, these variants corresponded to possible intermediates on evolutionary paths toward three distinct classes of methyltransferase proteins. The fact that these putative intermediates have in vivo activity shows that new methyltransferase topologies can have evolved gradually through multistep gene rearrangements. The authors point out that the majority of the selectants exhibit a substantial decrease in enzyme activity. In turn, they used random mutagenesis to demonstrate that point mutagenesis might compensate for changes in topology and loss of activity during the evolutionary process. In addition to the active sequences described above with N- and C-termini coinciding with natural methyltransferases, one set of selected variants did not correspond to any previously identified methyltransferase family. In these genes, the target-recognition domain was split in two; one part was located in the N-terminus and the remainder at the C-terminus. Homology searches identified previously unsuspected new methyltransferases in three different species, providing further indirect support for the permutation by duplication mechanism of evolution.

4 Conclusion

The examples given demonstrate the power of chimeragenesis as a tool for gaining insights into enzyme mechanism. In these examples, combining hybrid technology with other innovative methods, such as random mutagenesis or DNA shuffling, yielded information on structure–function relationships that was otherwise unattainable. The basic process in all these studies is the preparation and identification of hybrids exhibiting altered properties (such as selectivity, activity, pK_a, etc.) and the subsequent study of the hybrid sequence to deduce which domains are responsible for the changes. In many cases, chimeragenesis has shown that enzyme function is determined by discrete domains which can be interchanged between proteins of the same family. However, often factors such as domain–domain interactions, structural alterations, residue–residue coupling, etc. must be fine-tuned before domain-swapped hybrids are active.

4.1 Advantages and Disadvantages of Hybrid Approaches

One of the advantages of chimeragenesis in studying structure–function relationships is that large domains can be swapped at once, making it possible to rapidly scan large sequence space. This removes some of the guesswork from protein engineering that is necessary for rational point mutations. Another advantage of hybrid approaches is that they can be used for applications where other methods fall short. For example, they represent a rational approach for studying structure–function relationships when the three-dimensional structure of an enzyme is unavailable.

Hybrid approaches are also uniquely valuable in defining structure–function relationships that can be directly applied to the stabilization/humanization of engineered proteins for medical purposes (Clark 2000; Calvo and Rowinsky 2005). The work described in Sect. 3.1 by Benkovic, Iverson, Georgiou, and coworkers illustrates this: one of the hybrid proteins identified in this study (SCR23) can be considered a humanized enzyme because 82.9% of the sequence originates from the human GST sequence, yet the hybrid displays the activity of the rat parent.

A disadvantage of using hybrid proteins to understand structure–function relationships is that they frequently misfold locally or globally. It is difficult to interpret functional data because of a localized structural perturbation. As we have shown, if a segment of enzyme A is replaced by the complementary segment in enzyme B, it is likely that this segment would impart some of the catalytic character of enzyme B to the hybrid. However, it is also reasonable to expect that this replacement would alter the residue–residue interactions with the rest of the surrounding enzyme structure, resulting in unpredicted perturbations to functional and structural parameters. This issue derives from the fact that enzymes consist of residue–residue networks which cannot be disrupted without consequences. Computational methods, such as the SCHEMA or IPRO algorithms, can be used to minimize the introduction of unfavorable interactions (Bernhardt 2004; Saraf et al. 2006).

Thus, predicating or explaining the changes resulting from segment replacement represents a significant challenge. The difficulty of interpreting hybrid results, however, should not discourage an enzymologist. As illustrated in the examples given, the construction and analysis of hybrid enzymes offers a unique means to greatly further our understanding of enzyme chemistry.

4.2 Future Perspectives

In the future, we expect that the methods for producing hybrids will continue to evolve. One important area of research is the development of computational models to predict hybrid function. In many studies, hybrids are produced in a random fashion and efficient selections are required to sieve through the large libraries in order to identify active members. Thus, library-based random hybrid approaches depend on the development of novel and efficient selections and screening methods such as those highlighted in Sect. 3. As shown by Kim and coworkers in their work on recruiting a new activity into the GlyII scaffold (Sect. 3.2) (Park et al. 2006), innovative ways to combine chimeragenesis with other protein engineering methods such as random and rational mutagenesis, DNA shuffling, and in vitro evolution should be explored and are likely to result in impressive landmarks in protein engineering. The idea of discovering "generalist" or "primordial" enzymes and using them as starting points for engineering new activities is likely to be explored in the coming years. The examples presented herein show that preparation of hybrids, and their subsequent analysis, can be used to understand enzyme catalysis in a unique way that beautifully complements other structure–function studies. Clearly, a lack of understanding of structure–function relationships hinders rational protein design, and chimeragenesis is a useful tool in filling this void because what we learn about the interchangeability of domains can be used to generate new enzymes.

Acknowledgments We thank the National Institutes of Health (5 F32 GM072320-02) for supporting this work. We thank Todd Naumann for insightful discussions.

References

Adams J, Johnson K, Matthews R, Benkovic SJ (1989) Effects of distal point-site mutations on the binding and catalysis of dihydrofolate reductase from *Escherichia coli*. Biochemistry 28:6611–6618

Agarwal PK, Billeter SR, Rajagopalan PT, Benkovic SJ, Hammes-Schiffer S (2002) Network of coupled promoting motions in enzyme catalysis. Proc Natl Acad Sci USA 99:2794–2799

Appleman JR, Howell EE, Kraut J, Blakley RL (1990) Role of aspartate 27 of dihydrofolate reductase from *Escherichia coli* in interconversion of active and inactive enzyme conformers and binding of NADPH. J Biol Chem 265:5579–5584

Aravind L, Mazumder R, Vasudevan S, Koonin EV (2002) Trends in protein evolution inferred from sequence and structure analysis. Curr Opin Struct Biol 12:392–399

Armstrong R (1990) Structure–function relationships in enzymic catalysis can chimeric enzymes contribute? Chem Rev 90:1309–1325

Baird GS, Zacharias DA, Tsien RY (1999) Circular permutation and receptor insertion within green fluorescent proteins. Proc Natl Acad Sci USA 96:11241–11246

Benkovic SJ, Hammes-Schiffer S (2003) A perspective on enzyme catalysis. Science 301: 1196–1202

Bernhardt R (2004) Optimized chimeragenesis creating diverse P450 functions. Chem Biol 11:287–288

Betton J-M, Jacob JP, Hofnung M, Broome-Smith JK (1997) Creating a bifunctional protein by insertion of β-lactamase into the maltodextrin-binding protein. Nat Biotechnol 15:1276–1279

Brock BJ, Waterman MR (2000) The use of random chimeragenesis to study structure/function properties of rat and human P450c17. Arch Biochem Biophys 373:401–408

Bujnicki JM (2002) Sequence permutations in the molecular evolution of DNA methyltransferases. BMC Evol Biol 2:3

Calvo E, Rowinsky EK (2005) Approaches to optimize the use of monoclonal antibodies to epidermal growth factor receptor. Curr Oncol Rep 7:123–128

Cameron AD, Ridderstrom M, Olin B, Mannervik B (1999) Crystal structure of human glyoxalase II and its complex with a glutathione thiolester substrate analogue. Structure 7:1067–1078

Chandrasegaran S, Smith J (1999) Chimeric restriction enzymes. What is next? Biol Chem 380:841–848

Cheon YH, Park HS, Kim JH, Kim Y, Kim HS (2004) Manipulation of the active site loops of D-hydantoinase, a $(\beta/\alpha)_8$-barrel protein, for modulation of the substrate specificity. Biochemistry 43:7413–7420

Clark M (2000) Antibody humanization: a case of the 'Emperor's new clothes'? Immunol Today 21:397–402

Coco WM, Levinson WE, Crist MJ, Hektor HJ, Darzins A, Pienkos PT, Squires CH, Monticello DJ (2001) DNA shuffling method for generating highly recombined genes and evolved enzymes. Nat Biotechnol 19:354–359

Du Z, Tucker WC, Richter ML, Gromet-Elhanan Z (2001) Assembled F1-(αβ) and hybrid F1-$\alpha_3\beta_3\gamma$-ATPases from Rhodospirillum rubrum α, wild type or mutant β, and chloroplast γ subunits. Demonstration of Mg2+ versus Ca2+-induced differences in catalytic site structure and function. J Biol Chem 276:11517–11523

Eisenmesser EZ, Bosco DA, Akke M, Kern D (2002) Enzyme dynamics during catalysis. Science 295:1520–1523

Fierke CA, Johnson KA, Benkovic SJ (1987) Construction and evaluation of the kinetic scheme associated with dihydrofolate reductase from *Escherichia coli*. Biochemistry 26:4085–4092

Garcia-Viloca M, Gao J, Karplus M, Truhlar Donald G (2004) How enzymes work: analysis by modern rate theory and computer simulations. Science 303:186–195

Graf R, Schachman HK (1996) Random circular permutation of genes and expressed polypeptide chains: application of the method to the catalytic chains of aspartate transcarbamoylase. Proc Natl Acad Sci USA 93:11591–11596

Grandori R, Struck K, Giovanielli K, Carey J (1997) A three-step PCR protocol for construction of chimeric proteins. Protein Engin 10:1099–1100

Griswold KE, Kawarasaki Y, Ghoneim N, Benkovic SJ, Iverson BL, Georgiou G (2005) Evolution of highly active enzymes by homology-independent recombination. Proc Natl Acad Sci USA 102:10082–10087

Gurvitz A, Wabnegger L, Langer S, Hamilton B, Ruis H, Hartig A (2001) The tetratricopeptide repeat domains of human, tobacco, and nematode PEX5 proteins are functionally interchangeable with the analogous native domain for peroxisomal import of PTS1-terminated proteins in yeast. Mol Gen Genom 265:276–286

Hallet B, Sherratt DJ, Hayes F (1997) Pentapeptide scanning mutagenesis: Random insertion of a variable five amino acid cassette in a target protein. Nucleic Acids Res 25:1866–1867

Hayes JD, Flanagan JU, Jowsey IR (2005) Glutathione transferases. Annu Rev Pharmacol Toxicol 45:51–88

Heffron F, So M, McCarthy BJ (1978) In vitro mutagenesis of a circular DNA molecule by using synthetic restriction sites. Proc Natl Acad Sci USA 75:6012–6016

Hennecke J, Sebbel P, Glockshuber R (1999) Random circular permutation of DsbA reveals segments that are essential for protein folding and stability. J Mol Biol 286:1197–1215

Hopfner KP, Kopetzki E, Kresse GB, Bode W, Huber R, Engh RA (1998) New enzyme lineages by subdomain shuffling. Proc Natl Acad Sci USA 95:9813–9818

Horswill AR, Naumann TA, Benkovic SJ (2004) Using incremental truncation to create libraries of hybrid enzymes. Methods Enzymol 388:50–60

Horton RM, Hunt HD, Ho SN, Pullen JK, Pease LR (1989) Engineering hybrid genes without the use of restriction enzymes: Gene splicing by overlap extension. Gene 77:61–68

Jeltsch A (1999) Circular permutations in the molecular evolution of DNA methyltransferases. J Mol Evol 49:161–164

Karginov VA, Mamaev SV, An H, Van Cleve MD, Hecht SM, Komatsoulis GA, Abelson JN (1997) Probing the role of an active site aspartic acid in dihydrofolate reductase. J Am Chem Soc 119:8166–8176

Kawarasaki Y, Griswold KE, Stevenson JD, Selzer T, Benkovic SJ, Iverson BL, Georgiou G (2003) Enhanced crossover SCRATCHY: construction and high-throughput screening of a combinatorial library containing multiple non-homologous crossovers. Nucleic Acids Res 31: e126/121–e126/128

Ketterer B (2001) A bird's eye view of the glutathione transferase field. Chem Biol Interact 138:27–42

Kikuchi M, Ohnishi K, Harayama S (1999) Novel family shuffling methods for the in vitro evolution of enzymes. Gene 236:159–167

Koonin EV, Wolf YI, Karev GP (2002) The structure of the protein universe and genome evolution. Nature 420:218–223

Lee S-G, Lutz S, Benkovic SJ (2003) On the structural and functional modularity of glycinamide ribonucleotide formyltransferases. Protein Sci 12:2206–2214

Luckow B, Renkawitz R, Schuetz G (1987) A new method for constructing linker scanning mutants. Nucleic Acids Res 15:417–429

Lutz S, Benkovic SJ (2000) Homology-independent protein engineering. Curr Opin Biotechnol 11:319–324

Lutz S, Benkovic SJ (2002) Engineering protein evolution. Direct Mol Evol Proteins 177–213

Lutz S, Ostermeier M, Benkovic SJ (2001a) Rapid generation of incremental truncation libraries for protein engineering using a-phosphothioate nucleotides. Nucleic Acids Res 29: e16/11–e16/17

Lutz S, Ostermeier M, Moore GL, Maranas CD, Benkovic SJ (2001b) Creating multiple-crossover DNA libraries independent of sequence identity. Proc Natl Acad Sci USA 98:11248–11253

Mani M, Kandavelou K, Dy FJ, Durai S, Chandrasegaran S (2005) Design, engineering, and characterization of zinc finger nucleases. Biochem Biophys Res Commun 335:447–457

Manoil C, Bailey J (1997) A simple screen for permissive sites in proteins: analysis of *Escherichia coli* lac permease. J Mol Biol 267:250–263

Martin JL, McMillan FM (2002) SAM (dependent) I AM: the S-adenosylmethionine-dependent methyltransferase fold. Curr Opin Struct Biol 12:783–793

Mas MT, Chen CY, Hitzeman RA, Riggs AD (1986) Active human-yeast chimeric phosphoglycerate kinases engineered by domain interchange. Science 233:788–790

Mei H-C, Liaw Y-C, Li Y-C, Wang D-C, Takagi H, Tsai Y-C (1998) Engineering subtilisin YaB: restriction of substrate specificity by the substitution of Gly124 and Gly151 with Ala. Protein Eng 11:109–117

Mittermaier A, Kay LE (2006) New tools provide new insights in NMR studies of protein dynamics. Science 312:224–228

Nixon AE, Ostermeier M, Benkovic SJ (1998) Hybrid enzymes: manipulating enzyme design. Trends Biotechnol 16:258–264

Ostermeier M, Benkovic SJ (2001) Construction of hybrid gene libraries involving the circular permutation of DNA. Biotechnol Lett 23:303–310

Ostermeier M, Nixon AE, Benkovic SJ (1999a) Incremental truncation as a strategy in the engineering of novel biocatalysts. Bioorg Med Chem 7:2139–2144

Ostermeier M, Nixon AE, Shim JH, Benkovic SJ (1999b) Combinatorial protein engineering by incremental truncation. Proc Natl Acad Sci USA 96:3562–3567

Ostermeier M, Shim JH, Benkovic SJ (1999c) A combinatorial approach to hybrid enzymes independent of DNA homology. Nat Biotechnol 17:1205–1209

Park HS, Nam SH, Lee JK, Yoon CN, Mannervik B, Benkovic SJ, Kim HS (2006) Design and evolution of new catalytic activity with an existing protein scaffold. Science 311:535–538

Peisajovich SG, Rockah L, Tawfik DS (2006) Evolution of new protein topologies through multistep gene rearrangements. Nat Genet 38:168–174

Ponting CP, Russell RB (1995) Swaposins: circular permutations within genes encoding saposin homologs. Trends Biochem Sci 20:179–180

Rajagopalan PT, Lutz S, Benkovic SJ (2002) Coupling interactions of distal residues enhance dihydrofolate reductase catalysis: mutational effects on hydride transfer rates. Biochemistry 41:12618–12628

Rod TH, Brooks CL (2003) How dihydrofolate reductase facilitates protonation of dihydrofolate. J Am Chem Soc 125:8718–8719

Saraf MC, Horswill AR, Benkovic SJ, Maranas CD (2004) FamClash: a method for ranking the activity of engineered enzymes. Proc Natl Acad Sci USA 101:4142–4147

Saraf MC, Moore GL, Goodey NM, Cao VY, Benkovic Stephen J, Maranas CD (2006) IPRO: an iterative computational protein library redesign and optimization procedure. Biophys J 90:1–14

Schneider A, Stachelhaus T, Marahiel MA (1998) Targeted alteration of the substrate specificity of peptide synthetases by rational module swapping. Mol Gen Genet 257:308–318

Shao Z, Zhao H, Giver L, Arnold FH (1998) Random-priming in vitro recombination: an effective tool for directed evolution. Nucleic Acids Res 26:681–683

Short JM (1997) Recombinant approaches for accessing biodiversity. Nat Biotechnol 15: 1322–1323

Sieber V, Martinez CA, Arnold FH (2001) Libraries of hybrid proteins from distantly related sequences. Nat Biotechnol 19:456–460

Smith J, Berg JM, Chandrasegaran S (1999) A detailed study of the substrate specificity of a chimeric restriction enzyme. Nucleic Acids Res 27:674–681

Stemmer WP (1994a) DNA shuffling by random fragmentation and reassembly: in vitro recombination for molecular evolution. Proc Natl Acad Sci USA 91:10747–10751
Stemmer WP (1994b) Rapid evolution of a protein in vitro by DNA shuffling. Nature 370: 389–391
Stevenson JD, Benkovic SJ (2002) Combinatorial approaches to engineering hybrid enzymes. J Chem Soc Perkin Trans 2:1483–1493
Stevenson JD, Lutz S, Benkovic SJ (2001) Retracing enzyme evolution in the $(\beta\alpha)_8$-barrel scaffold. Angew Chem Int Ed Engl 40:1854–1856
Szomolanyi E, Kiss A, Venetianer P (1980) Cloning the modification methylase gene of Bacillus sphaericus R in *Escherichia coli*. Gene 10:219–225
Volkov AA, Shao Z, Arnold FH (2000) Random chimeragenesis by heteroduplex recombination. Methods Enzymol 328:456–463
Wang Z, Fast W, Benkovic SJ (1999) On the mechanism of the metallo-beta-lactamase from Bacteroides fragilis. Biochemistry 38:10013–10023
Wells JA, Vasser M, Powers DB (1985) Cassette mutagenesis: an efficient method for generation of multiple mutations at defined sites. Gene 34:315–323
Wu R, Ruben G, Siegel B, Jay E, Spielman P, Tu CP (1976) Synchronous digestion of SV40 DNA by exonuclease III. Biochemistry 15:734–740
Yanchak MP, Taylor RA, Crowder MW (2000) Mutational analysis of metallo-beta-lactamase CcrA from Bacteroides fragilis. Biochemistry 39:11330–11339
Zhao H, Giver L, Shao Z, Affholter JA, Arnold FH (1998) Molecular evolution by staggered extension process (StEP) in vitro recombination. Nat Biotechnol 16:258–261

Chemical Protein Engineering: Synthetic and Semisynthetic Peptides and Proteins

L. Merkel, L. Moroder, and N. Budisa(✉)

Contents

Abstract Chemical engineering of proteins provides a pool of various synthetic and semisynthetic methods. These involve techniques for the design of peptides and proteins equipped with chemical handles at defined positions in the target structures. Using water-compatible bioorthogonal chemical ligations gained immense importance in chemical and cellular biology for regioselective addressing to target structures. Beside classical protein modifications, we present examples not only of expressed protein modifications, but also of the attachment of small molecules by native and expressed chemical ligation, modified Staudinger ligation, and of the copper(I)-catalyzed Huisgen [3+2] cycloaddition of azides and alkynes. Recently, in addition to enzymatic methods, the use of organometallic chemistry, i.e. regioselective palladium-catalyzed C-C coupling (Sonogashira, Suzuki, and Mizoroki-Heck reactions), became available as a tool for tailored protein modifications. This chapter reviews all traditional and newly developed chemoselective modifications and ligation methods in the field.

N. Budisa
Max Planck Institute of Biochemistry, Am Klopferspitz 18, 82152 Martinsried, Germany,
e-mail: budisa@biochem.mpg.de

C. Köhrer and U.L. RajBhandary (eds.) *Protein Engineering*.
Nucleic Acids and Molecular Biology 22,

1 Introduction

Since the proteins participate in one way or another in all chemical processes in the living organisms, one may expect highly significant information for biological chemistry from the elucidation of their structure and their transformations. It is therefore no surprise that the study of these substances, from which chemists have largely withdrawn for more than a generation, because they found more worthwhile work in the development of synthetic methods or in the study of simpler natural compounds, has been cultivated by the physiologists in ever-increasing degree and with unmistakable success. Nevertheless, there was never any doubt that organic chemistry, whose cradle stood at the proteins, would eventually return to them. Whereas cautious colleagues fear that a rational study of this class of substances will encounter insuperable difficulties, because of their highly inconvenient physical properties, other optimistically inclined observers, among whom I number myself, believe that one should at least attempt to besiege the virgin fortress with all the present-day resources; since only through daring can the limits of the potentialities of our methods be determined (Fischer 1906).

Emil Fischer, the pioneer and founder of peptide chemistry, anticipated as long ago as 1906 that advances in both peptide and protein research would require the combined efforts of organic chemistry and biology (Fischer 1906). This cautiously optimistic prediction about the synthetic accessibility of peptides and proteins has been realized most successfully in the last few decades with the large and solid knowledge in chemical synthesis and semisynthesis of proteins accumulated by chemists and biochemists over the last century. The conceptual and methodological advances have provided many refined strategies for essentially routine synthesis in solution and on solid supports of small to medium-sized polypeptides. However, for peptides containing numerous sensitive amino acids or particular sequences and side chain modifications it is still neither a routine nor a trivial matter. Similarly, the synthesis of proteins in stepwise solid-phase mode or by optimized segment condensation strategies in solution and on resin, or by convergent techniques still represents a challenging task. However, with the advent of new synthetic methods such as the chemoselective ligation of synthetic or bioexpressed protein fragments, and with the fast developments for manipulation of the genetic code new ingenious and most efficient techniques have been devised for synthesizing proteins with native or tailored new functions. The intent of this chapter is to provide insights into emerging new synthetic tools available for protein/peptide functionalization mainly by using ligation methods and to a lesser extent by postsynthetic side chain modifications.

2 Synthesis of Peptides and Proteins

2.1 Synthesis in Solution

Two lectures at the 74th meeting of the Society of German Scientists and Physicians in Karlsbad in 1902 mark the emergence of the Fischer–Hofmeister theory on protein structure (Fischer 1902; Hofmeister 1902). The backbone structure consists of repeating peptide bond units and thus proteins are made up of α-amino acids that are linked from head to tail by amide bonds. Since then a plethora of chemical methods for peptide bond formation have been developed, among which only a restricted number fulfill all the basic requirements for efficient peptide synthesis, such as fast and quantitative coupling reactions that proceed without enantiomerization (generally termed "racemization") or other side reactions. In parallel to the chemistry of peptide bond formation, efficient orthogonal protection strategies have been devised for the synthesis both in solution and on resin of peptides and proteins, and the present state of the art in the field has been extensively reviewed in monographs (Benoiton 2006; Bodanszky 1993; Kates and Albericio 2000; Lloyd-Williams et al. 1997) and most comprehensively in a new Houben–Weyl treatise on the synthesis of peptides and peptidomimetics (Goodman et al. 2002).

Synthesis of peptides in solution was pioneered by du Vigneaud (1953) with the synthesis of oxytocin in 1953. The stepwise approach applied in the synthesis of larger peptides was soon replaced by fragment condensation procedures, as the major difficulty encountered in the synthesis of larger peptides was the poor solubility of the growing fully protected polypeptide chains in most of the organic solvents. An assembly of smaller fully protected fragments offered an efficient bypass to this problem, leading to the successful synthesis of longer polypeptide chains (Wünsch 1971). This synthetic strategy combined with optimized coupling reagents and protecting groups finally enabled the synthesis of ribonuclease A in solution (Yajima and Fujii 1981). As an alternative approach to the solubility problem, the strategy of minimum protection was developed, which, combined with the azide coupling procedure, led to the total synthesis of ribonuclease S by Hirschmann and associates (Denkewalter et al. 1969; Hirschmann et al. 1969; Veber et al. 1969). Other attempts to synthesize ribonuclease T1 (Storey et al. 1972; Yanaihara et al. 1969) and iso-1-cytochrome *c* (Moroder et al. 1972, 1975) failed because of solubility problems and particularly because of the low coupling efficiency of the azide method, which at that time was the only procedure compatible with this synthetic strategy. The concept of using minimally protected segments for the synthesis of proteins was further developed and improved by employing more efficient coupling reactions (Aimoto 1999; Aimoto et al. 1989; Blake 1981). Further progress of this strategy finally led to the most recent procedures based on chemoselective ligation of fully unprotected fragments into target polypeptide chains (see Sect. 2.5.2).

2.2 Synthesis on Solid Supports

The most innovative discovery in peptide chemistry is the ingenious development of *solid-phase peptide synthesis* by Merrifield (1963), which soon after allowed him to report the total synthesis of ribonuclease A (Gutte and Merrifield 1969). By this synthetic procedure the whole polypeptide chain is assembled in stepwise manner on a solid support to which the C-terminal amino acid residue is anchored covalently by a scissile bond that is cleaved concurrently with full deprotection of the target polypeptide chain or under selective conditions to produce fully protected peptide fragments except for the C-terminal carboxy group. Because of the insolubility of the peptide–resin conjugate, this technique allows at each synthetic step of the repetitive couplings of suitably protected amino acid residues with subsequent selective N^{α}-deprotection and neutralization, and exhaustive washings to remove excess reagents and soluble coproducts. Thereby most of the new advances in the chemistry of peptide synthesis in solution could be transferred to the new technique and were then further optimized in terms of protection strategies and coupling efficiencies to reach performances suitable for automatization. Indeed, nowadays standard synthesis of peptides is generally carried out using automated and commercially available instruments that perform most of the synthetic steps under the control of computers.

Despite these continuous improvements even the solid-phase peptide synthesis suffers from serious limitations related to the chemistry applied. In fact, by assuming that an n-membered peptide is generated with a particular coupling efficiency (r) per cycle, one can derive the percentage of correct molecules in the final product using the expression $(r/n)^{n-1} \times 100\%$. Correspondingly, for $r = 98\%$ the proportion of correct molecules in the synthesis of a 100-membered polypeptide is maximally 13%, whereas for a 300-membered chain it is reduced to 0.24%. Furthermore, if the coupling efficiency is lowered to 95%, the percentages of theoretically expected correct molecules are dramatically reduced to 0.62 and 0.00002%, respectively (Offord 1987). In addition to the problem of incomplete couplings, there are other serious shortcomings of the solid-phase peptide synthesis method which derive from loss of side chain protections and partial cleavage of the peptide from the resin during chain assembly as well as various side reactions in the chain elongation and final deprotection/resin-cleavage steps. In addition, sequence-dependent poor coupling yields are often encountered. These can partly be bypassed by the use of newly discovered coupling procedures such as the urethane-protected carboxyanhydrides (Fuller et al. 1990; Fuller and Yalamoori 2002), the coupling agent *N,N,N′,N′*-tetramethyl-*O*-(7-azabenzotriazol-1-yl)uronium hexafluorophosphate (HATU) (Bienert et al. 2002; Carpino 1993), or acyl fluorides (Beyermann et al. 2002; Carpino and Mansour 1992), difficulties arising from aggregation phenomena even on solid supports were at least partially circumvented by applying pseudoproline synthons (Haack and Mutter 1992; Mutter et al. 1995) or temporary depsipeptide structures (Sohma et al. 2004; Akaji et al. 1999; Carpino et al. 2004; Mutter et al. 1995, 2004) whenever possible. Despite these continuous new improvements, synthesis of peptides with lengths over 50 residues can by no means be classified as routine work. The successful accomplishment of such syntheses may become even more difficult with multiple-cysteine-containing peptides, where the production of correct disulfide connectivities by regioselective disulfide pairing procedures or by oxidative refolding often represents an additional serious challenge (Akaji and Kiso 2003; Annis et al. 1997; Kimura 2003; Moroder et al. 1996, 2005).

2.3 *Fragment Condensation Strategies*

Although various successful syntheses of smaller proteins by the stepwise chain elongation procedure on a solid support have been reported (Dawson and Kent

2000) since the pioneering synthesis of ribonuclease A by Gutte and Merrifield (1969), a shift from this linear to a convergent strategy relying on the assembly of peptide segments, in analogy to the fragment condensation in solution, has significantly advanced the access to smaller proteins (Lloyd-Williams et al. 1993). More appropriate, however, proved to be the assembly of the protected fragments in solution, i.e., by changing the phase. By this approach, various proteins were prepared in a homogeneous form (Sakakibara 1999). The crown of these efforts is certainly the total chemical synthesis of the 238-membered green fluorescent protein by Sakakibara's group (Nishiuchi et al. 1998). As shown in Fig. 1 fully protected fragments consisting of ten to 12 residues were synthesized on resin and purified to a degree of homogeneity suitable for their condensation in sequence order. The main drawback of these synthetic efforts was the low yield of the correctly folded protein, which additionally implies the spontaneous autocatalytic chromophore formation (Fig. 1).

Currently, there are still major shortcomings of this type of fragment condensation strategy which prevent a more general application for the synthesis of proteins. Fully protected intermediates with lengths of over 60 amino acid residues are usually poorly soluble or insoluble, especially when prone to β-sheet formation. Therefore, the

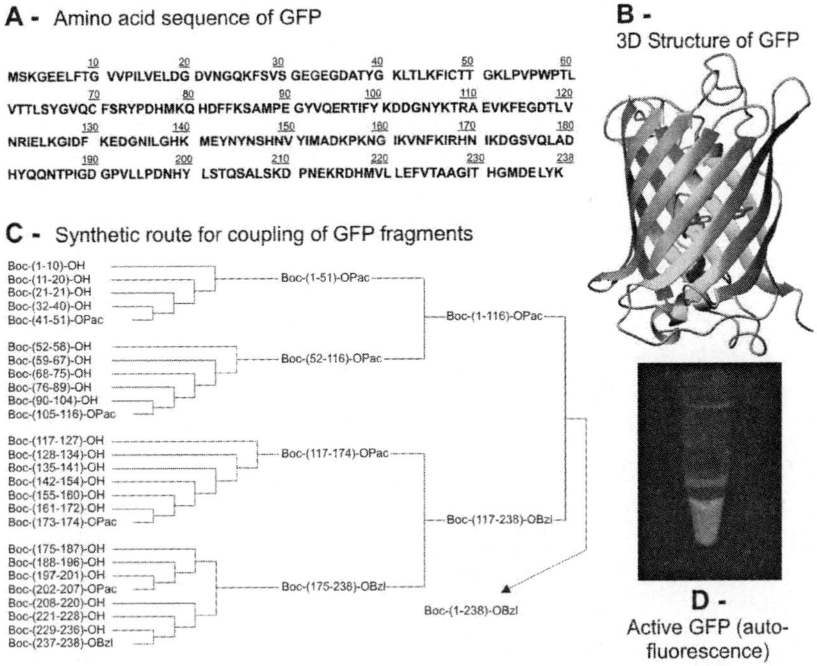

Fig. 1 Total chemical synthesis of the green fluorescent protein (*GFP*) by the *tert*-butyloxycarbonyl protection strategy in combination with OPac esters for the C-terminal carboxy groups. (**a**) Primary and (**b**) tertiary structure of GFP, and (**c**) scheme of fragment condensations. Upon folding of the fully deprotected polypeptide chain, only a small portion underwent spontaneous chromophore formation (**d**; Nishiuchi et al. 1998; Sakakibara 1999)

biggest contribution toward making the present methodology more widely applicable would be the finding of some powerful solvent systems suitable for dissolving the fully protected segments and reaction conditions for coupling segments effectively in such solvents (Sakakibara 1999).

Moreover, the homogeneity of the protected fragments is often not sufficient to allow efficient coupling reactions. Finally, there is always the risk of low yields deriving from improperly folded synthetic proteins. However, this problem is encountered independently whether the polypeptide chain is obtained by chemical synthesis or recombinant technology.

2.4 Chemoselective Ligation

The first attempts to assemble proteins by alternative procedures to the total chemical synthesis involved a combination of a synthetic with a biosynthetic portion by either noncovalent or covalent bonds. The biosynthetic protein fragment can be obtained by chemical or enzymatic fragmentation of natural proteins or by recombinant techniques. The noncovalent assembly implies tight complex formation between two protein fragments with the biological or enzymatic activities being fully retained (for a review see Offord 1990). Conversely, for the covalent assembly, the synthetic portion is designed with all the desired modifications, including the chemical functionality required for chemoselective ligation. The respective functionalization of the biosynthetic fragment is restricted to a few available reactions that can be performed on unprotected polypeptides. However, chemoselective ligation is often performed using only suitably modified unprotected synthetic peptides. The regio- and chemoselectivity is achieved with pairs of functional groups of complementary reactivity such as hydrazides or aminooxy groups, which readily react with aldehydes to form stable hydrazones or oxime bonds, while a thiol and a bromoacetyl group lead to thioester bonds. The advantage of these peptide bond replacements lies in the high and specific reactivity between two functional groups not found among the canonical amino acids (Offord 1991). However, the bioactivities of the protein constructs have to be retained despite the unnatural bond replacing the native peptide bond.

2.4.1 Hydrazone and Oxime Chemistries

Offord and Rose have pioneered the introduction of hydrazone and oxime bonds for regiospecific chemoselective ligation of unprotected fragments of proteins to generate macromolecular adducts as well as for conjugation of a variety of molecular probes to proteins (Gaertner et al. 1992; Gaertner and Offord 2003; Rose 1994). Because of the requirement of an aldehyde, the periodate oxidation of N-terminal serine or threonine residues of proteins was used to generate this functionality, while the hydrazide or aminooxy group is incorporated into the synthetic fragment.

These reactions can be performed in the presence of unprotected lysine residues in slightly acidic aqueous solution as the formation of the corresponding Schiff base is reversible, with the carbonyl form favored under such conditions. Conversely, aldehydes

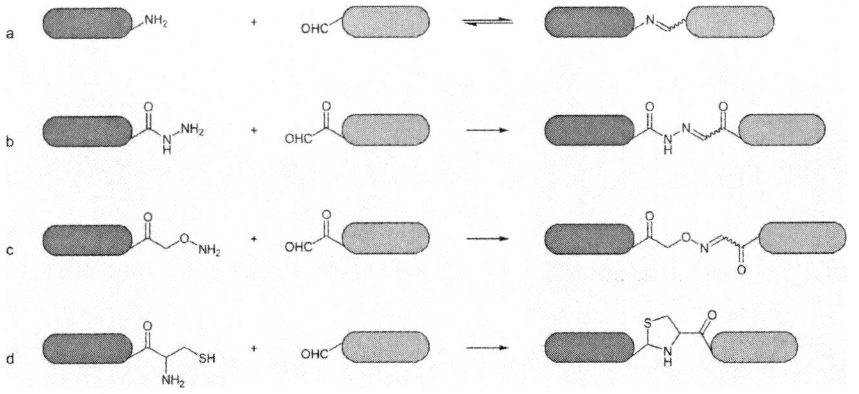

Fig. 2 Reaction of aldehydes with primary amines (**a**), hydrazide- and aminooxy-containing compounds (**b, c**), and cysteine (**d**; pseudoproline ligation)

react almost irreversibly with hydrazide- and aminooxy-containing compounds, yielding stable hydrazones and oximes under the same physiological conditions (Fig. 2).

2.4.2 Pseudoproline Chemistry

To achieve efficient condensation of relatively large peptide segments by bimolecular reactions, one can exploit a transient covalent linkage to bring two components together through a specific reaction of high efficiency (proximity principle; see Sect. 2.5.1). This principle is exploited in the pseudoproline ligation procedure (Liu and Tam 1994), where in a first reaction an aldehyde group in form of a glycoaldehyde ester at the C-terminus of one fragment forms a Schiff base with the amino group of the second fragment (Fig. 2). The Schiff base is converted into a five-membered ring involving the β-hydroxy of serine and threonine or the β-thiol function of cysteine with subsequent O→N or S→N acyl shift and generation of a pseudoproline residue at the ligation site (Fig. 3).

2.4.3 Thioester Bond Formation

A useful and highly selective chemical ligation can be achieved by reaction of a C-terminal carbothioate group with *N*-bromoacetyl peptides at pH < 4 to form a thioester bond (Schnölzer and Kent 1992). The chemoselectivity of this reaction derives from the strongly nucleophilic thioacid group with its pK_a of about 3. At this pH, no other groups of an unprotected peptide are sufficiently reactive to compete for the bromoacetyl group; in particular the cysteine thiol group has a pK_a of about 9. With the synthesis of a variety of protein targets, it has been demonstrated that the thioester ligation is chemoselective for all functional groups of the canonical amino acids (Baca et al. 1995). By this procedure a fully active HIV-protease analog was synthesized with a pseudo-Gly-Gly sequence at the ligation site (Fig. 4). However, even other pseudo-Xaa-Gly (Xaa is any amino acid) sequences can be mimicked by the thioester ligation using corresponding thioacids.

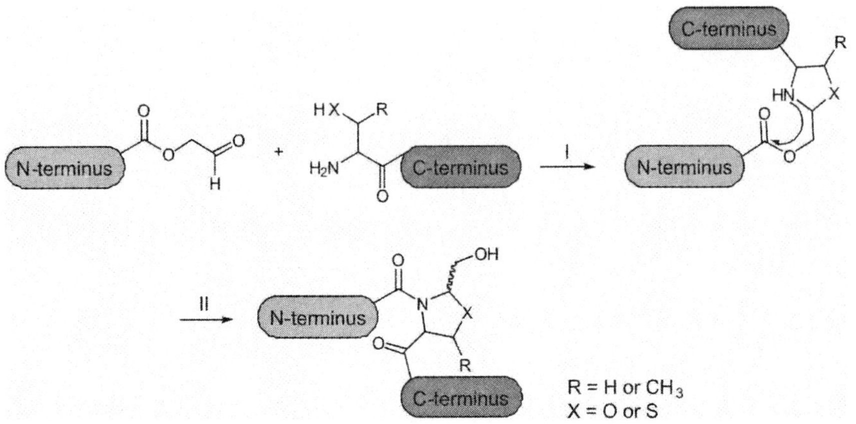

Fig. 3 Pseudoproline ligation via imine capture. A peptide containing a C-terminal glycoaldehyde ester reacts with a second peptide carrying cysteine, serine, or threonine as an N-terminal residue to form a pseudoproline analog (thiazolidine or oxazolidine)

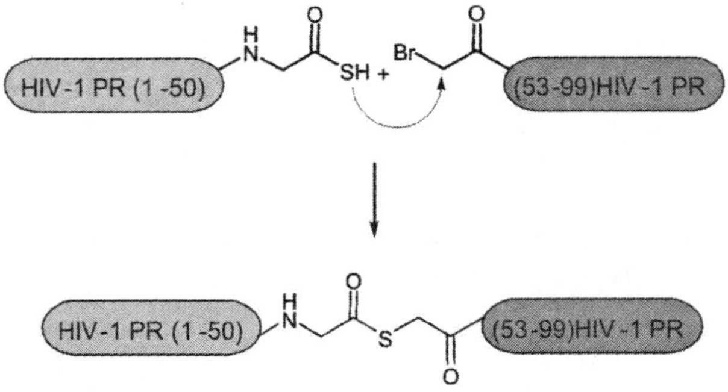

Fig. 4 Backbone engineering of HIV-protease via peptide bond replacement by a thioester bond (Schnölzer and Kent 1992)

2.5 Native Chemical Ligation

2.5.1 S→N Acyl Shift and the Proximity Rule

Wieland et al. (1953) showed in the early 1950s that *S*-aminoacyl cysteine derivatives undergo an intramolecular transacylation to form N^{α}-aminoacyl cysteine compounds via a five-membered transition state (Fig. 5). This reaction proceeds even under mild acidic conditions and, as expected, the rate of conversion increases with higher pH values. The potential of this reaction for the synthesis of peptides was obvious and indeed H-Val-Cys-OH was obtained in good yields from H-Val-SPh and H-Cys-OH. With this type of reaction, after the capture of the two components the peptide bond formation becomes a first-order reaction involving an intramolecular acyl transfer.

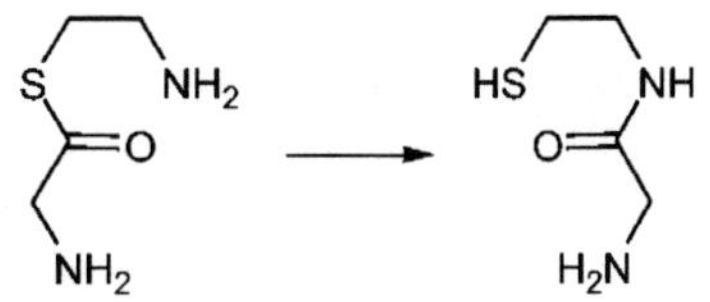

Fig. 5 Spontaneous intramolecular rearrangement of *S*-aminoacyl cysteamine via an S→N acyl shift (Wieland et al. 1953)

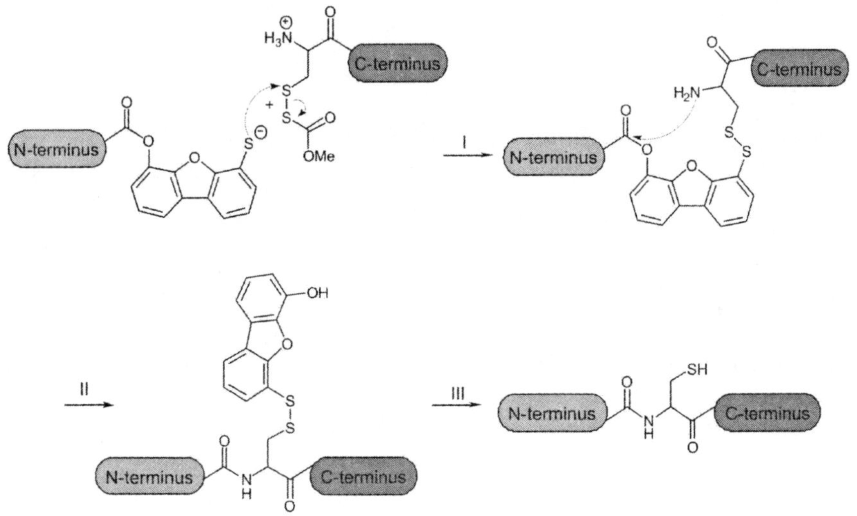

Fig. 6 Thiol capture ligation according to Kemp (1981)

These experiments led Brenner et al. (1957) to coin the concept of entropic activation, and to address the significance in the context of peptide bond formation via the proximity principle which leads to increased effective local concentration of the amine nucleophile and carboxyl electrophile if these two groups are brought into close proximity. This concept was later realized by Kemp (1981) by the "prior thiol capture strategy" making use of a mercaptodibenzofuranyl ester as a template to ligate chemoselectively two peptide fragments (Kemp and Carey 1993; Fig. 6). In the first step, a disulfide is formed via thiol–disulfide exchange between the activated disulfide of the cysteinyl peptide and the mercaptodibenzofuranyl ester of the N-terminal peptide in a bimolecular reaction. For this type of chemoselective reaction, protection of the peptide fragments is not required. After the first step, aminolysis of the ester occurs as an intramolecular reaction at high rates leading to the peptide bond formation. Final reduction of the disulfide releases the coupled peptide and the template molecule.

2.5.2 Native Chemical Ligation of Unprotected Peptides

The preliminary work of Wieland and Kemp was further developed by Tam by the use of glycoaldehyde ester (Liu and Tam 1994), and then almost simultaneously by

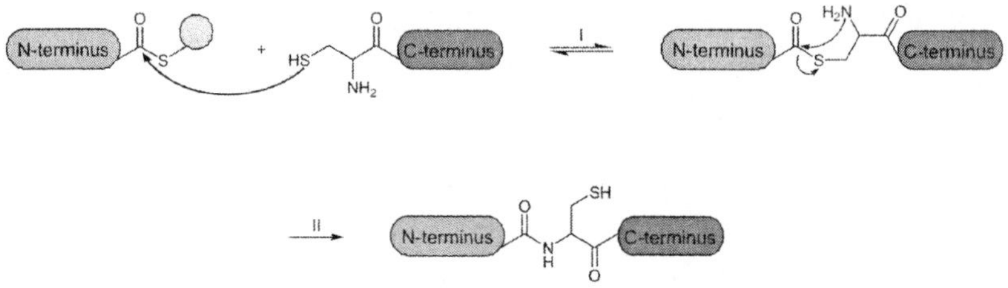

Fig. 7 Reaction mechanism of native chemical ligation

Kent (Dawson et al. 1994) and Tam (Tam et al. 1995) for the development of a highly chemoselective ligation of unprotected peptide fragments via peptide amide bonds to produce larger polypeptide chains and proteins. This strategy, termed "native chemical ligation" (NCL), exploits the reversible intermolecular transesterification between a peptide thioester and the nucleophilic thiol group of an *N*-cysteinyl peptide to produce the *S*-(peptidyl)-cysteinyl peptide as an intermediate (Fig. 7). In the absence of structurally favored proximity effects or of the assistance by functionalities in a correct architecture as present in the inteins for protein splicing (see Sect. 2.6), hydroxy groups and amines present in the unprotected peptide segments are too weak as nucleophiles to compete under neutral conditions with the transesterification by the thiol groups. The thioester itself is generally relatively unreactive to aminolysis; but the *S*-(peptidyl)-cysteinyl peptide formed as an intermediate in the first step rapidly undergoes the S→N acyl shift, through the favorable five-membered-ring intermediate, to produce the native peptide bond at the ligation site in an irreversible reaction step (Dawson and Kent 2000; Kent 2003).

Closely related to NCL above described is the acyl-initiated capture (Tam et al. 1995). In this case, a peptide with a C-terminal thioacid reacts with a second peptide bearing β-bromoalanine as an N-terminal residue. The product of this capture reaction is again the thioester that undergoes the S→N acyl shift to generate the native bond and a cysteine residue (Fig. 8). The limitation of this procedure is the use of bromoalanine, which can readily form aziridine or undergo β-elimination.

2.5.3 Potentials and Limitations of Native Chemical Ligation

The yields reported for the synthesis of proteins by NCL fully confirm the high potential of this procedure as an alternative or complementary approach to protein engineering methods based on ribosome-mediated protein expression. The method of choice will therefore be ultimately dictated by the envisaged practical applications (Dawson and Kent 2000). On the other hand, the synthesis of proteins by NCL currently has no rival in the experimental design of proteins, which includes sequential isotopic labeling, preparation of circular proteins, and insertion of non-native peptide fragments or nonpeptidic molecules at predefined sites to name just a few examples (Nilsson et al. 2005). The basic drawback of NCL is the obligatory requirement of a cysteine as the N-terminal residue in one ligating fragment.

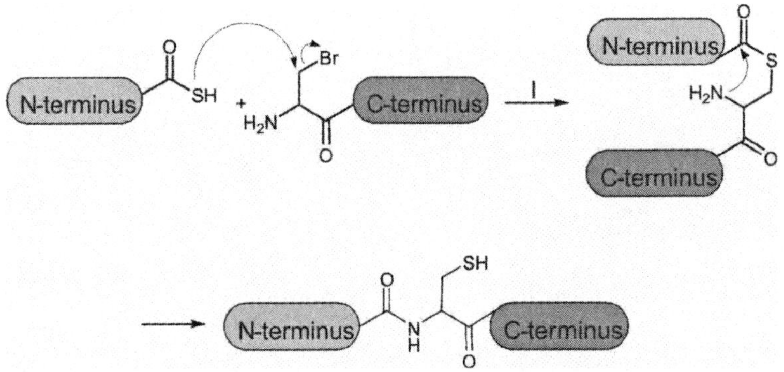

Fig. 8 Acyl-initiated capture for the thioester formation as the first step in native chemical ligation (Tam et al. 1995)

Additionally, it is important to prevent oxidation of the cysteine thiol group to the disulfide dimer (which is inactive in ligation). This is readily achieved by operating under reducing conditions. Denaturants such as guanidine hydrochloride or urea do not interfere with the ligation reaction and are generally added to enhance the solubility of the peptide segments, and thus their concentration. The reaction itself proceeds very fast, but reaction rates and yields do depend strongly on the amino acid residue at the thioester site (Dawson and Kent 2000).

2.5.4 Cysteine Mimetics for Native Chemical Ligation

In addition to cysteine, other nucleophiles such as serine, threonine, and even histidine and tryptophan which contain hydroxy and amine groups on their side chains in proximity of the α-amino group can act as nucleophilic partners in the capture reaction, although with lower efficiency (Tam et al. 2001). More efficient are cysteine mimetics such as homocysteine and selenocysteine (Sec). The homocysteine inserted synthetically as a replacement for methionine residues allows for regeneration of this residue after ligation by methylation with methyl *p*-nitrobenzenesulfonate (Tam and Yu 1998). On the other hand, a selenolate is more nucleophilic than a thiolate and the pK_a of the selenol group (5.2–5.7) is lower than that of a thiol (8.5). Correspondingly, the transesterification reaction should occur under acidic conditions at higher rates with peptide fragments containing a Sec residue as the N-terminal nucleophile compared with the cysteinyl peptides (Fig. 9). Indeed, at pH 5.0, the reactions with Sec are 10^3-fold faster than with cysteine (Gieselman et al. 2002). Moreover, ligation with N-Sec fragments could serve as a straightforward approach to generate Sec-containing proteins (Quaderer 2001). The fast and mild oxidative elimination of selenides to dehydroalanine could generate regioselectively a reactive handle for other chemoselective transformations (Gieselman et al. 2002).

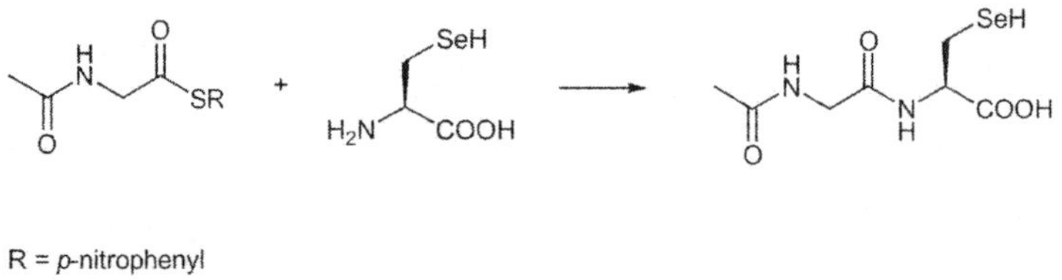

Fig. 9 Native chemical ligation with selenocysteine (Gieselman et al. 2002)

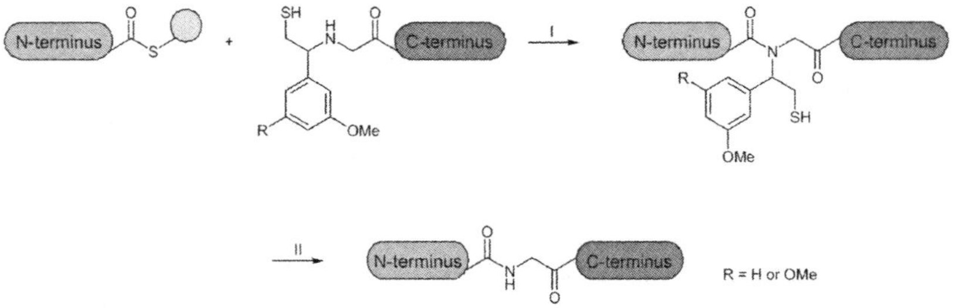

Fig. 10 An example of native chemical ligation with 1-phenyl-2-mercaptoethyl derivatives as auxiliaries (Botti et al. 2001). These auxiliaries are readily removed with trifluoroacetic acid after completion of the ligation reaction

2.5.5 Native Chemical Ligation with Removable Auxiliaries

In the absence of cysteine residues in suitable positions of the target peptide or protein, this residue can be introduced purposely for the ligation reaction as it can readily be transformed after completion of the reaction into amino acids with or without functionalities by alkylation reactions. For example, cysteine residues were introduced into synthetic protein fragments and afterwards alkylated with bromoacetic acid or bromoacetamide to generate the noncoded amino acid residues pseudoglutamate or pseudoglutamine, which are electronically and sterically similar to their natural counterparts glutamate/glutamine (Kochendoerfer et al. 2003; Clayton et al. 2004). Cysteine residues can also be converted to alanine by desulfuration with palladium or Raney nickel (Yan and Dawson 2001).

An alternative approach to bypass the absence of cysteines in the sequence of target proteins is the use of glycines N^{α}-substituted with oxyethanethiol, which can be removed by zinc to regenerate the glycine residue (Canne et al. 1996). The use of 2-mercaptobenzyl groups and 2-mercaptoethyl derivatives (Fig. 10) as second-generation auxiliaries provided an important advancement in the field (Botti et al. 2001; Kawakami et al. 2001). For additional auxiliaries developed over recent years the reader is directed to the recent comprehensive review of Nilsson et al. (2005).

2.5.6 Sequential Peptide Ligation

The synthesis of peptides and particularly proteins by Native chemical ligation NCL would significantly benefit from ligation of a larger number of smaller fragments more readily accessible in homogeneous form by synthetic means. These fragments should contain N-terminally a temporary protected cysteine residue and C-terminally the required thioester to allow their assembly into the target protein sequence in the C→N direction. Photolabile groups were proposed as readily removable cysteine protection (Hennard and Tam 1997). However, the most successful protection proved to be the conversion of cysteine into 1,3-thiazolidine-2-carboxylic acid, from which the unprotected cysteine is readily recovered by treatment with *O*-methylhydroxylamine (Thamm et al. 2002; Villain et al. 2001). An alternative N→C ligation of suitable fragments is feasible by kinetic control using the less reactive thioalkyl esters for intermediate C-terminal protection of the N^{α}-cysteinyl peptide while performing the transesterification reaction with the more reactive thioaryl ester of the N-terminal peptide fragment. A convergent approach based on the use of both ligation strategies, as exemplarily applied for the synthesis of crambin (Bang et al. 2006), should facilitate synthetic access to even larger proteins.

A theoretically interesting, but from a practical point of view probably less efficient sequential ligation relies on the use of a peptide mercaptoethyl ester which acts as the nucleophile for transesterification with the peptide thioester. Subsequent aminolysis of the thioester regenerates the mercaptoethyl ester in the two ligated peptides at the C-terminus for additional sequential ligations. However, the intramolecular aminolysis of the thioester, which in this case is not favored by the proximity principle, i.e., by an S→N acyl shift via a five-membered intermediate, proceeds at low rates under mild acidic conditions. The rate can be enhanced by addition of Ag^+ ions (Lu and Tam 2005).

2.6 Protein Splicing and Expressed Protein Ligation

Pure chemical ligation strategies are nowadays supplemented with or even replaced by biochemical approaches that involve molecular biology techniques that emerged from a burgeoning revolution in proteomics, which is fuelling the need for proteins with tailored modifications. In addition to the use of expressed protein fragments containing an N-terminal cysteine residue, mechanisms of autocatalytic post-translational modifications can be exploited to access expressed proteins or protein fragments as derivatives or variants suitable for orthogonal chemical ligations. An autocatalytic processing of expressed proteins that can be exploited for the preparation of expressed protein fragments as thioesters is protein splicing (Kane et al. 1990; Paulus 2000).

In this process an internal intein domain excises itself from a host protein, the extein. Thereby, the intein is split N- and C-terminally and splicing only occurs on reconstitution of the two extein fragments (Gogarten et al. 2002). All classes of inteins are characterized by several conserved sequence motifs containing the critical residues for catalyzing the splicing reaction (Perler 2002). With suitable mutations of the intein segment, the autolytic process can be stopped at the thioester intermediate, which can then be used to cleave the intein variant with excess of suitable mercaptanes to produce the polypeptide thioester as required for chemical ligation with synthetic *N*-cysteinyl peptides. This strategy, termed "expressed protein ligation" (Muir et al. 1998) or "intein mediated ligation" (Evans et al. 1998), represents an efficient extension of the purely synthetic Native chemical ligation NCL as it takes advantage of recombinant DNA technology to generate protein fragments via ribosomal synthesis. By this approach large proteins become accessible for chemoligation (Paulus 2000) and the tools for generation of C-terminal thioester-tagged proteins are now even commercially available (e.g., IMPACT™-system from New England Biolabs). The thioesters accessible by this recombinant technique may also serve for the synthesis of (diphenylphosphinyl) methane thioesters as intermediates in the traceless Staudinger ligation (Nilsson et al. 2003). Conversely, protein fragments that contain an N-terminal cysteine residue can be obtained directly by expression or by enzymatic cleavage of appropriate precursors containing suitable amino acid residues or sequences for selective cleavage at Xaa-Cys bonds (e.g., the endogenous methionyl-peptidases of *Escherichia coli* or proteases such as factor Xa; for more details see the chapter by Imperiali and Vogel Taylor, this volume).

2.7 Amide Bonds Generated by Decarboxylative Condensation

A decarboxylative condensation of α-keto carboxylic acids and *N*-alkylhydroxylamine derivatives may well represent a promising amide-forming reaction for ligation chemistry (Bode et al. 2006). The reaction tolerates several unprotected functional groups and is highly chemoselective. It proceeds in water without catalysts or other reagents, generating water and carbon dioxide as the only by-products (Fig. 11). Application of this reaction for condensation of model peptides has been reported (Carrillo et al. 2006). The great advantage of this ligation chemistry would be the absence of specifically required amino acids at the C- or N-termini of the peptide/protein fragments. Most importantly, epimerization of the ketoacid does not occur during the reaction (Bode et al. 2006).

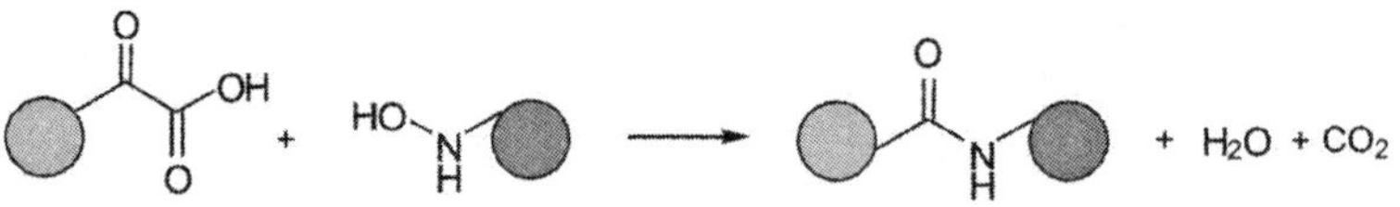

Fig. 11 Decarboxylative condensation of α-ketoacids and *N*-alkylhydroxylamine derivatives (Bode et al. 2006)

2.8 Staudinger Ligation

Staudinger und Meyer (1919) reported that azides react smoothly with triaryl phosphines under mild conditions. This reaction yields almost quantitatively aza-ylides without noticeable formation of side products (Fig. 12).

This classic organic reaction was recognized by Saxon and Bertozzi (2000) and Raines (Nilsson et al. 2000) as a tool for ligation of peptides via amide bond formation. The basic motivation for the application of this chemistry was the limits of the hydrazone ligation methods in vivo as hydrazine probes react readily with cellular ketones, whereas azides and phosphines are expected to represent bioorthogonal functionalities; these groups are abiotic and unreactive toward biomolecules in cells.

As the aza-ylides undergo spontaneous hydrolysis to form primary amines and phosphine oxide in aqueous environments, Saxon and Bertozzi (2000) introduced an ester moiety into the phosphine structure as an electrophilic trap. The ester moiety would then capture the nucleophilic aza-ylide by intramolecular cyclization, forming a stable amide before the competing aza-ylide hydrolysis can occur (Fig. 13). This reaction was successfully used for introduction of molecular probes and nonpeptidic molecules into proteins (Prescher and Bertozzi 2005).

On the basis of these first results, a traceless Staudinger reaction was developed in parallel efforts by the laboratories of Bertozzi (Saxon et al. 2000) and Raines (Nilsson et al. 2000), where the phosphine oxide moiety is released after completion of the reaction, as shown in Fig. 14. The key novelty of this reaction is the use of a phosphinothiol, which allows one in a first step to couple a peptide thioester to

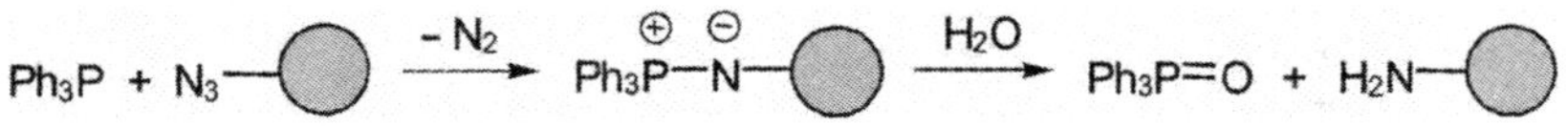

Fig. 12 Reaction of azides with phosphines to yield primary amines (Staudinger and Meyer 1919)

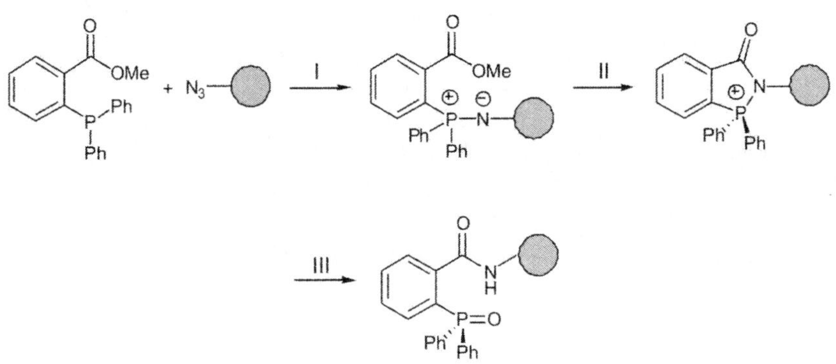

Fig. 13 Mechanism of the modified Staudinger reaction (Saxon and Bertozzi 2000)

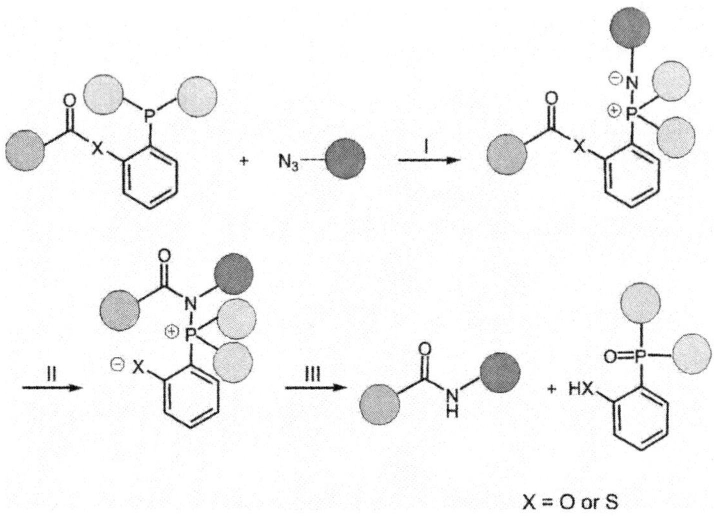

Fig. 14 Traceless Staudinger reaction for peptide ligation (Nilsson et al. 2000)

the phosphine via transthioesterification (Nilsson et al. 2000), while the peptidyl phosphine ester has to be prepared separately (Saxon et al. 2000). Reaction with an α-azidoacyl peptide as a second fragment leads to the aza-ylide, where the highly nucleophilic nitrogen attacks the carbonyl to produce the amidophosphonium salt that hydrolyzes to the amide product and phosphine oxide. To increase the relatively low reaction yields, various phosphines were analyzed and among these ligation with diphenylphosphinomethanethiol was found to be the most suitable reagent. This may derive from the five-membered transition state compared with the six-membered ring in the case of 2-diphenylphosphinylthiophenol or 2-(diphenylphosphinyl)phenol (Saxon et al. 2000).

The efficiency of the traceless Staudinger ligation has been well documented by the successful synthesis of the isotopically labeled ribonuclease A (Nilsson et al. 2003) using an expressed thioester fragment and a synthetic N^{α}-azido peptide. On the other hand, expression of a protein with azidohomoalanine as a replacement for methionine residues allowed for its regioselective modification with phosphines containing suitable molecular probes (Kiick et al. 2002). This latter approach appears highly promising even for in vivo protein modifications, a fact that would represent a great breakthrough in the field.

3 Chemical Modification of Proteins

There are only a few proteins in nature which exhibit a final covalent structure corresponding to the accurate translation of the related messenger RNA. Obviously, nature has a good reason to expand the inventory of the side chains available to functional proteins. Various post-translational covalent modifications enable the regulation of interactions with other proteins and small molecules, and in this way influence, modulate, or change their properties and functions. For example, recent sequencing of the human chromosomes revealed that there are 20,000–25,000 protein-coding genes. On the other hand, it is expected that there are at least 300,000 to millions of functionally active protein variants in living cells, which are the result of existing pathways for molecular variations between a gene and its corresponding active protein product. Not surprisingly, between 1,000 and 2,000 genes, representing more than 5% of the human genome, are believed to encode enzymes dedicated to protein post-translational modifications (Walsh 2006).

It should always be kept in mind that despite the vast diversity of post-translational modifications in nature, there is a rather small number of basic chemical principles behind them. For example, chromophore maturation of green fluorescent protein and protein splicing rely in fact on the same chemical principle – the generation of a linear ester bond.

3.1 Chemical versus Ribosomal Synthesis

Despite the general limitation of the chemical synthesis of proteins to polypeptide chains of about 10 kDa, it allows the most different structures to be incorporated into the polypeptide chain as well as the access to suitable precursors for postsynthetic modifications at will. Conversely, the synthesis of proteins on RNA templates is subject to strict constraints in terms of stereochemistry (L-amino acids) and side chain variability (only 20 residues). Intensive research is therefore addressed to reprogram the protein translation machinery in the context of an expanded genetic code (see the chapters by Beatty and Tirrell, Mascarenhas et al., Köhrer and RajBhandary, Dougherty, Hecht, Hirao et al., and Leconte and Romesberg, this volume).

The ability and capacity of living cells to synthesize functional proteins is superior to all known synthetic methods and approaches (Budisa 2004b). From the point of view of a synthetic chemist, two basic features, i.e., regiospecificity and stereospecificity, clearly demonstrate the superior power and versatility of a ribosome-mediated protein synthesis over the chemical synthesis. Furthermore, the capacity of templated synthesis to generate polypeptides of almost unlimited length with structural homogeneity and precisely defined stereochemical composition is unrivaled (Scheme 1). For these reasons, it is often easier to achieve desired modifications by operating directly on bioexpressed proteins, domains, or subdomains which are already functionally optimized by evolution than by total chemical synthesis. In other words, such natural proteins or related fragments can serve as ready-made intermediates for tailored chemical modifications to produce, e.g., enzymes with altered or new catalytic properties, to alter protein–protein interactions and to dissect structure–function relationships.

Chemical modifications of solvent-accessible reactive side chains have a long history in protein chemistry (Means and Feeney 1990) (for more details see the

CHEMICAL SYNTHESIS OF POLYPEPTIDES	RIBOSOMAL PROTEIN SYNTHESIS
➢ Chain elongation modus: C-terminal ⇒ N-terminal	➢ Chain elongation modus: N-terminal ⇒ C-terminal
➢ Protection/deprotection chemistries obligatory; high reaction yields in each step necessary	➢ Living cells assign ~ 35-45% of the genome for translational machinery
➢ Automated synthesizers; for good coupling efficiency suitable agents are necessary; problem of racemisation	➢ Living cells invest ~ 30-50% of the energy (ATP, GTP) for protein translation
➢ Decrease in coupling with the increase of polypeptide chain (typical length limit: 50 – 100 amino acids)	➢ High fidelity of translation; invariant chemistries (only 20 amino acids!); no limitations in chain size
➢ Expanded amino acid repertoire; no stereochemical restrictions, alternative backbone chemistries	➢ Strict control of regio-, and stereoselectivity (L-amino acids); co-, and post-translational processing
➢ Sequence dependence: synthesis of "difficult" proteins and peptides are still a formidable chemical effort	➢ All known protein/peptide structural types are efficiently synthesized at ribosome

Scheme 1 Basic characteristics of ribosomal and chemical peptide synthesis. Ribosomal polypeptide synthesis proceeds stepwise from the N-terminus to the C-terminus by reading the messenger RNA template in 5′→3′ direction. Peptide bond formation is catalyzed by ribosomal RNA (23S rRNA); sequence fidelity is ensured by codon–anticodon interactions. After the proteins/peptides emerge from the ribosomal biosynthetic assembly line, the polypeptides are usually post-translationally modified to increase the side chain diversity beyond the 20 standard amino acids. Deviations from standard decoding rules allow cotranslational incorporation of special chemical groups (e.g., formylmethionine, selenocysteine, pyrrolysine)

chapter by van Kasteren et al., this volume). Indeed, covalent chemical modifications were the main tool for classical enzymologists to identify residues involved in catalysis or binding, to introduce reporter groups such as spectroscopic probes, or to determine protein topologies by chemical cross-linking experiments (Govardhan 1999). In this technique, reactive functions such as carboxy, imidazole, indole, thiol, amino, and hydroxy groups can be modified with more or less selective monofunctional or bifunctional reagents (Lundblad and Bradshaw 1997). Indeed numerous monographs have addressed these specific aspects of protein chemistry (Hermanson 1996) and commercial kits are nowadays available for a large number of group-specific reactions, such as acylations, amidations, reductive alkylations, lipid coatings, cross-linking, caging, adsorption, gel entrapment, molecular imprinting, and immobilization.

3.2 *Side Chain Modifications*

Orthogonally reactive amino acid side chains are especially useful for regiospecific conjugation of unprotected polypeptide chains with probes of desired properties in aqueous solutions (Narayan et al. 2005). Such reactive side chains that do not occur naturally in proteins and peptides might also be used to restrict their conformational flexibility, to enhance their metabolic stability, and to affect receptor binding affinities as well as to study steric requirements in distinct secondary structure formation. Although peptide synthesis combined with chemical ligation procedures offers the most straightforward access to such functionalized polypeptide chains, recombinant DNA technology can also be used for such purposes in the context of a ribosomal synthesis with an expanded amino acid repertoire (Budisa et al. 1999).

3.2.1 Isosteric Replacements: From “Chemical” to “Atomic” Mutations

The first methods available for altering the functional properties of proteins were purely chemical in nature (Means and Feeney 1990). The basic intention was to change as much as possible the catalytic properties of enzymes by inducing predictable perturbations into the protein structures. This concept assumes that no major changes in the conformational properties of protein molecules take place upon defined chemical transformations. Curiously, the “cysteine–alanine” (“Raney nickel”) experiment, performed in 1962 in the laboratories of Lipmann, von Ehrenstein and Benzer, followed exactly these principles (Chapeville et al. 1962). This was a crucial experiment for the elucidation of the flow of genetic information. In particular, a cysteine attached enzymatically to its cognate transfer RNA ($tRNA^{Cys}$) was transformed into alanine by the reduction with Raney nickel without affecting its covalent attachment to $tRNA^{Cys}$. Such a hybrid or misacylated Ala-$tRNA^{Cys}$ was efficiently translated into the target sequence and the resulting peptide chain was found to contain alanine residues at the positions coded for cysteine. Later, Bender (Polgar and Bender 1966) and Koschland (Neet and Koshland 1966) introduced the concept of "chemical mutation" and paved the way towards site-selective chemical modifications. In this way, they could show how subtle changes in the active-site geometry of enzymes might dramatically affect the original, but generate novel catalytic activities. In particular, by chemical modification of a single amino acid residue in the active site of the serine protease subtilisin (Ser→Cys), an enzyme variant was generated that does not show any peptidase activity but which hydrolyses active esters such as *p*-nitrophenyl acetate (Neet et al. 1968). This selective modification of the active-site serine by reaction with arylsulfonyl fluoride and subsequent displacement with sodium sulfide was subsequently applied by Wu and Hilvert (1989) to transform subtilisin into an artificial selenoenzyme by chemical modification of its catalytic serine into a Sec. The Sec enzyme lost the peptidase activity, but was shown to mimic the antioxidant enzyme glutathione peroxidase by catalyzing the reduction of hydroperoxides with 3-carboxy-4-nitrobenzenethiol (Hilvert 2001). Similarly, also trypsin has been converted chemically into selenotrypsin, leading to a variant devoid of proteolytic activity, which however exhibited good glutathione peroxidase activity (Liu et al. 1998).

These examples clearly demonstrate that tailored functions can be delivered into particular protein scaffolds by generating new reactive centers or, more generally, by direct chemical modifications (Tann et al. 2001). It is now well documented that nature itself inserts Sec into some proteins in the context of alternative reading or recoding of the UGA stop codon of their genes (Böck et al. 2002). The chemical rationale behind the context-dependent translational integration of Sec into proteins is easy to understand as there are significant physicochemical differences between cysteine and its analog Sec such as the pK_a(Sec) of 5.2 compared with the pK_a(Cys) of 8.3 (Moroder 2005). While at the physiological pH the cysteine thiol function is not deprotonated, Sec exists almost exclusively as selenolate, which is a better nucleophile than thiolate. This explains the evolutionary conservation of Sec at the active site of a few enzymes such as formate dehydrogenase; the Sec→Cys replacement in this enzyme (Axley et al. 1991) lowers its turnover number by 2 orders of magnitude (Fig. 15). Sec is found in a small number of proteins in archaea, eubacteria, and eukaryotes and often is referred as the 21st proteinogenic amino acid, whereas its analog tellurocysteine has not yet been identified as a naturally occurring amino acid (Moroder 2005).

In spite of a wide range of possible chemical transformations to generate bespoke proteins, this approach is hampered by difficulties arising from the lack of chemo- and regiospecificity of the reactions applied, which can thus lead to either heterogeneous or hardly reproducible protein product mixtures (deSantis and Jones 1999). This is the major reason why direct protein modifications are presently not so popular anymore. This method has been largely replaced by site-directed amino acid mutagenesis (Hutchison et al. 1978) using genetic methods which provide very precise tools for dissecting and designing protein functions in the frame of the standard repertoire of 20 amino acids as prescribed by the genetic code. For example, in the absence of precise three-dimensional structures, alanine (or glycine)

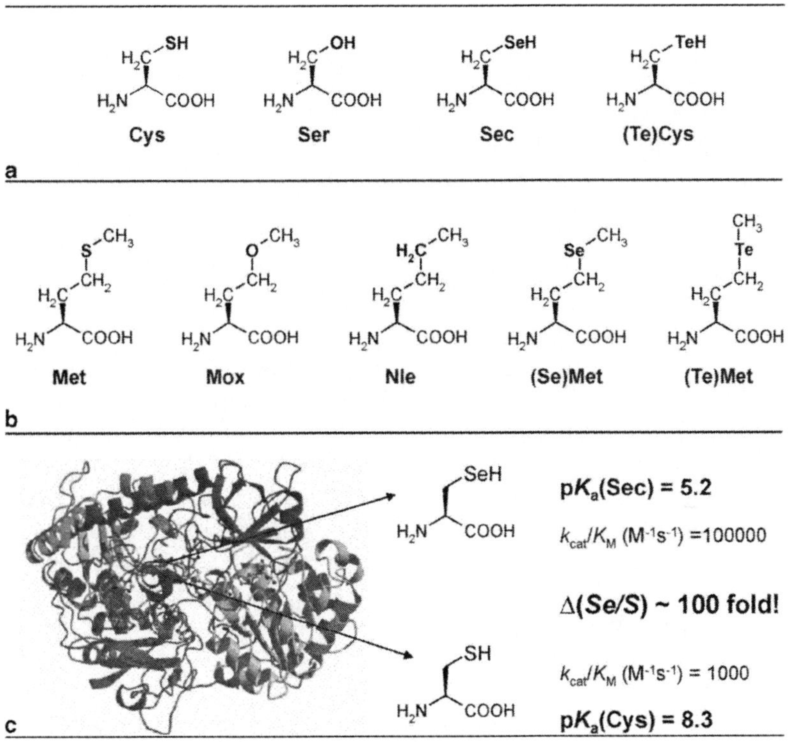

Fig. 15 (**a**) L-Cysteine (*Cys*) and the isosteric analogs serine (*Ser*), selenocysteine (*Sec*), and tellurocysteine [(*Te*)*Cys*]. (**b**) Methionine (*Met*) is compared with the related isosteric analogs 2-aminohexanoic acid (*Nle*), methoxinine (*Mox*), selenomethionine [(*Se*)*Met*], and telluromethionine [(*Te*)*Met*]. (**c**) Formate dehydrogenase contains a selenocysteine residue in the active site whose replacement with cysteine reduces 100-fold the catalytic efficiency of the enzyme (Axley et al. 1991)

scanning mutagenesis (Lefèvre et al. 1997) is well suited for systematic substitutions of all residues in (or around) putative active centers of proteins and thus for identification of "essential" functional groups.

On the other hand, it is often desirable to generate isosteric changes that are not disruptive. Although in the frame of the standard amino acid repertoire replacements such as Ser/Ala/Cys, Thr/Val, Glu/Gln, Asp/Asn, and Tyr/Phe are feasible in protein structures, there are no such possibilities for the other canonical amino acids such as tryptophan, methionine, proline, or histidine. It is also not possible to introduce Sec into proteins by using such approaches, i.e., mutagenesis techniques are limited to the exchange of one canonical amino acid for one of the 19 other canonical amino acids. The recent advances in the genetically encoded incorporation of noncanonical amino acids into proteins are highly promising in this context (see the chapters by Beatty and Tirrell, Mascarenhas et al., Köhrer and RajBhandary, Dougherty, Hecht, Hirao et al., and Leconte and Romesberg, this volume). Such genetically encoded replacements at the level of single atoms such as H/F and CH_2/S/Se/Te (Fig. 15) are known as "atomic mutations" (Budisa et al. 1998).

The best example assessing the utility of such replacements is the Met→SeMet isosteric replacement for phase calculation in protein X-ray crystallography (Budisa et al. 1995). The sulfur of the methionine side chain can also be replaced with methylene (Nle), oxygen (Mox), and even tellurium [(Te)Met]. All these methionine analogs are substrates for the protein biosynthetic machinery and can be translated into target sequences as a response to the methionine coding triplet AUG (Budisa 2004b; Budisa et al. 1999; Moroder 2005). The requirements for isosteric replacements are especially pronounced in studies of protein interiors to redesign or rationally manipulate protein structures. In fact, the core of the protein structures is probably the most tightly packed form of organic matter in nature. It is believed that the conformational specificity of protein cores is dictated by a stereochemical code that discriminates between the native and other conceivable chain folds (Rose and Wolfenden 1993). One of the possible ways to crack this code may be the use of isosteric analogs of core-building amino acid residues (Budisa et al. 1998). With site-directed mutagenesis, usually several sets of interactions are affected by the introduction of new side chains, whereas with atomic mutations, the altered protein properties arise solely from exchanges of single atoms, i.e., in favorable cases the impact of a single interaction can be analyzed (Minks et al. 1999).

3.3.2 Nonisosteric Chemical Modifications for Enhanced Functionality

A design of proteins with novel properties generally is not limited to isosteric chemical changes, i.e., to atomic mutations. For example, Kaiser (1988) and associates have generated new enzymatic activities via attachment of the flavin cofactor to the active site of the protease papain. The resulting chimera exhibited oxidoreductase activities with an increase in reaction rate by a factor of about 1,000 over the corresponding uncatalyzed reaction. Even nature itself uses the nonisosteric and bulky lysine-surrogate pyrrolysine to improve catalytic performances of vital enzymes involved in the methane-producing metabolism of some archaebacteria (Srinivasan et al. 2002). However, the cotranslational incorporation of this amino acid into proteins (as a response to a stop codon) is possible only when special requirements are fulfilled which cannot be met by simple site-directed mutagenesis protocols (Schimmel and Beebe 2004).

The recent explosion in commercial and synthetic applications of enzymes with enhanced stability and functionality has caused a renaissance of the chemical modification procedures (Hermanson 1996). Glutharylaldehyde-mediated cross-linking and poly(ethylene glycol) (PEG) modifications of enzyme surfaces are two of many well-known examples. Pharmaceutical companies have shown particular interest in PEGylation of proteins, which usually increases the therapeutic potential of proteins by reducing their toxicity and immunogenicity, increasing their half-life in serum and bioavailability (Lundblad and Bradshaw 1997). Other widely used modifications include (1) N-terminal acetylation and C-terminal amidation, (2) biotinylation or fluorophore-conjugation in N-, C-, or internal positions generally exploiting amino

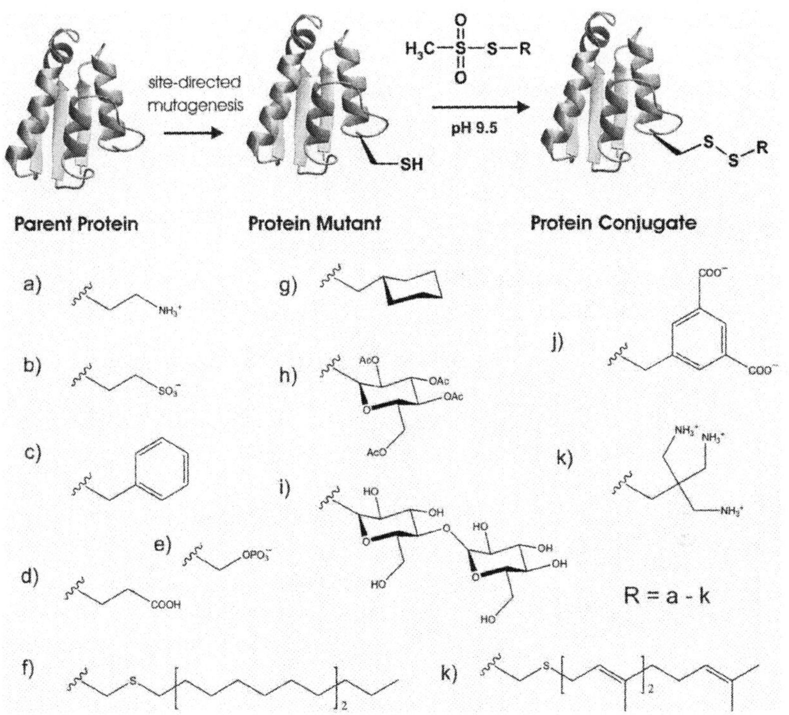

Fig. 16 Site-specific chemical modification: the combination of site-directed mutagenesis with covalent attachment of chemical functionalities. A unique cysteine residue is introduced into the target protein by site-directed mutagenesis. Upon subsequent reaction of the thiol group with the methanethiosulfonate reagent thiol–disulfide exchange reactions lead to conjugation with a variety of interesting functional groups and moieties. In this way, quantitative site-specific chemical modifications of the target protein with the desired–R groups (**a–l**) is achieved (deSantis and Jones 2002)

functions, (3) phosphorylation (phosphoserine, phosphothreonine, and phosphotyrosine), and (4) cyclization via cystine disulfides (Hermanson 1996).

The cysteine thiol function represents an attractive target for site-specific modification of proteins because of their rare natural abundance in many proteins and the relative ease of their selective chemical modification. Indeed cysteine residues are readily modified chemoselectively with a variety of reagents as shown in Fig. 16. The strategy employed includes the introduction of a cysteine residue at a key protein position via site-directed mutagenesis; subsequently this residue can be modified at will. For example, Gloss and Kirsch (1995) generated γ-thialysine with bromoethylamine in a Lys258Cys mutant of the wild-type aspartate aminotransferase. The modified enzyme exhibited an acidic shift of 1.3 pH units in the pK_a and the k_{cat}/K_M values for substrates were decreased by 1 order of magnitude. Additionally, cysteine residues react with a variety of interesting compounds that might serve as markers. In the same manner, a nitroxide group, which contains an unpaired electron (spin label), can be attached to proteins for electron paramagnetic resonance studies (Hubbell et al. 1998). Such examples illustrate how site-directed mutagenesis combined with chemoselective modifications may represent a versatile strategy to move beyond the structural limitations of the 20 canonical amino acids side chains in protein engineering (deSantis and Jones 2002). With no doubt classical chemical transformations will further play an important and general role in protein and cellular chemistry along with biotechnology; but the potential of the new approaches offered by site-directed mutagenesis will be further amplified by new developments for incorporation of noncanonical amino acids with particular functionalities into proteins leading to the new strategies of genetically encoded modifications (Budisa 2004b).

3.3.3 C–C Coupling Chemistries for Chemical Modification of Proteins

Schmidtchen and Dibowski (1998a) reported an in vitro method for the regioselective coupling of peptide fragments containing stable iodoarenes and alkynes as orthogonal reaction partners. Their C–C cross-coupling is catalyzed by palladium complexes under very mild aqueous conditions, thus enabling minimal interferences with fragile biologically active tertiary structures (Dibowski and Schmidtchen 1998b). These earlier studies were further optimized by Kodama et al. (2006), who developed a regioselective palladium chemistry for protein modifications under near-physiological conditions. The essential novelty was the efficient coupling of the expanded genetic code with organometallic chemistry. In particular, a *p*-iodophenylalanine was incorporated in vitro into a single site of the Ras protein. The unique aryl iodide of the modified Ras protein provided the reactive moiety for a regioselective palladium-catalyzed reaction with vinylated/alkenylated (Mizoroki–Heck reaction) and alkynylated/propargylated biotin derivatives (Sonogashira reaction; Fig. 17).

It should always be kept in mind that proteins are in general only marginally stable and thus allow for modification reactions to be performed only under mild conditions in aqueous environments at neutral pH values. In view of these limitations, the classical palladium chemistry is rather delicate (absence of oxygen, long reaction time, +4°C, and high catalyst concentrations – approximately 0.5 mM), and complicated and requires patient and skillful experimentation (King and Yasuda 2004). Therefore, it is not surprising that the reported reaction yields were rather low and that further optimization is needed as well as alternative chemistries. For example, proteins with pre-installed alkyne and alkene functionalities were expressed at high levels. Improved protocols for the attachment of halogenated ligands in both Sonogashira and Mizoroki-Heck reactions should result in functionally active protein conjugates after the Pd-reaction. Incorporating boronated amino acid into the target protein sequence would enhance the repertoire of protein conjugation techniques with the Suzuki-reaction as well.

3.3.4 "Click Chemistry" for Protein Modification

The concept of click chemistry was introduced by Sharpless and associates in 2001 and is based on the cycloaddition reaction between azide and alkyne units (Kolb

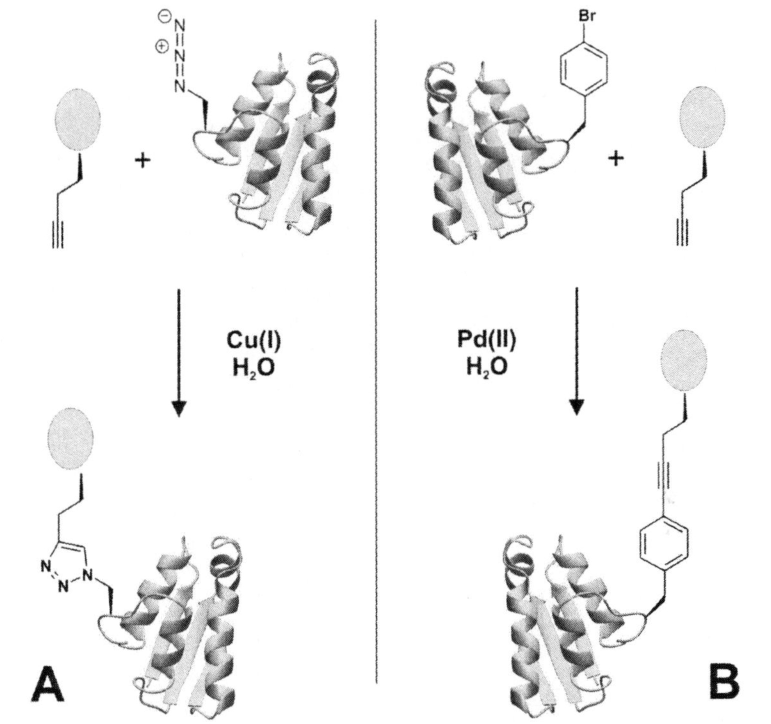

Fig. 17 Protein conjugation via Cu(I)-catalyzed cycloaddition. "Click chemistry" (**a**) and Pd-catalyzed C–C conjugation chemistry (a variation of the Heck reaction) (**b**) Cycloaddition is a pericyclic chemical reaction where two π bonds are lost and two σ bonds are gained. Two major conjugation types of Pd(II)-catalyzed C–C coupling are currently applied in proteins, i.e., vinylated/alkenylated (Mizoroki–Heck reaction) and alkynylated/propargylated (Sonogashira reaction) ligands (King and Yasuda 2004)

et al. 2001). The connecting units consist of carbon-heteroatom bonds C–X–C rather than C–C bonds as in the Heck reaction. Huisgen (1968) was the first to understand the versatility of this cycloaddition reaction and its utility for synthetic purposes. In fact, azides and alkynes are inert to most chemical functionalities and are stable in a wide range of solvents, temperatures, ion strengths, and pH values. The reaction generates the two regioisomers 1,2-triazole and 1,4-triazole. Almost simultaneously, Meldal and Sharpless discovered that the regioselective cycloaddition with formation of 1,4-triazole as the preferred regioisomer can be achieved with Cu(I) as a catalyst (Rostovtsev et al. 2002; Tornoe et al. 2002). In addition, in the presence of copper the reaction is 10^6-fold faster than in the absence of the catalyst and is 25 times faster than the Staudinger ligation (Prescher and Bertozzi 2005). For these reasons the azide/alkyne Huisgen cycloaddition is considered as *the cream of the crop* of click chemistry (Wang et al. 2005). Not surprisingly, this reaction has found increasing applications for bioconjugations.

Tirrell and coworkers were the first to demonstrate that the cell surface of *E. coli* can be labeled with biotin by exploiting the Cu(I)-catalyzed cycloaddition reaction (Fig. 17; Link et al. 2004; see also the chapter by Beatty and Tirrell, this volume). For this purpose, cells expressing a membrane protein with the methionine analog azidohomoalanine were biotinylated, subsequently stained by avidin–AlexaFluor 488, and subjected to flow cytometry analyses. Similarly, the methionine and phenylalanine residues of histidine-tagged barstar as the model protein were efficiently replaced by the nonnatural amino acids homopropargylglycine and ethylphenylalanine in expression experiments. Once the proteins containing these amino acids had been expressed, the *E. coli* cell cultures were allowed to react with an azide derivative of coumarin in the presence of Cu(I) catalyst (Beatty et al. 2005). In a parallel effort, Schultz and coworkers expressed human superoxide dismutase-1 with a *p*-azidophenylalanine residue which then allowed PEGylatation of the protein (Deiters et al. 2004). By this procedure, a homogeneous population of the highly active enzyme with improved properties was obtained (increased metabolic stability, protease resistance, and lowered immune response). Recently, click chemistry was also successfully applied for enzyme profiling and in drug discovery (Kolb and Sharpless 2003). Furthermore, Tirrell, Bertozzi, and coworkers reported the possibility of a Huisgen cycloaddition on living cells without the need for Cu(I) as the catalyst (Link et al. 2006).

3.3.5 Backbone Modifications by Surrogate Peptide Bonds: Peptidomimetics

Replacement of an amide bond in a peptide backbone is widely used in peptidomimetics. The peptide primary structure is mimicked by amide bond isosteres and/or modification of the native peptide backbone, including chain extension or incorporation of heteroatoms. Peptides modified in this way are often protease-resistant, and may have reduced immunogenicity and improved bioavailability compared with the parent peptides. In addition to mimicking the peptidic structure, such backbone modifications can induce and stabilize well-defined secondary structural elements such as helices, turns, and small, sheet-like structures. Examples of types of peptide backbone modifications are presented in Scheme 2. These include exchange of single chain units (–CO– or –NH–) by oxygen (depsipeptide), a sulfur atom (thioester), or a $-CH_2-$ group (ketomethylene ester). The C^{α} moiety (–CH–) can be replaced with nitrogen (azapeptide) or boron (boropeptide), or converted to a quaternary carbon by addition of another alkyl group. The carbonyl group in the peptide bond can be replaced by a thiocarbonyl group (endothiopeptide) or by sulfonamides. The peptide bond itself can be stereochemically inverted (at C^{α} atom) or isosterically replaced by alkane/alkene structures. Finally, the peptide chain can be extended by one atom or atom groups such as aminoxy, hydrazine, or β-amino acids (Ahn et al. 2002).

All these backbone modifications are accessible by synthesis; and by employing the new procedures of chemical ligations, one can even introduce related peptide fragments at selected positions into proteins.

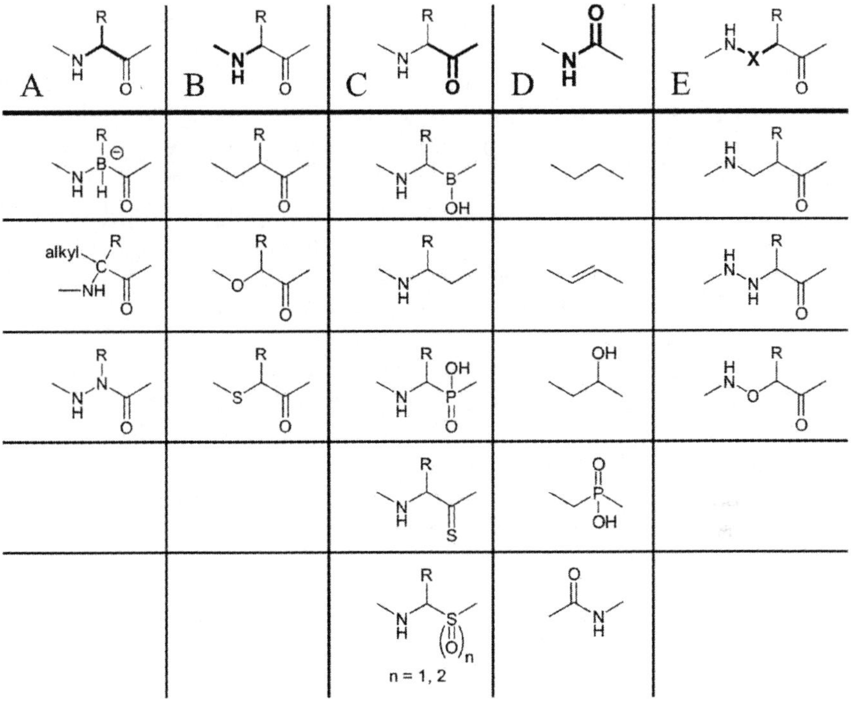

Scheme 2 Peptide mimicry: some selected types of peptide backbone modifications. Although the mimetic structures are often not stable in acidic or basic aqueous solutions, such backbone modifications differ substantially from the peptide amide structure and thus may affect local conformational geometries. The modifications include exchange of single chain units (*A, B, C*), double exchange at the peptide bond (*D*), and extension of the peptide chain (*E*). (Adapted from similar compilations in Sewald and Jakubke 2002 and Toniolo 2003)

4 Enzyme-Mediated Peptide Bond Formation

Nature uses ribosomes or multienzyme complexes for ribosomal and nonribosomal polypeptide synthesis, respectively. These complex machineries are not amenable to simple laboratory routine. However, nature offers a plethora of proteolytic enzymes which could possibly be forced to shift the equilibrium in favor of peptide bond formation by the currently available techniques of directed evolution. This would generate biocatalysts with features that nature did not develop during evolution. So far, reversal of the proteolytic activity of proteases can only be achieved by thermodynamic or kinetic manipulation of the enzymatic equilibrium reactions (Bordusa and Jakubke 2002).

In 1938 Max Bergmann, a pupil of Emil Fischer, reported on the enzymatic synthesis of peptide bonds, describing the papain-catalyzed synthesis of benzoylleucyl-leucine anilide by manipulating experimentally the kinetic equilibria of the catalyzed reaction (Bergmann and Fraenkel-Conrat 1938). This seminal report has forced increased research on enzymatic synthesis mainly based on peptide chain elongation via transamidation reactions (Fruton 1982). This led more recently to the development of the substrate mimetic approach (Bordusa 2002), which was successfully applied in protease-mediated ligation of a protein fragment thioester substrate mimetic, produced by the recombinant intein technique, with a synthetic peptide containing an N-terminal serine residue using V8, the Glu/Asp-specific serine protease from *Staphylococcus aureus* (Fig. 18a) (Machova et al. 2003).

The concept of modifying the active site of proteases for reduction of the proteolytic activity in favor of synthetase properties was first realized by Kaiser et al. (1985) with the successful conversion of the protease subtilisin into a ligase. Chemical exchange of the active-site serine with cysteine (see Sect. 3.2.1) led to thiosubtilisin, which reacts rapidly with an activated peptide to form an acylated enzyme and then transfers the peptide onto another peptide, completing the ligation reaction (Nakatsuka et al. 1987). Since the enzyme is a poor peptidase, it does not attack the resulting product rapidly. For such a kinetically controlled system the ratio between aminolysis and hydrolysis is of crucial importance for efficient peptide

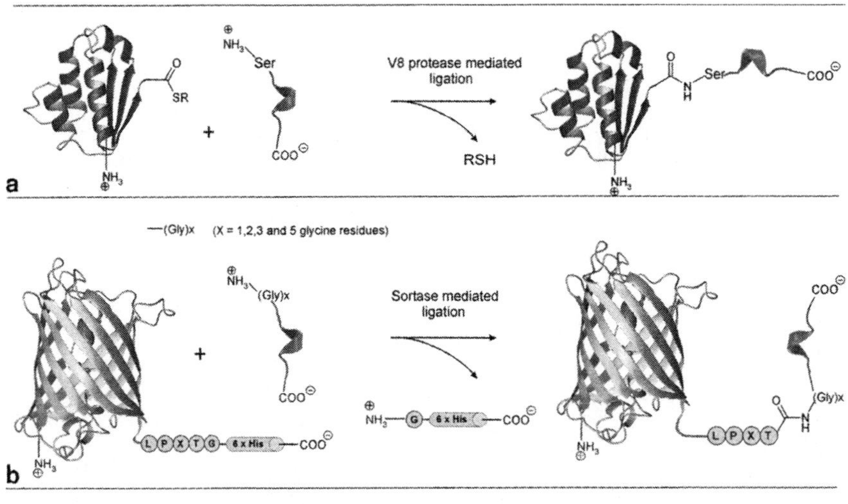

Fig. 18 Enzyme-catalyzed peptide bond formation. Selected examples: (**a**) Expressed enzyme ligation with specific thioesters as a substrate mimetic for *Staphylococcus aureus* V8 serine-protease (Machova et al. 2003). A serine residue is required at the N-terminus of the donor peptide fragment for the enzyme to act as an inverse protease. (**b**) Sortase from *S. aureus*, a membrane-anchored transpeptidase, cleaves any polypeptide provided with a C-terminal sorting signal, between the threonine (*T*) and glycine (*G*) of the LPXTG motif. It then catalyzes the formation of an amide bond between the carboxy group of threonine and the amino group of pentaglycine from cell wall peptidoglycan. The extension of this strategy to tagged GFP (Nt-GFP-LPXTG-6His-Ct) enables its successful conjugation with various donor molecules containing two or more N-terminal glycines (D- and L-peptides, nonpeptide fragments, and even other GFPs) (Mao et al. 2004)

bond synthesis. These early findings with subtilisin provided a solid base for engineering an active site of this enzyme (Jackson et al. 1994) that efficiently catalyzes the ligation of peptide fragments (Chang et al. 1994). The enzyme variant subtiligase exhibits a largely reduced proteolytic activity and it is functionally active as an acyltransferase (Braisted et al. 1997). This property was exploited for enzymatic condensation of six peptide fragments into a ribonuclease A analog containing 4-fluorohistidine in the active site (Jackson et al. 1994).

Most recently, a new enzyme (sortase) with its highly selective catalytic properties has entered the arena of enzyme-mediated protein modifications (Mao et al. 2004). Sortases are bacterial enzymes that are responsible for the covalent attachment of specific proteins to the cell wall of Gram-positive bacteria (Mazmanian et al. 1999). Proteins that are substrates of sortase possess a "sorting signal" at the C-terminus consisting of the LPXTG motif (where X can be any amino acid), a hydrophobic region, and a tail of charged residues. Sortase is capable of catalyzing a two-step transpeptidation reaction, either in vivo or in vitro. First, the LPXTG motif is cleaved between threonine and glycine. Then the threonine is covalently attached to the amino group of a pentaglycine segment of the cell wall peptidoglycan, resulting in a cell-wall-attached protein (Huang et al. 2003). With use of a similar experimental set-up with sortase from *S. aureus*, selective protein–peptide and protein–protein ligations were achieved, confirming the efficiency of this enzyme (Mao et al. 2004). Furthermore, it was shown that nonnative peptide fragments including D-peptides and nonpeptide derivatives (e.g., folate) of glycine, diglycine, or triglycine are efficiently conjugated by sortase to the acceptor protein such as the green fluorescent protein containing the LPXTG motif (Fig. 18b). With this enzyme, the arsenal of tools for the production of proteins with tailored properties is enriched by an additional procedure with great potential to become part of routine recipes.

A continuously increasing number of enzymes that excel for highly specific protein modification properties are being discovered and applied with success for such purposes. For example, the *E. coli* enzyme biotin ligase links in sequence-specific mode biotin to a 15-membered acceptor peptide (Chen et al. 2005). Since the enzyme accepts as substrates also ketone-containing isosteric biotin derivatives (and ketones are absent at native cell surfaces), the acceptor peptide fused with recombinant cell-surface proteins can be tagged not only with biotin, but also with the biotin–ketone moiety that allows for further transformations by the hydrazide or hydroxylamine chemistries (see Fig. 2).

Similar surprising findings will most probably be made by genome and proteome mapping among the different kingdoms of life. In fact, there are enough documented cases where the discovery of particular biological processes during studies of mechanisms of microbial pathogenicity or of unique biological phenomena in rather obscure microorganisms might serve as direct input for the development of novel technologies (Budisa 2004a). A typical example is protein splicing (see Sect. 2.6; Gogarten et al. 2002; Paulus 2000).

Acknowledgment This work was supported by the BioFuture Program of the Federal Ministry of Education and Research of Germany.

References

Ahn JM, Boyle NA, MacDonald MT, Janda KD (2002) Peptidomimetics and peptide backbone modifications. Mini Rev Med Chem 2:463–473

Aimoto S (1999) Polypeptide synthesis by the thioester method. Biopolymers 51:247–265

Aimoto S, Mizoguchi N, Hojo H, Yoshimura S (1989) Development of a facile method for polypeptide synthesis. Synthesis of bovine pancreatic trypsin inhibitor (BPTI). Bull Chem Soc Jpn 62:524–531

Akaji K, Kiso Y (2003) Synthesis of cystine peptides. In: Goodman M, Felix A, Moroder L, Toniolo C (eds) Houben-Weyl, Methods of organic chemistry: Synthesis of peptides and peptidomimetics, vol E22b. Georg Thieme Verlag, Stuttgart, pp 101–141

Akaji K, Hayashi Y, Kiso Y, Kuriyama N (1999) Convergent synthesis of dolastatin 15 by solid phase coupling of an N-methylamino acid. J Org Chem 64:405–411

Annis I, Hargittai B, Barany G (1997) Disulfide bond formation in peptides. Method Enzymol 289:198–221

Anthony-Cahill SJ, Magliery TJ (2002) Expanding the natural repertoire of protein structure and function. Curr Pharm Biotechnol 3:299–315

Axley MJ, Böck A, Stadtman TC (1991) Catalytic properties of an *Escherichia coli* formate dehydrogenase mutant in which sulfur replaces selenium. Proc Natl Acad Sci USA 88:8450–8454

Baca M, Muir TW, Schnölzer M, Kent SBH (1995) Chemical ligation of cysteine-containing peptides: Synthesis of a 22kDa tethered dimer of HIV-1 protease. J Am Chem Soc 117:1881–1887

Bang D, Pentelute BL, Kent SBH (2006) Kinetically controlled ligation for the convergent chemical synthesis of proteins. Angew Chem Int Ed 45:3985–3988

Beatty KE, Xie F, Wang Q, Tirrell DA (2005) Selective dye-labeling of newly synthesized proteins in bacterial cells. J Am Chem Soc 127:14150–14151

Benoiton NL (2006) Chemistry of peptide synthesis. CRC Press, Taylor & Francis Group, Boca Raton, Florida

Bergmann M, Fraenkel-Conrat H (1938) The enzymatic synthesis of peptide bonds. J Biol Chem 124:1–6

Beyermann M, Bienert M, Carpino LA (2002) Acid halides. In: Goodman M, Felix A, Moroder L, Toniolo C (eds) Houben–Weyl, Methods of organic chemistry: Synthesis of peptides and peptidomimetics, vol E22a. Georg Thieme Verlag, Stuttgart, pp 475–494

Bienert M, Henklein P, Beyermann M, Carpino LA (2002) Uronium/guanidinium salts. In: Goodman M, Felix A, Moroder L, Toniolo C (eds) Houben–Weyl, Methods in organic chemistry: Synthesis of peptides and peptidomimetics, vol E22a. Georg Thieme Verlag, Stuttgart, pp 555–580

Blake J (1981) Peptide segment coupling in aqueous medium – Silver ion activation of the thiolcarboxyl group. Int J Pept Prot Res 17:273–274

Böck A, Thanbichler M, Rother M, Resch A (2002) Selenocysteine. In: Ibba M, Cusack S, Francklyn C (eds) Aminoacyl-tRNA synthetases. Landes Bioscience, Austin, Texas, pp 320–328

Bodanszky M (1993) Principles of peptide synthesis. Springer-Verlag, Berlin

Bode JW, Fox RM, Baucom KD (2006) Chemoselective amide ligations by decarboxylative condensations of N-alkylhydroxylamines and α-ketoacids. Angew Chem Int Ed 45:1248–1252

Bordusa F (2002) Proteases in organic synthesis. Chem Rev 102:4817–4867

Bordusa F, Jakubke HD (2002) Enzymatic synthesis. In: Goodman M, Felix A, Moroder L, Toniolo C (eds) Houben–Weyl, Methods of organic chemistry: Synthesis of peptides and peptidomimetics, vol E22a. Georg Thieme Verlag, Stuttgart, pp 642–664

Botti P, Carrasco MR, Kent SBH (2001) Native chemical ligation using removable N-α-(1-phenyl-2-mercaptoethyl) auxiliaries. Tetrahedron Lett 42:1831–1833

Braisted AC, Judice JK, Wells JA (1997) Synthesis of proteins by subtiligase. Methods Enzymol 289:298–313

Brenner M, Zimmermann JP, Wehrmüller J, Quitt P, Hartmann A, Schneider W, Beglinger U (1957) Aminoacyl-einlagerung. 1. Definition, Übersicht und Beziehung zur Peptidsynthese. Helv Chim Acta 40:1497–1517

Budisa N (2004a) Adding new tools to the arsenal of expressed protein ligation. ChemBioChem 5:1176–1179

Budisa N (2004b) Prolegomena to future experimental efforts on genetic code engineering by expanding its amino acid repertoire. Angew Chem Int Ed 43:6426–6463

Budisa N, Steipe B, Demange P, Eckerskorn C, Kellermann J, Huber R (1995) High-level biosynthetic substitution of methionine in proteins by its analogs 2-aminohexanoic acid, selenomethionine, telluromethionine and ethionine in *Escherichia coli*. Eur J Biochem 230:788–796

Budisa N, Huber R, Golbik R, Minks C, Weyher E, Moroder L (1998) Atomic mutations in annexin V – Thermodynamic studies of isomorphous protein variants. Eur J Biochem 253:1–9

Budisa N, Minks C, Alefelder S, Wenger W, Dong FM, Moroder L, Huber R (1999) Toward the experimental codon reassignment in vivo: Protein building with an expanded amino acid repertoire. FASEB J 13:41–51

Canne LE, Bark SJ, Kent SBH (1996) Extending the applicability of native chemical ligation. J Am Chem Soc 118:5891–5896

Carpino LA (1993) 1-Hydroxy-7-azabenzotriazole. An efficient peptide coupling additive. J Am Chem Soc 115:4397–4398

Carpino LA, Mansour ESME (1992) Protected β- and γ-aspartic and -glutamic acid fluorides. J Org Chem 57:6371–6373

Carpino LA, Krause E, Sferdean CD, Schümann M, Fabian H, Bienert M, Beyermann M (2004) Synthesis of "difficult" peptide sequences: Application of a depsipeptide technique to the Jung-Redemann 10- and 26-mers and the amyloid peptide Aβ(1–42). Tetrahedron Lett 45:7519–7523

Carrillo N, Davalos EA, Russak JA, Bode JW (2006) Iterative, aqueous synthesis of β(3)-oligopeptides without coupling reagents. J Am Chem Soc 128:1452–1453

Chang TK, Jackson DY, Burnier JP, Wells JA (1994) Subtiligase - A tool for semisynthesis of proteins. Proc Natl Acad Sci USA 91:12544–12548

Chapeville F, Lipmann F, von Ehrenstein G, Weisblum B, Ray WJ, Benzer S (1962) On role of soluble ribonucleic acid in coding for amino acids. Proc Natl Acad Sci USA 48:1086–1092

Chen I, Howarth M, Lin W, Ting AY (2005) Site-specific labeling of cell surface proteins with biophysical probes using biotin ligase. Nat Methods 2:99–104

Clayton D, Shapovalov G, Maurer JA, Dougherty DA, Lester HA, Kochendoerfer GG (2004) Total chemical synthesis and electrophysiological characterization of mechanosensitive channels from *Escherichia coli* and *Mycobacterium tuberculosis*. Proc Natl Acad Sci USA 101:4764–4769

Dawson PE, Kent SBH (2000) Synthesis of native proteins by chemical ligation. Annu Rev Biochem 69:923–960

Dawson PE, Muir TW, Clark-Lewis I, Kent SBH (1994) Synthesis of proteins by native chemical ligation. Science 266:776–779

deSantis G, Jones JB (1999) Chemical modification of enzymes for enhanced functionality. Curr Opin Biotechnol 10:324–330

deSantis G, Jones JB (2002) Combining site-specific chemical modification with site-directed mutagenesis: A versatile strategy to move beyond the structural limitations of the 20 natural amino acids side chains in protein engineering. In: Braman J (ed) Methods in Molecular Biology: In vitro mutagenesis protocols, vol 182. Humana Press, Totowa, New Jersey, pp 55–65

Deiters A, Cropp TA, Summerer D, Mukherji M, Schultz PG (2004) Site-specific PEGylation of proteins containing unnatural amino acids. Bioorg Med Chem Lett 14:5743–5745

Denkewalter RG, Veber DF, Holly FW, Hirschmann R (1969) Studies on total synthesis of an enzyme. I. Objective and strategy. J Am Chem Soc 91:502–503

Dibowski H, Schmidtchen FP (1998a) Bioconjugation of peptides by Palladium catalyzed C–C cross-coupling in water. Angew Chem Int Ed 37:476–478

Dibowski H, Schmidtchen FP (1998b) Sonogashira cross-couplings using biocompatible conditions in water. Tetrahedron Lett 39:525–528

du Vigneaud V, Ressler C, Swan CJM, Roberts CW, Katsoyannis PG, Gordon S (1953) The synthesis of an octapeptide amide with the hormonal activity of oxytocin. J Am Chem Soc 75:4879–4880

Evans TC, Benner J, Xu MQ (1998) Semisynthesis of cytotoxic proteins using a modified protein splicing element. Protein Sci 7:2256–2264

Fischer E (1902) Über die Hydrolyse der Proteinstoffe. Chemiker-Zeitung 26:939–940

Fischer E (1906) Untersuchungen über Aminosäuren, Polypeptide und Proteine Ber Chem Ges 39:530–610 (Translated by Fruton JS (1972) In: The nature of proteins, molecules and life. Wiley, Chichester, pp 112–113)

Fruton JS (1982) Proteinase-catalyzed synthesis of peptide bonds. Adv Enzymol 53:239–306

Fuller WD, Yalamoori VV (2002) Acid anhydrides. In: Goodman M, Felix A, Moroder L, Toniolo C (eds) Houben–Weyl, Methods of organic chemistry: Synthesis of peptides and peptidomimetics, vol E22a. Georg Thieme Verlag, Stuttgart, pp 495–516

Fuller WD, Cohen MP, Shabankareh M, Blair RK, Goodman M, Naider FR (1990) Urethane protected amino acid N-carboxy anhydrides and their use in peptide synthesis. J Am Chem Soc 112:7414–7416

Gaertner HF, Offord RE (2003) Semisynthesis. In: Goodman M, Felix A, Moroder L, Toniolo C (eds) Hobuben–Weyl, Methods of organic chemistry: synthesis of peptides and peptidomimetics, vol E22b. Georg Thieme Verlag, Stuttgart, pp 81–100

Gaertner HF, Rose K, Cotton R, Timms D, Camble R, Offord RE (1992) Construction of protein analogues by site-specific condensation of unprotected fragments. Bioconjugate Chem 3: 262–268

Gieselman MD, Zhu YT, Zhou H, Galonic D, van der Donk WA (2002) Selenocysteine derivatives for chemoselective ligations. ChemBioChem 3:709–716

Gloss LM, Kirsch JF (1995) Decreasing the basicity of the active site base, Lys-258, of *Escherichia coli* aspartate aminotransferase by replacement with γ-thialysine. Biochemistry 34: 3990–3998

Gogarten JP, Senejani AG, Zhaxybayeva O, Olendzenski L, Hilario E (2002) Inteins: structure, function, and evolution. Annu Rev Microbiol 56:263–287

Goodman M, Felix A, Moroder L, Toniolo C (eds) (2002) Houben–Weyl, Methods of organic chemistry: Synthesis of peptides and peptidomimetics. vol E22a–E22e, Georg Thieme Verlag, Stuttgart

Govardhan CP (1999) Crosslinking of enzymes for improved stability and performance. Curr Opin Biotechnol 10:331–335

Gutte B, Merrifield RB (1969) The total synthesis of an enzyme with ribonuclease A activity. J Am Chem Soc 91:501–502

Haack T, Mutter M (1992) Serine derived oxazolidines as secondary structure disrupting, solubilizing building blocks in peptide synthesis. Tetrahedron Lett 33:1589–1592

Hennard C, Tam JP (1997) Sequential orthogonal coupling strategy for the synthesis of biotin tagged β-defensin. In: Peptides, frontiers in peptide science (Proceedings of the 15th Am Peptide Symposium, Nashville, Tennessee). Kluwer, New York

Hermanson GT (1996) Bioconjugate techniques. Academic Press, San Diego, California

Hilvert D (2001) Enzyme engineering. Chimia 55:867–869

Hirschmann R, Nutt RF, Veber DF, Vitali RA, Varga SL, Jacob TA, Holly FW, Denkewalter RG Studies on the total synthesis of an enzyme. V. The preparation of enzymatically active material. J Am Chem Soc 91:507–508

Hofmeister F (1902) Ueber den Bau des Eiweissmoleküls. Naturwiss Rundschau 17:529–545

Huang XY, Aulabaugh A, Ding WD, Kapoor B, Alksne L, Tabei K, Ellestad G (2003) Kinetic mechanism of *Staphylococcus aureus* sortase SrtA. Biochemistry 42:11307–11315

Hubbell WL, Gross A, Langen R, Lietzow MA (1998) Recent advances in site-directed spin labeling of proteins. Curr Opin Struct Biol 8:649–656

Huisgen R (1968) Cycloadditions – definition, classification, and characterization. Angew Chem Int Ed 7:321–328

Hutchison CA, Phillips S, Edgell MH, Gillam S, Jahnke P, Smith M (1978) Mutagenesis at a specific position in a DNA sequence. J Biol Chem 253:6551–6560

Jackson DY, Burnier J, Quan C, Stanley M, Tom J, Wells JA (1994) A designed peptide ligase for total synthesis of ribonuclease A with unnatural catalytic residues. Science 266:243–247

Kaiser ET (1988) Catalytic activity of enzymes altered at their active sites. Angew Chem Int Ed 27:913–922

Kaiser ET, Lawrence DS, Rokita SE (1985) The chemical modification of enzymatic specificity. Annu Rev Biochem 54:565–595

Kane PM, Yamashiro CT, Wolczyk DF, Neff N, Goebl M, Stevens TH (1990) Protein splicing converts the yeast TFP1 gene product to the 69-kD subunit of the vacuolar H+ -adenosine triphosphatase. Science 250:651–657

Kates SA, Albericio F (2000) Solid-phase synthesis. A practical guide. Marcel Dekker, New York

Kawakami T, Akaji K, Aimoto S (2001) Peptide bond formation mediated by 4, 5-dimethoxy-2-mercaptobenzylamine after periodate oxidation of the N-terminal serine residue. Org Lett 3:1403–1405

Kemp DS (1981) The amine capture strategy for peptide bond formation – An outline of progress. Biopolymers 20:1793–1804

Kemp DS, Carey RI (1993) Synthesis of a 39-peptide and a 25-peptide by thiol capture ligations: Observation of a 40-fold rate acceleration of the intramolecular O,N-acyl-transfer reaction between peptide fragments bearing only cysteine protective groups. J Org Chem 58:2216–2222

Kent SBH (2003) Total chemical synthesis of enzymes. J Pept Sci 9:574–593

Kiick KL, Saxon E, Tirrell DA, Bertozzi CR (2002) Incorporation of azides into recombinant proteins for chemoselective modification by the Staudinger ligation. Proc Natl Acad Sci USA 99:19–24

Kimura T (2003) Oxidative refolding of multiple-cystine peptides. In: Goodman M, Felix A, Moroder L, Toniolo C (eds) Hobuben–Weyl, Methods of organic chemistry: Synthesis of peptides and peptidomimetics, vol E22b. Georg Thieme Verlag, Stuttgart, pp 142–161

King AO, Yasuda N (2004) Palladium-catalyzed cross-coupling reactions in the synthesis of pharmaceuticals. In: Larsen RD (ed) Topics in organometallic chemistry: Organometallics in process chemistry, vol 6. Springer, Berlin, pp 205–245

Kochendoerfer GG, Chen SY, Mao F, Cressman S, Traviglia S, Shao HY, Hunter CL, Low DW, Cagle EN, Carnevali M, Gueriguian V, Keogh PJ, Porter H, Stratton SM, Wiedeke MC, Wilken J, Tang J, Levy JJ, Miranda LP, Crnogorac MM, Kalbag S, Botti P, Schindler-Horvat J, Savatski L, Adamson JW, Kung A, Kent SBH, Bradburne JA (2003) Design and chemical synthesis of a homogeneous polymer-modified erythropoiesis protein. Science 299:884–887

Kodama K, Fukuzawa S, Nakayama H, Kigawa T, Sakamoto K, Yabuki T, Matsuda N, Shirouzu M, Takio K, Tachibana K, Yokoyama S (2006) Regioselective carbon-carbon bond formation in proteins with Palladium catalysis; New protein chemistry by organometallic chemistry. ChemBioChem 7:134–139

Kolb HC, Sharpless KB (2003) The growing impact of click chemistry on drug discovery. Drug Discov Today 8:1128–1137

Kolb HC, Finn MG, Sharpless KB (2001) Click chemistry: Diverse chemical function from a few good reactions. Angew Chem Int Ed 40:2004–2021

Lefèvre F, Rémy MH, Masson JM (1997) Alanine-stretch scanning mutagenesis: A simple and efficient method to probe protein structure and function. Nucleic Acids Res 25:447–448

Link AJ, Vink MK, Tirrell DA (2004) Presentation and detection of azide functionality in bacterial cell surface proteins. J Am Chem Soc 126:10598–10602

Link AJ, Vink MK, Agard NJ, Prescher JA, Bertozzi CR, Tirrell DA (2006) Discovery of aminoacyl-t RNA synthetase activity through cell-surface display of noncanonical amino acids. Proc Natl Acad Sci USA 103:10180–10185

Liu C, Tam JP (1994) Peptide segment ligation strategy without use of protecting groups. Proc Natl Acad Sci USA 91:6584–6588

Liu JQ, Jiang MS, Luo GM, Yan GL, Shen JC (1998) Conversion of trypsin into a selenium-containing enzyme by using chemical mutation. Biotechnol Lett 20:693–696

Lloyd-Williams P, Albericio F, Giralt E (1993) Convergent solid-phase peptide synthesis. Tetrahedron 49:11065–11133

Lloyd-Williams P, Albericio F, Giralt E (1997) Chemical approaches to the synthesis of peptides and proteins. CRC Press, Boca Raton, Florida

Lu YA, Tam JP (2005) Peptide ligation by a reversible and reusable C-terminal thiol handle. Org Lett 7:5003–5006

Lundblad RL, Bradshaw RA (1997) Applications of site-specific chemical modification in the manufacture of biopharmaceuticals. Biotechnol Appl Biochem 26:143–151

Machova Z, von Eggelkraut-Gottanka R, Wehofsky N, Bordusa F, Beck-Sickinger AG (2003) Expressed enzymatic ligation for the semisynthesis of chemically modified proteins. Angew Chem Int Ed 42:4916–4918

Mao HY, Hart SA, Schink A, Pollok BA (2004) Sortase-mediated protein ligation: A new method for protein engineering. J Am Chem Soc 126:2670–2671

Mazmanian SK, Liu G, Hung TT, Schneewind O (1999) *Staphylococcus aureus* sortase, an enzyme that anchors surface proteins to the cell wall. Science 285:760–763

Means GE, Feeney RE (1990) Chemical modifications of proteins: History and applications. Bioconjug Chem 1:2–12

Merrifield RB (1963) Solid phase peptide synthesis. J Am Chem Soc 85:2149–2154

Minks C, Huber R, Moroder L, Budisa N (1999) Atomic mutations at the single tryptophan residue of human recombinant annexin V: Effects on structure, stability, and activity. Biochemistry 38:10649–10659

Moroder L (2005) Isosteric replacement of sulfur with other chalcogens in peptides and proteins. J Pept Sci 11:187–214

Moroder L, Borin G, Marchiori F, Scoffone E (1972) Synthetic peptides related to the entire sequence of yeast iso-I-cytochrome *c*. Biopolymers 11:2191–2194

Moroder L, Filippi B, Borin G, Marchiori F (1975) Studies on cytochrome *c*. X. Synthesis of N-benzyloxycarbonyl-(Thr-107)-dotetracontapeptide (sequence 67–108) of baker's yeast iso-I-cytochrome *c*. Biopolymers 14:2061–2074

Moroder L, Besse D, Musiol HJ, Rudolph-Böhner S, Siedler F (1996) Oxidative folding of cystine-rich peptides vs. regioselective cysteine pairing strategies. Biopolymers 40:207–234

Moroder L, Musiol HJ, Götz M, Renner C (2005) Synthesis of single- and multiple-stranded cystine-rich peptides. Biopolymers 80:85–97

Muir TW, Sondhi D, Cole PA (1998) Expressed protein ligation: A general method for protein engineering. Proc Natl Acad Sci USA 95:6705–6710

Mutter M, Nefzi A, Sato T, Sun X, Wahl F, Wöhr T (1995) Pseudo-prolines (psi-Pro) for accessing inaccessible peptides. Pept Res 8:145–153

Mutter M, Chandravarkar A, Boyat C, Lopez J, Dos Santos S, Mandal B, Mimna R, Murat K, Patiny L, Saucéde L, Tuchscherer G (2004) Switch peptides in statu nascendi: Induction of conformational transitions relevant to degenerative diseases. Angew Chem Int Ed 43:4172–4178

Nakatsuka T, Sasaki T, Kaiser ET (1987) Peptide segment coupling catalyzed by the semisynthetic enzyme thiolsubtilisin. J Am Chem Soc 109:3808–3810

Narayan S, Muldoon J, Finn MG, Fokin VV, Kolb HC, Sharpless KB (2005) "On water": Unique reactivity of organic compounds in aqueous suspension. Angew Chem Int Ed 44:3275–3279

Neet KE, Koshland DE (1966) The conversion of serine at the active site of subtilisin to cysteine: A chemical mutation. Proc Natl Acad Sci USA 56:1606–1611

Neet KE, Nanci A, Koshland DE (1968) Properties of thiol-subtilisin. The consequences of converting the active serine residue to cysteine in a serine protease. J Biol Chem 243: 6392–6401

Nilsson BL, Kiessling LL, Raines RT (2000) Staudinger ligation: A peptide from a thioester and azide. Org Lett 2:1939–1941

Nilsson BL, Hondal RJ, Soellner MB, Raines RT (2003) Protein assembly by orthogonal chemical ligation methods. J Am Chem Soc 125:5268–5269

Nilsson BL, Soellner MB, Raines RT (2005) Chemical synthesis of proteins. Annu Rev Biophys Biomol Struct 34:91–118

Nishiuchi Y, Inui T, Nishio H, Bódi J, Kimura T, Tsuji FI, Sakakibara S (1998) Chemical synthesis of the precursor molecule of the Aequorea green fluorescent protein, subsequent folding, and development of fluorescence. Proc Natl Acad Sci USA 95:13549–13554

Offord RE (1987) Protein engineering by chemical means? Protein Eng 1:151–157

Offord RE (1990) Chemical approaches to protein engineering. In: Hook JB, Poste GD (eds) Protein design and the development of new therapeutics and vaccines. Plenum Press, New York, pp 253–281
Offord RE (1991) Going beyond the code. Protein Eng 4:709–710
Paulus H (2000) Protein splicing and related forms of protein autoprocessing. Annu Rev Biochem 69:447–496
Perler FB (2002) InBase: The intein database. Nucleic Acids Res 30:383–384
Polgar L, Bender ML (1966) A new enzyme containing a synthetically formed active site. Thiosubtilisin. J Am Chem Soc 88:3153–3154
Prescher JA, Bertozzi CR (2005) Chemistry in living systems. Nat Chem Biol 1:13–21
Quaderer R, Sewing A, Hilvert D (2001) Selenocysteine-mediated native chemical ligation. Helv Chim Acta 84:1197–1206
Radzicka A, Wolfenden R (1988) Comparing the polarities of the amino acids: Side-chain distribution coefficients between the vapor phase, cyclohexane, 1-octanol, and neutral aqueous solution. Biochemistry 27:1664–1670
Rose GD, Wolfenden R (1993) Hydrogen bonding, hydrophobicity, packing, and protein folding. Annu Rev Biophys Biomol Struct 22:381–415
Rose K (1994) Facile synthesis of homogeneous artificial proteins. J Am Chem Soc 116:30–33
Rostovtsev VV, Green LG, Fokin VV, Sharpless KB (2002) A stepwise Huisgen cycloaddition process: Copper(I)-catalyzed regioselective "ligation" of azides and terminal alkynes. Angew Chem Int Ed 41:2596–2599
Sakakibara S (1999) Chemical synthesis of proteins in solution. Biopolymers 51:279–296
Saxon E, Bertozzi CR (2000) Cell surface engineering by a modified Staudinger reaction. Science 287:2007–2010
Saxon E, Armstrong JI, Bertozzi CR (2000) A "traceless" Staudinger ligation for the chemoselective synthesis of amide bonds. Org Lett 2:2141–2143
Schimmel P, Beebe K (2004) Genetic code sizes pyrrolysine. Nature 431:257–258
Schnölzer M, Kent SBH (1992) Constructing proteins by dovetailing unprotected synthetic peptides: Backbone-engineered HIV protease. Science 256:221–225
Sewald N, Jakubke HD (2002) Peptides: chemistry and biology. Wiley-VCH, Weinheim
Sohma Y, Sasaki M, Hayashi Y, Kimura T, Kiso Y (2004) Novel and efficient synthesis of difficult sequence-containing peptides through O–N intramolecular acyl migration reaction of O-acyl isopeptides. Chem Commun 124–125
Srinivasan G, James CM, Krzycki JA (2002) Pyrrolysine encoded by UAG in archaea: Charging of a UAG-decoding specialized tRNA. Science 296:1459–1462
Staudinger H, Meyer J (1919) Über neue organische Phosphorverbindungen III. Phosphinmethylenderivate und Phosphinimine. Helv Chim Acta 2:635–646
Storey HT, Beacham J, Cernosek SF, Finn FM, Yanaihara C, Hofmann K (1972) Studies on polypeptides. 51. Application of S-ethylcarbamoylcysteine to the synthesis of a protected heptatetracontapeptide related to the primary sequence of ribonuclease T1. J Am Chem Soc 94:6170–6178
Tam JP, Lu Y, Liu C, Shao J (1995) Peptide synthesis using unprotected peptides through orthogonal coupling methods. Proc Natl Acad Sci USA 92:12485–12489
Tam JP, Xu J, Eom KD (2001) Methods and strategies of peptide ligation. Biopolymers 60:194–205
Tam JP, Yu Q (1998) Methionine ligation strategy in the biomimetic synthesis of parathyroid hormones. Biopolymers 46:319–327
Tann CM, Qi D, Distefano MD (2001) Enzyme design by chemical modification of protein scaffolds. Curr Opin Chem Biol 5:696–704
Thamm P, Kolbeck W, Musiol HJ, Moroder L (2002) Amide group – "Pseudoproline" protection. In: Goodman M, Felix A, Moroder L, Toniolo C (eds) Hobuben–Weyl, Methods of organic chemistry: Synthesis of peptides and peptidomimetics, vol E22a. Georg Thieme Verlag, Stuttgart, pp 267–268
Toniolo C (2003) Main-chain-modified peptides. In: Goodman M, Felix A, Moroder L, Toniolo C (eds) Houben–Weyl, Methods of organic chemistry: Synthesis of peptides and peptidomimetics, vol E22c. Georg Thieme Verlag, Stuttgart, pp 213–633

Tornoe CW, Christensen C, Meldal M (2002) Peptidotriazoles on solid phase: [1,2,3]-triazoles by regiospecific copper(I)-catalyzed 1,3-dipolar cycloadditions of terminal alkynes to azides. J Org Chem 67:3057–3064

Veber DF, Varga SL, Milkowski JD, Joshua H, Conn JB, Hirschmann R, Denkewalter RG (1969) Studies on the total synthesis of an enzyme. IV. Some factors affecting the conversion of protected S-protein to ribonuclease S'. J Am Chem Soc 91:506–507

Villain M, Vizzavona J, Rose K (2001) Covalent capture: A new tool for the purification of synthetic and recombinant polypeptides. Chem Biol 8:673–679

Walsh CT (2006) Posttranslational modifications of proteins: Expanding nature's inventory. Roberts, Co., Englewood, Colorado

Wang Q, Chittaboina S, Barnhill HN (2005) Advances in 1,3-dipolar cycloaddition reaction of azides and alkynes - a prototype of "click" chemistry. Lett Org Chem 2:293–301

Wieland T, Bokelman E, Bauer L, Lang HU, Lau H (1953) Über Peptidsynthesen: Bildung von S-haltigen Peptiden durch intramolekulare Wanderung von Aminoacylresten. Liebigs Ann Chem 583:129–149

Wu ZP, Hilvert D (1989) Conversion of protease into an acyl transferase: Selenosubtilisin. J Am Chem Soc 111:4513–4514

Wünsch E (1971) Synthesis of naturally occurring polypeptides, problems of current research. Angew Chem Int Ed 10:786–795

Yajima H, Fujii N (1981) Totally synthetic crystalline ribonuclease A. Biopolymers 20: 1859–1867

Yan LZ, Dawson PE (2001) Synthesis of peptides and proteins without cysteine residues by native chemical ligation combined with desulfurization. J Am Chem Soc 123:526–533

Yanaihara N, Yanaihara C, Dupuis G, Beacham J, Camble R, Hofmann K (1969) Studies on polypeptides. XLII. Synthesis and characterization of seven fragments spanning the entire sequence of ribonuclease T1. J Am Chem Soc 91:2184–2185

Native Chemical Ligation: SemiSynthesis of Post-translationally Modified Proteins and Biological Probes

E. Vogel Taylor and B. Imperiali(✉)

Contents

Abstract Native chemical ligation (NCL) is a chemoselective reaction that joins synthetic or recombinant peptide and protein fragments through a native peptide bond. Advances in generating the required peptide and protein components for NCL have enabled the homogeneous incorporation of unnatural amino acids, fluorescent probes, and other modifications into protein domains and full-length proteins. The resulting semisynthetic proteins, which can be made in multi-milligram quantities, may be used in biophysical and biochemical studies to address challenging biological problems not accessible with traditional methods. This chapter provides an overview of NCL,

B. Imperiali
Department of Chemistry and Department of Biology, Massachusetts Institute of Technology, 77 Massachusetts Avenue, Cambridge, MA 02139, USA, e-mail: imper@mit.edu

C. Köhrer and U.L. RajBhandary (eds.) *Protein Engineering.*
Nucleic Acids and Molecular Biology 22,

including relevant design considerations and technological advances, and a discussion of recent applications that apply protein-based probes accessed through NCL.

1 Introduction

Methods for the modification of proteins to introduce novel chemical functionalities and for the preparation of homogenous samples of post-translationally modified proteins are critical for probing protein structure/function relationships and protein/protein interactions. *Native chemical ligation* (NCL) is a chemoselective reaction that joins synthetic or biologically expressed protein fragments via a native amide bond, enabling the construction of modified proteins in multi-milligram quantities sufficient for biophysical and biochemical studies (Dawson et al. 1994). NCL can be carried out with fully deprotected peptide and protein fragments in neutral aqueous media, enabling modified peptides to be incorporated into target peptides or proteins. Required for the reaction are an N-terminal fragment containing a C-terminal α-thioester and a C-terminal fragment with an N-terminal cysteine residue. The general reaction is shown in Fig. 1a. The transformation begins with a reversible transthioesterification to associate the two fragments via a thioester bond involving the cysteine residue of the C-terminal fragment. An energetically favorable acyl rearrangement involving a five-membered-ring transition state then occurs to form a stable amide bond. Since the thioester-linked intermediates have never been observed, it is presumed that the rate-limiting step of NCL is transthioesterifi-

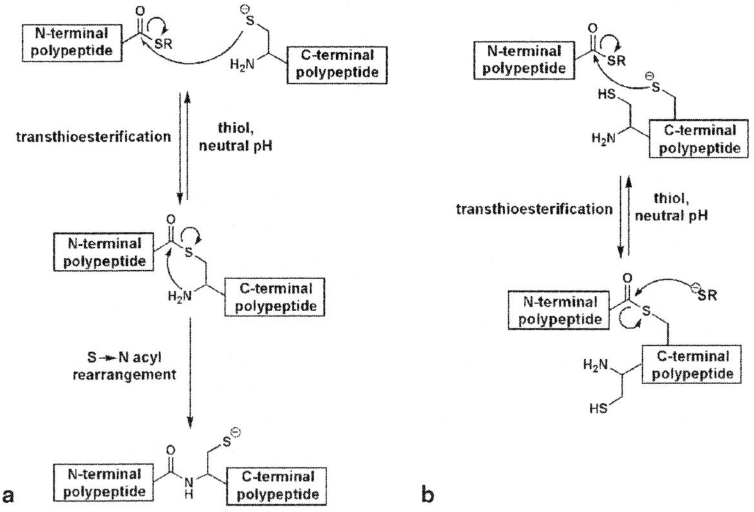

Fig. 1 a Native chemical ligation (NCL) of unprotected polypeptide fragments. **b** Reversible transthioesterification with internal cysteine residues cation. Typically, the reaction is run in the presence of thiol additives, which both suppress cysteine oxidation and catalyze the reaction by generating more reactive thioesters (Dawson et al. 1997). Importantly, NCL is compatible with the presence of fully deprotected internal cysteine residues in either protein fragment (Fig. 1b). Thioester exchange is fully reversible, and it is only upon reaction with the terminal cysteine that the S→N acyl rearrangement can occur, resulting in the thermodynamically stable product and accounting for the chemoselective nature of the reaction.

NCL was first introduced in 1994 with the synthesis of a 72-residue polypeptide comprising entirely native peptide bonds, thus overcoming the practical limits of solid-phase peptide synthesis (SPPS) through the ligation of two fully deprotected peptide fragments (Dawson et al. 1994). While NCL was originally limited to ligations involving α-thioesters generated by chemical synthesis, the scope of the reaction was advantageously extended to proteins of any size through the application of *expressed protein ligation* (EPL), in which α-thioester or N-terminal cysteine-containing fragments that are generated via recombinant methods are used for the semisynthesis of protein domains and full-length proteins (Erlanson et al. 1996; Muir et al. 1998; Severinov and Muir 1998). EPL was originally defined specifically as the extension of NCL that involves recombinant α-thioester peptides, but the term is now used more generally to designate NCL processes involving any recombinant fragments, either thioester or N-terminal cysteine-containing (Muir 2003).

2 Engineering Design Considerations for NCL

2.1 The Ligation Junction

An important consideration in the design of semisynthetic proteins is the designation of a ligation junction, Xaa-Cys (Xaa is any amino acid). Since the practical limit of SPPS is between 40 and 60 residues, junction sites for ligations involving a synthetic peptide must fall within those distances from the N- or C-terminus of the protein target. In some cases naturally occurring cysteine residues are present in the protein of interest at a suitable position for the ligation. However, in the absence of a suitably placed cysteine, one may be introduced in the place of a nonessential residue. Appropriate residues may be selected on the basis of structural knowledge or previous experiments that demonstrate that the residues are not involved in protein function. The ease of site-directed mutagenesis allows for further confirmation that the cysteine mutation does not significantly alter protein structure or function. In general, ideal ligation sites are in segments of a protein that lack secondary structure, such as in terminal regions, loop regions, or linker regions between two domains. If the reaction is carried out under nondenaturing conditions and if the protein will not be subjected to refolding, it is particularly important that the folded state of the protein is not disturbed by the ligation site, regardless of whether a naturally occurring or an introduced cysteine residue is used as the Xaa-Cys junction.

While standard NCL dictates that a cysteine residue be present at the N-terminus of the C-terminal fragment, it is also important to consider the effect of the C-terminal residue of the α-thioester fragment, –Xaa–SR, on the ligation. It has been reported that α-thioester peptides containing any of the 20 encoded amino acid residues at the terminal position are compatible with NCL, but the kinetics of ligation differ dramatically depending on the properties of that residue (Hackeng et al. 1999). In general, the reaction proceeds most rapidly when the C-terminal position is occupied by the sterically unhindered glycine residue, or by histidine or cysteine residues, which are hypothesized to increase the rate of thioester exchange by catalysis with the imidazole or thiol side chain functionalities. Other terminal residues shown to result in rapid product formation are phenylalanine, methionine, tyrosine, alanine, and tryptophan. Ligation has been found to be prohibitively slow only with two of the β-branched amino acids (valine and isoleucine) and proline, although it may be possible to overcome these residue limitations using highly activated thioesters.

In another study, ligations involving aspartic or glutamic acid residues in the C-terminal position of an α-thioester were shown to result in significant side product formation (Villain et al. 2003). In this case, the proximity of the carboxyl functionality in the amino acid side chains to the activated thioester can result in the formation of backbone isomers, in which a β-amide or a γ-amide bond forms between the cysteine-containing peptide and the aspartate or glutamate residues, respectively (Fig. 2a). Side product formation can be avoided by orthogonal protection of the carboxyl side chains. However, when possible, selecting an Xaa-Cys ligation site that avoids sluggish ligations or side reactions is beneficial to ensure maximal product formation.

2.2 *Thiol Additives Used in NCL*

Another factor affecting the efficiency of NCL is the presence of thiol additives (Dawson et al. 1997), which potentially combat difficult ligation reactions. The alkylthioesters commonly used in NCL react slowly because of the poor leaving group properties of the corresponding alkyl thiols, requiring thiol–thioester exchange with the thiol additive to promote ligation with the cysteine-containing fragment. To minimize confusion, this exogenous thiol replacement will be referred to throughout the chapter as *thiol–thioester exchange* to distinguish it from the attack of a cysteine residue in the first step of NCL, which will be referred to as *transthioesterification*. Rapid and complete ligation is best facilitated by thiols that are both good nucleophiles, to promote the in situ formation of a more reactive thioester, and good leaving groups, to favor the transthioesterification (Fig. 2b). For NCL involving peptide fragments, a benzylmercaptan/thiophenol mixture is commonly employed in an aqueous/organic buffer (Fig. 2c). Thiophenol promotes rapid thiol–thioester exchange and

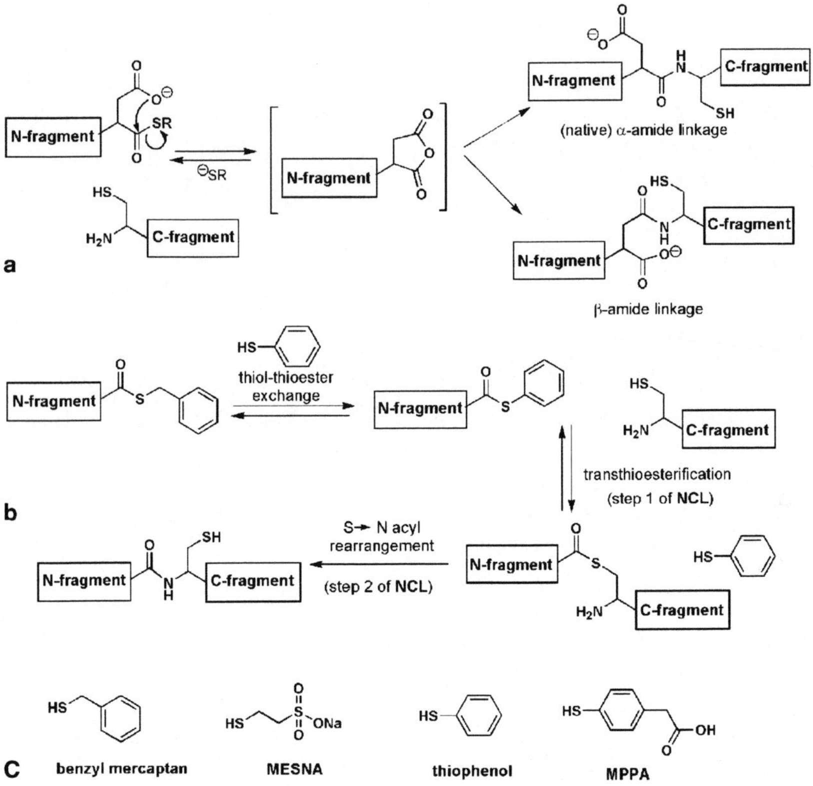

Fig. 2 **a** Generation of a β-amide-linked side product for NCL with an Asp-Cys ligation junction. **b** Mechanism for NCL with thiol catalysis. An ideal thiol additive results in rapid thiol–thioester exchange of the thioester fragment and provides a good leaving group for transthioesterification with the cysteine fragment. **c** Thiols used for NCL

serves as a good leaving group, but has poor aqueous solubility and therefore can only be used at very low concentrations for ligations involving proteins that cannot tolerate the addition of organic solvent. 2-Mercaptoethanesulfonic acid (MESNA) has excellent solubility in water and is a popular choice for ligations using recombinant protein fragments. However, a recent investigation of a number of thiol additives revealed that MESNA, an alkanethiol with pK_a of 9.2, shows rapid thiol–thioester exchange, but has poor leaving group properties (Johnson and Kent 2006). In the case of MESNA, the rate-limiting step for NCL is transthioesterification, whereas with certain other thiols, the thiol–thioester exchange is rate limiting. In general, alkylthioesters, such as those generated by MESNA or benzyl mercaptan, were found to be less reactive than phenylthioesters. The investigators found (4-carboxymethyl)thiophenol (MPPA) to be a superior catalyst with peptide-based test ligations and recommend use of MPPA for protein ligations as well, predicting that it may be more effective than MESNA for

EPL. It will be interesting to see if MESNA is eventually replaced as the predominantly used thiol for EPL by MPPA or another aryl thiol catalyst. Another option for rapid ligation is the use of preformed aryl thioesters (Dawson et al. 1994; von Eggelkraut-Gottanka et al. 2003; Bang et al. 2006), which eliminates the thiol–thioester exchange step that is rate limiting with certain thiol additives.

3 Thioester Generation

3.1 *Synthetic Thioester Peptides*

The synthesis of α-thioester peptides for NCL can be accomplished using a variety of SPPS-based methods. The modular nature of SPPS allows facile replacement of any residue of the target peptide thioester with nonencoded amino acids, as well as the site-specific incorporation of tags and probes. Initially most investigators employed acid-labile *tert*-butyloxycarbonyl (Boc)-protection strategies for SPPS (Dawson et al. 1994; Canne et al. 1995), since the thioester linkage is not stable to the basic deprotection treatment with piperidine used in N^{α}-(9-fluorenylmethoxycarbonyl) (Fmoc)-based SPPS. However, a number of approaches have been developed to circumvent the sulfur–carbonyl cleavage associated with Fmoc deprotection, allowing the generation of thioester peptides without the harsh cleavage conditions employed in Boc-based SPPS. This use of milder conditions is particularly important for the synthesis of glycopeptide and phosphopeptide thioesters, since glycoside and phosphoryl linkages are labile to the anhydrous hydrofluoric acid required for peptide cleavage in Boc-based SPPS.

In one Fmoc-based approach, deblocking agents were developed as alternatives to piperidine to effectively remove the Fmoc protecting group without nucleophilic cleavage of the thioester linkage (Li et al. 1998; Bu et al. 2002). In an alternative strategy, a backbone amide linkage is used to anchor the penultimate residue, which contains an orthogonally protected C-terminus, to the solid support (Alsina et al. 1999). The peptide can be elaborated using standard Fmoc-based SPPS protocols, and subsequently deprotected at the C-terminus prior to being coupled to a thioester amino acid and finally fully deprotected and cleaved from the resin (Fig. 3a). Additional backbone amide linkage and side chain linkage based approaches have been reported (Gross et al. 2005), including an example that masks the terminal glycine thioester as a trithioorthoester and releases the deprotected thioester by acid treatment (Brask et al. 2003).

A more commonly applied strategy utilizes an alkanesulfonamide “safety catch” linker, in which a peptide is assembled on a solid support linked via the C-terminus by a sulfonamide bond. This sulfonamide linkage is stable to the repeated piperidine deprotection treatments in Fmoc-based SPPS. Following peptide synthesis, the sulfonamide is activated by N-alkylation and subsequently cleaved from the solid support by nucleophilic attack of a thiol additive. The

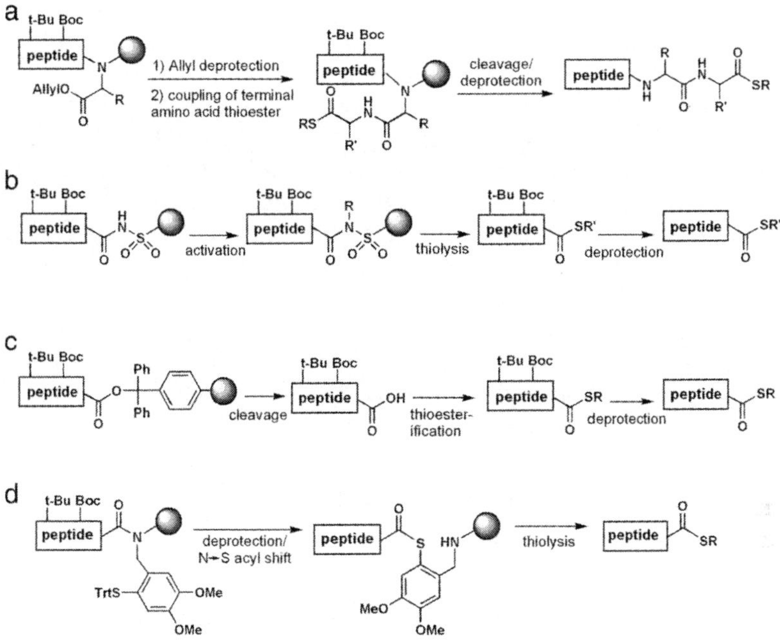

Fig. 3 N^{α}-(9-Fluorenylmethoxycarbonyl) (Fmoc)-based solid-phase peptide synthesis methods for peptide α-thioester synthesis include **a** the backbone amide linkage strategy, **b** the "safety catch" linker strategy, **c** in situ thioesterification of a protected peptide, and **d** thiolysis following a thiol-auxiliary mediated N→S acyl shift. *Boc tert*-butyloxycarbonyl

thioester bond is stable in trifluoroacetic acid (TFA), and the peptide can therefore be deprotected with a standard TFA cleavage cocktail (Fig. 3b) (Ingenito et al. 1999; Shin et al. 1999). A similar method employs an aryl hydrazide linker that can be activated following peptide synthesis by mild oxidation (Camarero et al. 2004a, b).

Arguably the most straightforward Fmoc-based SPPS approach for generating thioesters involves peptide synthesis on highly acid labile resin, such as commercially available 2-chlorotrityl or TGT resin, followed by cleavage of the fully protected peptide from solid support with mild acid treatment. The acid releases the peptide as a C-terminal carboxylic acid, but does not affect the Fmoc-compatible side chain protecting groups. The protected peptide is then derivatized to a C-terminal thioester in situ by treatment with the desired thiol and standard peptide-coupling activating agents (Futaki et al. 1997; Mezo et al. 2001). Side chain deprotection affords the corresponding free thioester (Fig. 3c). An evaluation of activating agents for this strategy has identified conditions, specifically phosphonium salt based activating agents, that result in high yields and low levels of epimerization (less than 1.4%) (Von Eggelkraut-Gottanka et al. 2003).

While still in a preliminary stage of development, a final method of α-thioester formation is worth mentioning, both for its creative approach and for its similarity to the protein splicing mechanism that is exploited for the generation of recombinant protein α-thioesters (Sect. 3.2). This method involves the generation of a thioester-linked peptide by utilizing an N→S acyl shift facilitated by a protected thiol-containing auxiliary attached to the peptide backbone (Fig. 4d) (Kawakami et al. 2005). Following Fmoc-based SPPS of a peptide linked to resin by an amide bond, deprotection of the peptide side chains with a TFA cleavage cocktail concurrently deprotects the thiol moiety of the auxiliary, initiating an acyl shift that results in a thioester bond-linkage of the peptide to solid support. Subsequent thiolysis releases the corresponding peptide thioester.

A recent review of methods for thioester synthesis provides more information on these and other approaches (Camarero and Mitchell 2005).

3.2 *Recombinant Thioester Fragments*

While direct synthesis is suitable for peptide fragments under approximately 60 residues in length, the scope of NCL was initially limited by the lack of recombinant methods to generate longer peptide or protein fragments containing C-terminal α-thioesters. This limitation was overcome in 1998 with the introduction of EPL, which applies a modified protein splicing mechanism to generate thioesters suitable for ligation (Muir et al. 1998; Severinov and Muir 1998). To place the technology developed for recombinant protein thioesters in context, a brief explanation of protein splicing follows.

3.2.1 Protein Splicing

Protein splicing is a protein processing event that results in the extrusion of an internal protein segment, termed an "intein," and the concomitant joining of two flanking regions, termed "N- and C-exteins," through an amide bond (Paulus 2000). Inteins, which are functionally analogous to self-splicing RNA introns, are protein segments that catalyze the intramolecular protein rearrangement that mediates their excision (Fig. 4a). A number of conserved or key residues in inteins contribute to the splicing event through structural or electronic influences (David et al. 2004). In general, inteins contain a cysteine or serine residue at the N-terminal position (termed the 1 position), a conserved asparagine residue at the C-terminal position, and a cysteine, serine, or threonine residue at the first position of their flanking C-extein (termed the +1 position). In the initial step of standard protein splicing, an N→S (or O) acyl shift occurs, involving the cysteine (or serine) at position 1, transferring the N-extein from the backbone to the side chain of residue 1 of the intein. While this step appears thermodynamically unfavorable, it is favored by the conformation of the intein, which is thought to distort the scissile amide bond into a higher-energy confor-

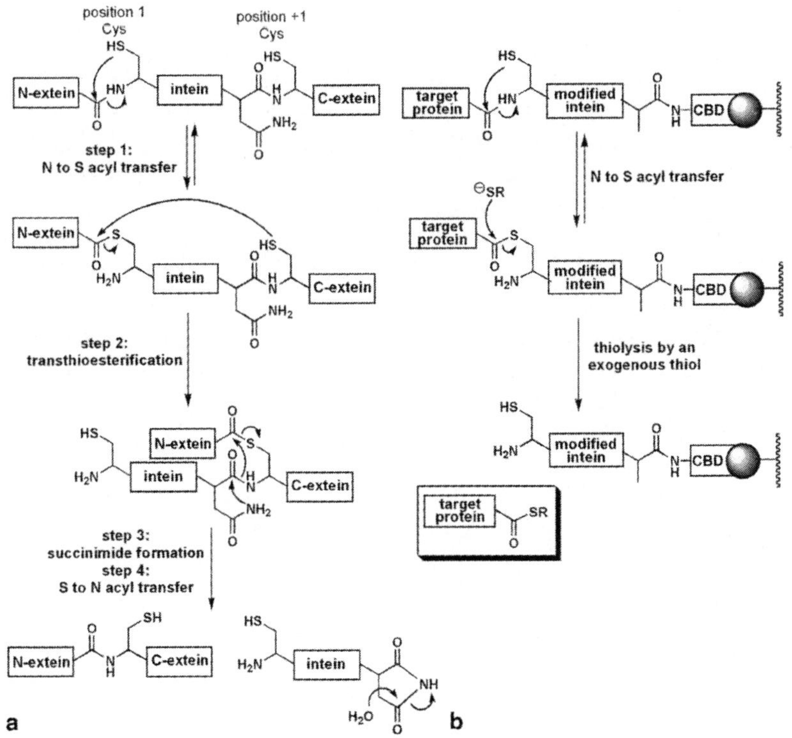

Fig. 4 **a** Mechanism of protein splicing. Splicing results in the joining of two exteins through a native peptide bond and the extrusion of the intein segment. **b** Recombinant generation of an α-thioester using a modified intein system. The target protein is N-terminal to the intein, and a chitin binding domain (*CBD*) is C-terminal to the intein, immobilizing the construct on solid support. An asparagine to alanine mutation prevents succinimide formation. Thiolysis results in release of the α-thioester target protein

mation. In the next step, transesterification involving the cysteine (or serine or threonine) at position +1 joins the two exteins through a thioester bond. In the third step, the asparagine residue cyclizes, resulting in cleavage at the C-terminal splice junction to liberate the intein as a C-terminal succinimide. In the final step, which is the sole reaction that does not require catalysis by the intein, a spontaneous acyl rearrangement produces the spliced exteins with an amide linkage (Muir 2003).

3.2.2 Mutant (Asparagine to Alanine) Inteins To Generate Recombinant Thioesters

Protein splicing has been exploited with great success for the generation of recombinant thioesters. Inteins have been engineered with an asparagine to alanine mutation, which permits the initial step of the protein splicing mechanism involving an N→S acyl shift to produce a thioester linkage, but prevents subsequent succinimide formation (Chong et al. 1996). Addition of exogenous thiols results in release of the N-extein as the corresponding C-terminal thioester. The isolation of recombinant thioesters from a resin-bound intein system has been commercialized by New England Biolabs as the IMPACT™ (intein-mediated purification with an affinity chitin binding tag) system. In this system, a target gene, functioning as the N-extein, is cloned immediately N-terminal to a genetically modified (asparagine to alanine) intein gene. A chitin binding domain is cloned C-terminal to the intein, functioning as the C-extein and facilitating immobilization of the resulting protein construct on chitin beads. Thus, the expressed three-segment (target protein–intein–chitin binding domain) construct can be isolated from all other cellular proteins by immobilization and washing, and the target protein thioester can be released by subsequent thiolysis (Fig. 4b) (Chong and Xu 2005).

EPL can be performed with the eluted thioester or directly on the chitin beads (Evans et al. 1998; Muir et al. 1998; Severinov and Muir 1998). For solid-phase ligation, the thiol- and cysteine-containing peptide/protein fragment can be simultaneously incubated with the resin-bound protein fusion, enabling concurrent thiolysis and ligation.

4 N-Terminal Cysteine Fragments

The synthesis of peptides containing an N-terminal cysteine residue is straightforward using SPPS and requires no additional manipulations. For longer fragments, there are several approaches for accessing biochemically expressed proteins with an N-terminal cysteine. The most commonly applied methods involve the generation of a precursor protein designed with a cysteine residue immediately C-terminal to a protease cleavage site (Fig. 5, method A). The first example of affinity cleavage for NCL used factor Xa, a protease that cuts C-terminal to its Ile-Glu-Gly-Arg recognition sequence (Erlanson et al. 1996). Other C-terminal cleaving proteases include enterokinase, ubiquitin C-terminal hydrolase, and furin. The single disadvantage with employing proteases is the possibility for undesired cleavage at secondary sites. For instance, factor Xa can cleave after Gly-Arg pairs or other basic residues in a target protein.

Recently *Tobacco etch virus* (TEV) protease, a highly specific cysteine protease with a seven-residue recognition sequence, was applied for the generation of N-terminal cysteine fragments (Tolbert and Wong 2002). TEV protease typically recognizes a serine or glycine residue in the P1′ site, but will also tolerate a cysteine in that position. TEV protease demonstrates high sequence selectivity and overexpresses well on a large scale, making it an ideal protease for EPL applications.

Endogenous methionine aminopeptidases have also been utilized to access N-terminal cysteine proteins from the corresponding Met-Cys-containing precursor proteins (Fig. 5, method B). The resulting cysteine polypeptides can be

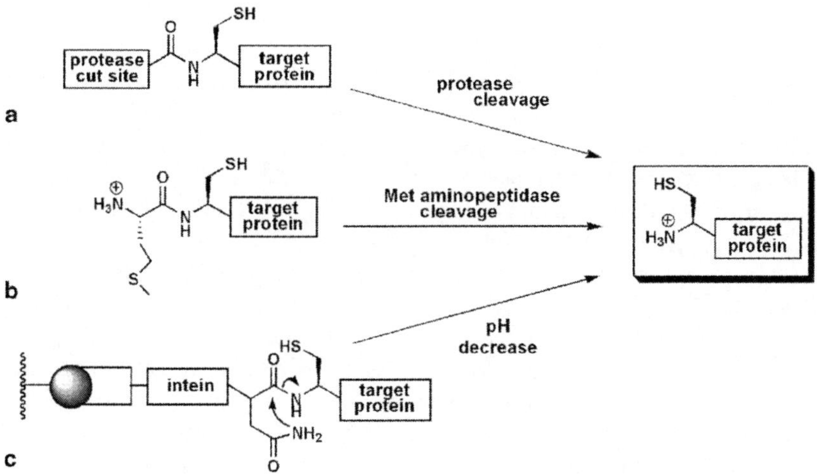

Fig. 5 Recombinant methods for generating N-terminal cysteine-containing protein fragments using exogenous protease **(A)**, endogenous methionine aminopeptidase **(B)**, and a genetically modified intein system **(C)**

isolated from cell lysate using aldehyde-functionalized resin (Villain et al. 2001), or reacted directly for in vivo NCL (Camarero et al. 2001). For exogenous cleavage of a Met-Cys junction, cyanogen bromide was successfully applied to access an N-terminal cysteine in a recombinant glycoprotein that was insoluble in buffers compatible with the commonly employed proteases (Macmillan and Arham 2004).

An intein-based strategy can also be applied to generate N-terminal cysteine proteins (Fig. 5, method C). Several commercially available expression vectors contain genetically modified inteins that lack the conserved cysteine (or serine) residue at the N-terminus (1 position) of the intein for this purpose. Cleavage of the intein by succinimide formation, induced by lowering the pH and increasing temperature, simultaneously releases the N-terminal cysteine protein (Chong and Xu 2005). The limitation to this method is the possibility for spontaneous intein cleavage.

5 Extensions of NCL

5.1 Sequential NCL

Although many NCL applications require the ligation of only two fragments, the technique is not limited to a single ligation step. Multiple modifications throughout a protein or modification in the middle of a large protein can be accomplished by sequential ligations (Fig. 6a) (Cotton et al. 1999). In sequential NCL, an initial ligation is carried out between an N-terminal cysteine-containing fragment and a second fragment containing a thioester moiety and a protected N-terminal cysteine residue, masked to prevent the thioester fragment from undergoing intra- or intermolecular ligations. Strategies used to mask the cysteine residue include the incorporation of the factor Xa prosequence immediately N-terminal to the cysteine residue (Cotton et al. 1999), acetamidomethyl protection of the side chain thiol (Macmillan and Bertozzi 2004), thiazolidine cysteine protection (Bang and Kent 2004), and photolabile protection of the thiol or amino group (Ueda et al. 2005). Following the initial ligation, the cysteine is deprotected and a second ligation is carried out. Theoretically there is no limit to the number of fragments that can be ligated in this fashion. Segmental ligation has been performed on a solid support, in a strategy analogous to SPPS (Cotton and Muir 2000), as well as in situ, either in "one-pot" reactions (Bang and Kent 2005; Ueda et al. 2005) or with high-performance liquid chromatograpy or affinity-tag purification of the ligated intermediates (Bang and Kent 2005).

5.2 *Accessing Xaa-Ala and Xaa-Gly Ligation Junctions*

The only major limitation of NCL is the general requirement for a cysteine residue at the ligation site. For specific applications, the requisite N-terminal cysteine residue can be replaced by another nucleophilic residue, such as selenocysteine (Gieselman et al. 2001). Alternatively, NCL can be combined with postligation desulfurization with palladium or H_2/Raney nickel to convert the resultant cysteine to an alanine, and thereby access peptides or proteins with an Xaa-Ala ligation junction (Yan and Dawson 2001). This method has been applied to several proteins that lack cysteine residues (He et al. 2003; Bang et al. 2005), but cannot be used for proteins that contain a cysteine anywhere in the protein sequence, since all thiols will be reduced by the desulfurization step.

In a more generally applicable strategy, NCL can be performed with a removable cysteine mimic to ultimately produce an Xaa-Gly ligation junction (Fig. 6b) (Canne et al. 1996). In this approach the C-terminal ligation fragment is linked via the N-terminal amine to a thiol-containing auxiliary that facilitates thioester exchange with the α-thioester fragment. Following transthioesterification, an S→N acyl shift results, and subsequent cleavage of the N-linked auxiliary yields a native amide bond at the ligation site. Rapid ligation is best promoted by auxiliaries that proceed through a five-membered-ring acyl rearrangement intermediate, and most of these utilize a 2-aryl mercaptoethyl group as the cysteine mimic. Depending on the aryl substituents, removal of the auxiliary is accomplished with strong acid treatment (Botti et al. 2001), milder TFA treatment (Macmillan and Anderson 2004), or photolysis (Marinzi et al. 2004).

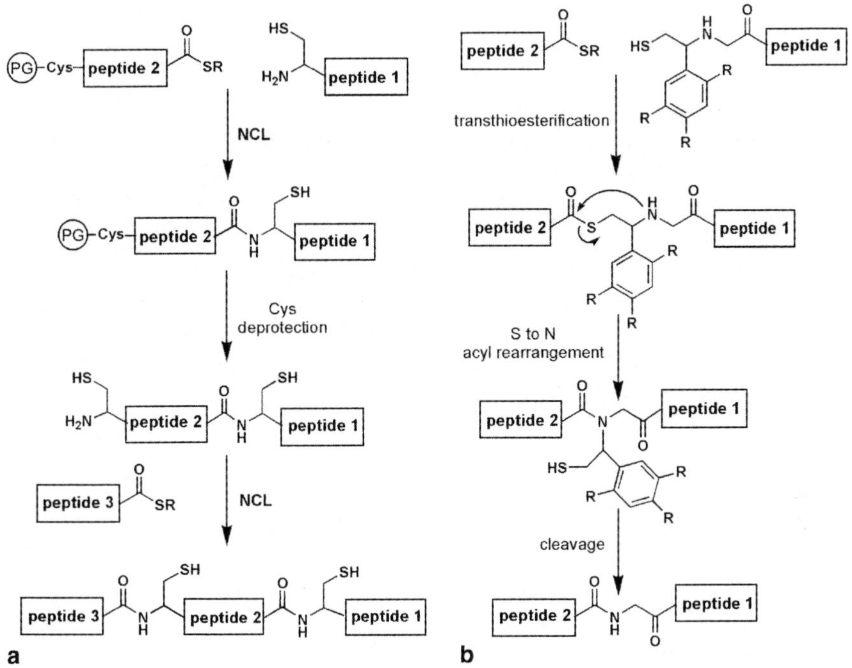

Fig. 6 **a** Sequential NCL. **b** NCL with a 2-aryl mercaptoethyl ligation auxiliary to produce an Xaa-Gly ligation junction. *PG* cysteine protecting group

6 Applications

The strongest testament to the utility and generality of NCL is the impressive number of applications that have been reported for the study of complex biological systems using ligation methods to access homogenous samples of native or modified proteins or protein analogs. Owing to the vast number of applications reported since NCL was first introduced, only a fraction of these can be covered with adequate detail in this chapter. Since a number of excellent reviews are available (Muir 2003; David et al. 2004; Schwarzer and Cole 2005), particular focus is on reports from the past 3 years.

6.1 Post-translational Modifications

Post-translational (and cotranslational) modification of proteins increases the diversity of the proteome by more than an order of magnitude beyond that programmed by the genetic code (Walsh 2006). Direct characterization of the impact of post-translational modifications, including phosphorylation, glycosylation, lipidation, and acetylation, on the structure and function of proteins can be prohibitively complex owing to the difficulty of obtaining homogenously modified protein in high yields. The nature of this difficulty stems from the lack of genetic encoding for these modifications, the heterogeneity of biological samples, and the inherent nonspecificity in the enzymes that catalyze post-translational modifications. NCL and EPL allow the generation of modified proteins in milligram quantities by the reaction of a biologically expressed non-modified fragment with a synthetic fragment containing the desired modification (see also the chapter by van Kasteren et al., this volume). Many post-translational modifications occur within the N- or C-terminus of proteins, making the corresponding modified proteins ideal targets for NCL. However, segmental ligation can be used to incorporate modifications within the central regions of a protein as well.

6.1.1 Phosphorylation

Phosphorylation is the most prevalent post-translational modification, affecting an estimated one third of human proteins (Walsh 2006). Protein kinases, enzymes which catalyze transfer of the γ-phosphoryl from ATP to a serine, threonine, or tyrosine side chain within a peptide or protein, can have tens or hundreds of substrates, and may modify multiple sites in a single protein, making the production of discretely phosphorylated proteins challenging. NCL has facilitated the preparation of homogeneous samples of phosphoproteins, enabling investigators to isolate the specific roles of phosphorylation on protein function in vitro.

In one example, synthetic variants of a Cys_2His_2 zinc finger protein were prepared by NCL in various phosphorylated forms and used to study the effects of linker phosphorylation on DNA binding (Jantz and Berg 2004). Cys_2His_2 zinc finger proteins are a class of transcription factors comprising zinc-binding domains joined by highly conserved linker regions, which each include a single threonine or serine residue. These linker regions are known to be phosphorylated, but previous studies on the effects of phosphorylation were limited to using partially purified protein isolated from cell lysates, prohibiting detailed quantitative evaluation. NCL enabled the preparation of pure phosphorylated variants of a representative 86-residue protein containing three zinc finger domains joined by two linker regions. The proteins were synthesized by sequential ligation of peptide fragments to access variants with a single phosphothreonine residue in the N-terminal or C-terminal linker, phosphothreonine residues in both linkers, or no linker phosphorylation (Fig. 7, semisynthesis A). Direct comparison of the DNA binding affinity of the three phosphorylated variants and the nonphosphorylated protein using a fluorescence-anisotropy-based assay indicated that phosphorylation of either linker resulted in an approximately 40-fold decrease in DNA binding affinity, and that phosphorylation of both linkers resulted in a 130-fold loss. These quantitative measurements, made possible by pure preparations of phosphorylated variants, provide strong support for a model the authors set out to evaluate, in which coordinated regulation of zinc

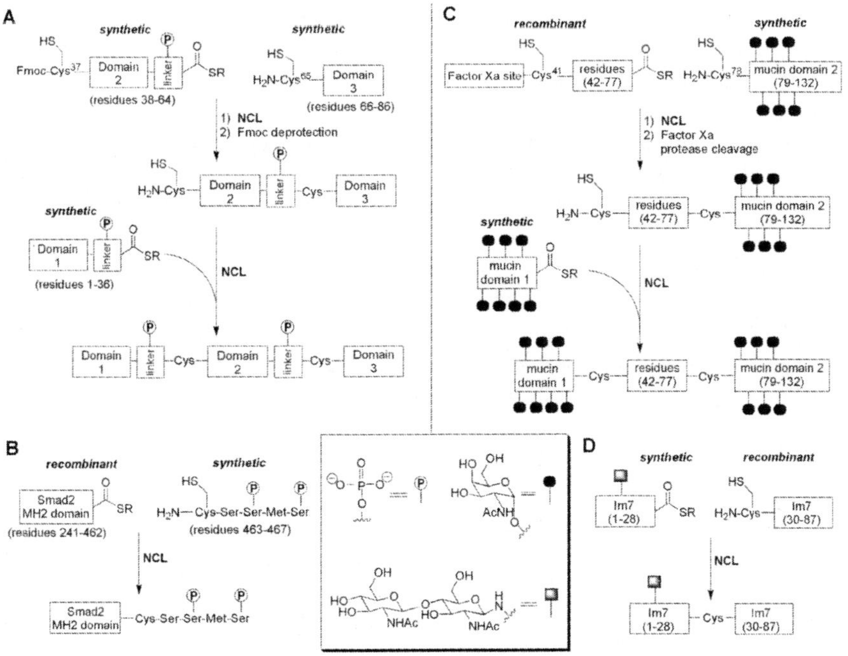

Fig. 7 Semisynthesis of a dually phosphorylated zinc finger protein **(A)**, Smad2 phosphorylated on residues 465 and 467 of its C-terminal tail **(B)**, GlyCAM-1 with 13 *N*-acetylgalactosamine modifications on two mucin-like domains **(C)**, and an N-linked glycoprotein variant of Im7 **(D)**

finger transcription factors is achieved by cellular phosphorylation of the linker regions.

In another study, singly and dually phosphorylated variants of a signaling protein, Smad2, were prepared using EPL, allowing the investigators to deconvolute the impact of each phosphoryl group on protein oligomerization and on further Smad2 phosphorylation by an upstream kinase (Ottesen et al. 2004). Phosphorylation of receptor-activated Smad (R-Smad) proteins, such as Smad2, occurs on the final two serine residues of a C-terminal phosphorylation sequence. Phosphorylation of this C-terminus results in the dissociation of the R-Smad from the membrane-bound receptor complex, and the formation of a new heteromeric complex between the R-Smad and a related co-Smad protein, which subsequently translocates to the nucleus and regulates gene expression. Phosphorylated R-Smads can also form homotrimers in vitro. With the serine residues under investigation located at the extreme C-terminus, Smad2 and the three possible phosphorylated variants were ideal candidates for EPL and were prepared by ligation of an expressed α-thioester corresponding to Smad2 (residues 241–462) with one of four synthetic pentapeptides containing a phosphoserine residue at either position 465 or 467, at positions 465 and 467, or at neither position (Fig. 7, semisynthesis B). Biophysical studies on the oligomerization state of the variants, including analytical ultracentrifugation, revealed that stable Smad2 oligomer formation requires phosphorylation at both serine residues, but that phosphorylation of Ser465 provides the driving force for oligomerization. In addition, phosphorylation of Smad2 at Ser467 proceeds more rapidly when the protein is already phosphorylated at Ser465, while the rate of phosphorylation of Smad2 at Ser465 does not significantly increase by prephosphorylation at Ser467. Since Smad2 is enzymatically phosphorylated on both sites by the same kinase, a semisynthetic strategy was critical to enable access to pure samples of singly phosphorylated Smad2 for these studies. Interestingly, the receptor kinase that phosphorylates Smad2, TβRI, is itself activated by phosphorylation, and a tetraphosphorylated variant of this kinase was among the first applications of NCL to produce phosphoproteins (Huse et al. 2000; Flavell et al. 2002) and was utilized in the Smad2 phosphorylation assay.

6.1.2 Glycosylation

Homogenous samples of glycosylated proteins are particularly challenging to access because cellular glycoproteins exist as complex mixtures of glycoforms, which are difficult to purify or even characterize (Macmillan and Bertozzi 2000). Similar to phosphorylation, enzymatic glycosylation results in heterogeneous samples owing to modification of multiple residues, and glycoprotein mixtures may be further complicated by the diversity of oligosaccharides and possible sugar linkages. Tremendous effort has been devoted to the chemical and chemoenzymatic synthesis of glycopeptides (Grogan et al. 2002), and these advances have been applied to the synthesis and semisynthesis of homogenous glycoproteins by NCL (Shin et al. 1999; Marcaurelle et al. 2001; Hojo et al. 2003; Miller et al. 2003; Tolbert et al. 2005). For the preparation of glycopeptide α-thioesters, Fmoc-based SPPS strategies are exclusively applied, since glycosidic linkages are not stable to the repetitive acid treatments required in Boc-based SPPS.

In an impressive synthetic undertaking, a 132-residue mucin-like glycoprotein, GlyCAM-1, was semisynthesized in three distinct glycoforms containing up to 13 glycans for future investigations on the effect of glycosylation on GlyCAM-1 structure and function (Macmillan and Bertozzi 2004). GlyCAM-1 consists of a central unglycosylated domain flanked by two mucin domains, characterized by dense groups of α-O-linked *N*-acetylgalactosamine on serine and threonine residues. All glycans were introduced as the monosaccharide *N*-acetylgalactosamine derivative, which, in principle, can be elaborated enzymatically to produce fully active glycoprotein. The semisynthetic glycoforms generated in this study include GlyCAM-1 variants glycosylated exclusively in either the N-terminal or the C-terminal mucin domain and a variant glycosylated in both mucin domains. The authors applied three different NCL approaches to achieve all the glycoforms, including a sequential strategy using two ligations and involving a recombinant α-thioester with a masked N-terminal cysteine to generate the GlyCAM-1 variant glycosylated on both mucin domains (Fig. 7, semisynthesis C).

NCL has also been applied to the semisynthesis of an N-linked glycoprotein variant of a well-studied bacterial protein, Im7, to investigate the influence of glycosylation on protein folding (Hackenberger et al. 2005). N-linked glycosylation is a cotranslational event that is thought to assist in the correct folding of expressed proteins. The four-helix Im7 protein, which is not naturally glycosylated, served as a tractable model for folding studies owing to the significant quantity of data available on the kinetics and thermodynamics of its three-state folding mechanism. The glycoprotein analog was prepared with an asparagine-linked chitobiose building block at residue 13 on helix I by the ligation of a glycosylated α-thioester to an expressed C-terminal fragment comprising Im7 residues 29–87 (Fig. 7, semisynthesis D). Biophysical analysis of the glycosylated and nonglycosylated semisynthetic Im7 variants revealed that glycosylation at position 13 had minimal effect on protein folding. Investigations of other Im7 glycovariants are currently underway to probe the effect of glycosylation at other sites.

6.1.3 Lipidation

The post-translational lipidation of proteins is involved in regulating function by targeting modified proteins to specific membranes. The covalent addition of lipid anchors can be divided into four main classes: N-terminal myristoylation, C-terminal addition of a glycosyl phosphatidylinositol (GPI), acetylation, and prenylation (Walsh 2006). As with other post-translationally modified proteins, lipoproteins are challenging to access by genetic or enzymatic methods and have become exciting targets for NCL.

Prenylation involves the addition of a farnesyl or a geranylgeranyl group to one or two cysteines at the C-terminus of a protein. The C-terminal location of the modified cysteine residues renders prenylated proteins well suited for semisynthesis by EPL. A successful application has been the semisynthesis of fluorescently labeled mono- and diprenylated variants of Rab7, a Rab guanosine triphosphatase (GTPase) of the Ras-GTPase superfamily (Durek et al. 2004). An expressed α-thioester corresponding to Rab7 (residues 1–201) was reacted with fluorescently labeled and prenylated hexapeptides to generate Rab7 variants with geranylgeranyl modifications on either Cys205 or Cys207, or on both Cys205 and Cys207 of the C-terminus (Fig. 8a). Semisynthesis involving the hydrophobic lipopeptides required several modifications to the standard EPL protocol, including the addition of specific detergents and an organic extraction of unreacted peptide, which binds to the protein noncovalently, from precipitated protein. In addition, a Rab chaperone protein was necessary to stabilize the denatured Rab7 variants during refolding. A novel fluorescence-based prenylation assay, utilizing the environment-sensitive properties of the dansyl fluorophore, was developed to probe the mechanism of diprenylation. The results of the prenylation assay support a proposed random sequential mechanism of prenylation. The straightforward assay, facilitated by the creative application of a fluorescent tag, was possible because the researchers were able to generate homogenous monoprenylated protein.

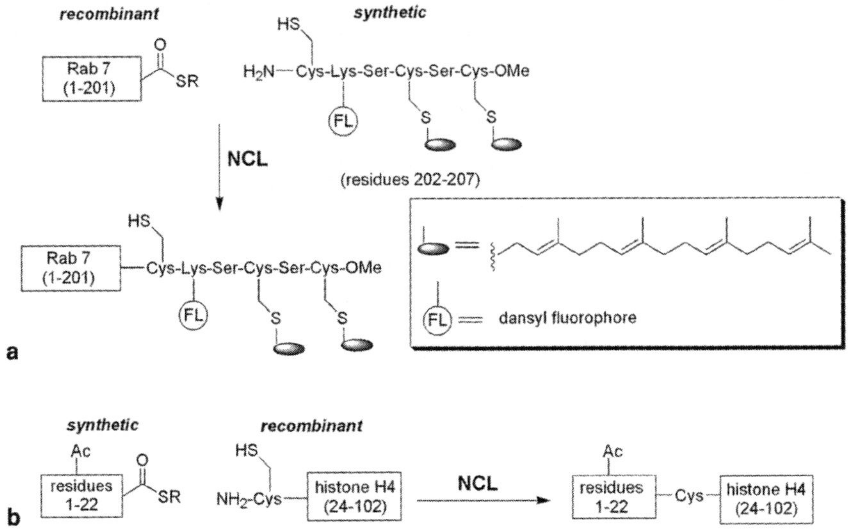

Fig. 8 Semisynthesis of **a** fluorescently labeled and diprenylated Rab7 and **b** monoacetylated histone H4

As with glycosylation, a primary challenge in the EPL of prenylated proteins is the synthesis of modified peptides. Comparison of a number of solution-phase and solid-phase methods for synthesizing prenylated peptides revealed the strength of a hydrazide-linker-based solid-phase approach that was used to incorporate several prenylated and fluorescently labeled peptides onto the α-thioester fragment of Rab7 (Brunsveld et al. 2005). The extensive semisynthetic work with Rab7 was recently extended to other Ras-type GTPases: those in the Ras subfamily (Gottlieb et al. 2006). Prenylated Ras proteins pose additional semisynthetic challenges because they are not known to interact with a chaperone, the use of which was essential for the stabilization and purification of Rab7. This challenge was successfully addressed by the use of polybasic prenylated peptides, which eliminated nonspecific peptide/protein aggregation and enabled ligation and purification in nondenaturing conditions.

Progress has also been made in the preparation of proteins with GPI anchors. GPI proteins are modified via a C-terminal amide linkage with a lipopentasaccharide anchor. With use of EPL, lipidated analogs of green fluorescent protein (GFP) were created with a simplified GPI anchor, a phospholipid without glycans, to demonstrate a flexible strategy for generating proteins lipidated at the C-terminus (Grogan et al. 2005). The lipidated GFP variants were shown to incorporate stably into supported membranes, and quantification of their lateral fluidity was achieved by fluorescence imaging. This strategy can therefore be applied to the semisynthesis of naturally lipidated proteins and their study in lipid bilayers.

6.1.4 Acetylation

Reversible acetylation involves modification of proteins, notably histones and transcription factors, on the ε-amino group of lysine residues. In core histones, which comprise the octomeric protein core of nucleosomes, post-translational modification of the N-terminus alters histone–DNA interactions and is implicated in regulating gene transcription (Walsh 2006). NCL has been employed to generate pure samples of modified histones, acetylated or methylated on the N-terminal tail (He et al. 2003; Shogren-Knaak and Peterson 2004). In a recent example, a homogenous monoacetylated variant of histone H4 was prepared by NCL to characterize the structural and functional effects of acetylation of Lys16 (Shogren-Knaak et al. 2006). Toward this end, a synthesized peptide thioester acetylated at Lys16 and corresponding to residues 1–22 of histone H4 was ligated to a recombinant C-terminal fragment (residues 23–102) (Fig. 8b). The H4 variant was incorporated into nucleosomal arrays and found to inhibit the formation of higher-order chromatin structures and prevent the functional interaction of histones with a specific chromatin-associated protein. This characterization of a selectively acetylated variant complements previous peptide competition studies and provides direct evidence not accessible with truncated or randomly hyperacetylated histone derivatives.

6.2 Fluorescent Probes

Genetically encoded fluorophores are widely used for imaging proteins in live cells. GFP and the ever-increasing number of GFP variants with improved properties and a spectrum of excitation and emission wavelengths are invaluable probes for imaging protein localization and examining intramolecular and intermolecular protein interactions. One powerful application of fluorophore-labeled proteins is fluorescence resonance energy transfer (FRET). FRET results when two fluorophores are within close proximity and the emission spectrum of the "donor fluorophore" overlaps with the excitation spectrum of the "acceptor fluorophore." The proximity and spectral overlap enable a transfer of energy and a corresponding increase in the intensity of the acceptor fluorophore emission, allowing for quantification of the distance between the two fluorophores. The major drawback of using genetically encoded fluorophores for FRET and other fluorescence-based imaging is the significant size of the fluorophores (27 kDa for GFP) appended onto the protein of interest, potentially altering native interactions and localization. Organic fluorophores are significantly smaller (less than 1 kDa) and often possess superior photophysical properties, such as higher extinction coefficients and greater resistance to photobleaching, but can be challenging to incorporate into proteins in a chemoselective manner.

NCL has been used to chemoselectively install a donor–accepter pair of organic fluorophores into proteins to study both intramolecular conformational changes (Cotton and Muir 2000) and intermolecular interactions (Scheibner et al. 2003) using FRET. In an example of the latter, variants of serotonin *N*-acetyltransferase

(AANAT), a circadian rhythm enzyme, were constructed via EPL with a fluorescein- or rhodamine-containing peptide at the C-terminus (Scheibner et al. 2003). Since standard methods for determining oligomerization state, such as size-exclusion chromatography and dynamic light scattering, proved inconclusive with AANAT, FRET was employed to probe for homooligomerization (Fig. 9a). Fluorescence studies showed a significant increase in FRET upon incubation of the donor (fluorescein)- and acceptor (rhodamine)-containing AANAT variants, indicating a preference for AANAT to oligomerize.

Along with FRET, there have been numerous applications reported for fluorescent probes incorporated into proteins by NCL, some of which are discussed in the context of other sections in this chapter. Examples include the C-terminal labeling of proteins for anisotropy-based binding studies (Maag and Lorsch 2003), the domain-specific replacement of tryptophan residues with a redshifted and environment-sensitive tryptophan analog to probe domain function in the context of a full-length protein (Muralidharan et al. 2004), and the site-specific incorporation of an environment-sensitive fluorophore into an effector domain to monitor protein-domain interactions (Becker et al. 2003). Fluorescent labeling of proteins has also been accomplished in vivo (Yeo et al. 2003). A cell-permeable fluorescent thioester was introduced into cells expressing a target protein with an N-terminal cysteine, generated by intein-mediated splicing, resulting in NCL to generate a protein variant labeled at the C-terminus (Fig. 9b).

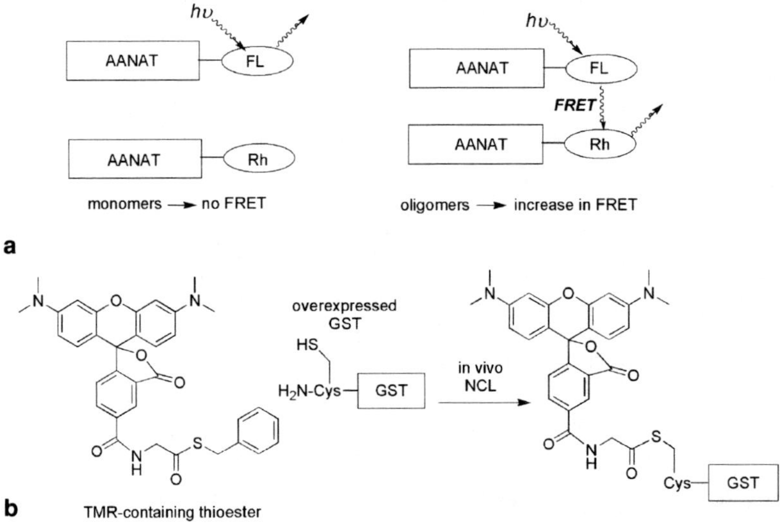

Fig. 9 a Serotonin *N*-acetyltransferase (*AANAT*) labeled with fluorescein (*FL*) and rhodamine (*Rh*). Expected fluorescence energy resonance transfer (*FRET*) outcome for AANAT as a monomer (no FRET) and as an oligomer (FRET). **b** In vivo labeling of glutathione *S*-transferase (*GST*; containing an N-terminal cysteine residue) with the membrane-permeable tetramethyl rhodamine (TMR) thioester

6.3 Unnatural Amino Acids

The power of NCL is even more pronounced in its application to the semisynthesis of protein domains or full-length proteins containing unnatural amino acids. Amino acid analogs chemoselectively introduced by NCL can be used to probe specific aspects of amino acid or protein function. This approach can involve introducing residues that differ from the natural amino acid at a position of interest in a single aspect, such as side chain geometry, steric effects, or electronic effects. In one example, variants of Src, a substrate of the kinase Crk, were prepared, in which a tyrosine residue known to be phosphorylated was replaced with one of five tyrosine analogs to dissect the contribution of individual factors affecting tyrosine phosphorylation (Fig. 10a) (Wang and Cole 2001). Evaluation of the semisynthetic substrates in a radioactivity-based kinase assay led the authors to conclude that the phenolic hydroxyl of tyrosine is not involved in ground-state Src/Crk interactions and that stabilization of tyrosine conformers increases the efficiency of phosphorylation. In another study, EPL was used to generate analogs of the blue copper protein azurin, an electron transfer protein, to study the role of the copper-coordinating methionine residue on reduction potential (Fig. 10b) (Berry et al. 2003). The methionine residue was replaced with either a norleucine or a selenomethionine using NCL; subsequent spectroscopic and biochemical studies of the semisynthetic variants revealed hydrophobicity to be the most significant factor affecting reduction potential.

Introduction of unnatural amino acids has also been used to confer proteins with new properties or functions. For example, a variant of the enzyme ribonuclease A was engineered with a β-peptide module in place of two natural amino acids, generating an enzyme with increased conformational stability without compromising catalytic activity (Fig. 10c) (Arnold et al. 2002). NCL has similarly been used to create proteins with more potent binding properties. Src, a protein substrate of Csk, was prepared by EPL with an ATPγS conjugate in place of the tyrosine at the site of phosphorylation, creating a Csk inhibitor (Fig. 10d) (Shen and Cole 2003). That report demonstrates a strategy that could potentially be used to identify unknown kinases through pull-down assays, or to generate tightly bound kinase/substrate pairs for X-ray crystallography. In another example in which unnatural amino acid introduction leads to modulation of protein binding, a DNA-binding zinc finger protein was tuned to recognize a specific DNA sequence by incorporating citrulline, an amino acid which combines the side chain length of arginine with an alternative functional group (Fig. 10e) (Jantz and Berg 2003). While tandem zinc finger domains provide a flexible framework for creating discriminating DNA-binding proteins, the domains almost exclusively require a guanosine at the 5′-end of the DNA for specific binding. Replacement of the DNA-interacting arginine residue with citrulline, which provides both a hydrogen-bond donor and a hydrogen-bond acceptor, created a variant that binds preferentially to adenosine at the 5′-end of the DNA, a function not possible with zinc finger domain variants restricted to the 20 encoded amino acids.

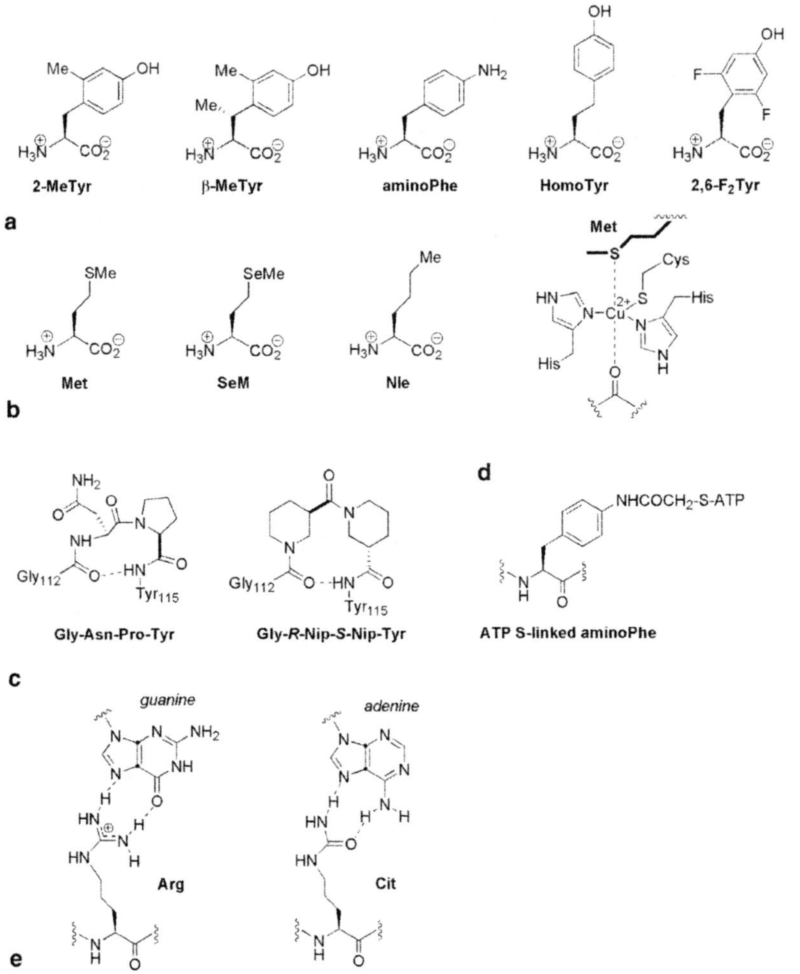

Fig. 10 **a** Unnatural tyrosine analogs incorporated into Src in place of a tyrosine residue known to be phosphorylated. **b** Unnatural methionine analogs used to replace the active-site methionine in azurin variants. **c** Naturally occurring residues (Gly-Asn-Pro-Tyr) forming a type VI reverse turn in ribonuclease A and the replacement β-peptide-containing moiety [Gly–(*R*)-nipecotic acid–(*S*)-nipecoic acid]. **d** ATPγS-linked aminophenylalanine residue incorporated into Src to replace the tyrosine residue phosphorylated by the kinase Csk. **e** Arginine–guanosine interaction and the hypothesized citrulline–adenosine interaction

The reports described above represent the use of unnatural amino acids to modulate or probe protein function but are by no means comprehensive. Several particularly creative applications that utilize NCL for the incorporation of unnatural amino acids are covered in detail in the following sections.

6.3.1 Phosphatase-Resistant Phosphoamino Acid Analogs

Characterizing the role of protein phosphorylation in vivo is complicated by the presence of protein phosphatases, which are promiscuous enzymes that catalyze the removal of phosphoryl groups from phosphoproteins. Phosphatases can cause unanticipated and undetected hydrolysis of the phosphoprotein being studied. The phosphatases themselves can be modulated by phosphorylation, creating phosphoproteins challenging to study both in vivo and in vitro, since phosphatases have the propensity to autodephosphorylate. The introduction by EPL of nonhydrolyzable phosphotyrosine analogs into full-length protein has been used to great effect in the study of protein tyrosine phosphatase SHP-2 (Fig. 11a) (Lu et al. 2001; Lu et al. 2003). Two phosphatase-resistant phosphotyrosine analogs were incorporated into the C-terminal tail (at either residue 542 or 580, or at both 542 and 580) of SHP-2: phosphonomethylene phenylalanine, which is commercially available as the N-Boc derivative, and difluoromethylene phosphonate, which is a superior mimic of phosphotyrosine but requires synthesis of the amino acid for introduction into peptides. Biochemical investigations of the SHP-2 variants revealed that phosphonates at Tyr542 and Tyr580 bind to N-terminal and C-terminal SH2 domains, respectively, within SHP-2, disrupting basal inhibition and thereby increasing phosphatase activity. Near-additive effects were observed for the doubly phosphonylated mutant. Similar incorporation of phosphonomethylene phenylalanine into a related phosphatase, SHP-1, provided equally illuminating mechanistic insight into the role of tyrosine phosphorylation on phosphatase regulation (Zhang et al. 2003). More recently, single and double phosphonate substitutions were incorporated by EPL into the low molecular weight protein tyrosine phosphatase and were shown to inhibit dephosphorylation of phosphopeptide substrates, representing the first example of negative regulation of a tyrosine phosphatase by phosphorylation (Schwarzer et al. 2006). Also noteworthy, the effect of phosphorylation on cellular stability and localization was investigated following the microinjection of the phosphonate variants into cells, an experiment that would not be possible with native phosphoproteins, which are susceptible to cellular phosphatases.

Phosphonate isosteres of phosphothreonine and phosphoserine, phosphonomethylene alanine and phosphonodifluoromethylene alanine, have also been introduced into a full-length protein by NCL (Fig. 11b) (Zheng et al. 2003, 2005). Unlike phosphotyrosine, which has no suitable surrogate within the encodable amino acids, glutamic and aspartic acid are commonly used as stable phosphothreonine or phosphoserine mimics; however, these residues are not always successful mimics, as was reported with replacement of phosphothreonine and phosphoserine residues in the protein AANAT, a regulatory enzyme involved in the production and secretion of melatonin. In contrast, the replacement of an N-terminal Thr31 or a C-terminal Ser205 with phosphonomethylene alanine by NCL enabled direct analysis of the role of phosphorylation on AANAT stability. Microinjection of the phosphonate mutants into cells provided direct support of increased AANAT stability by intermolecular binding facilitated by phosphorylation of Thr31 and Ser205.

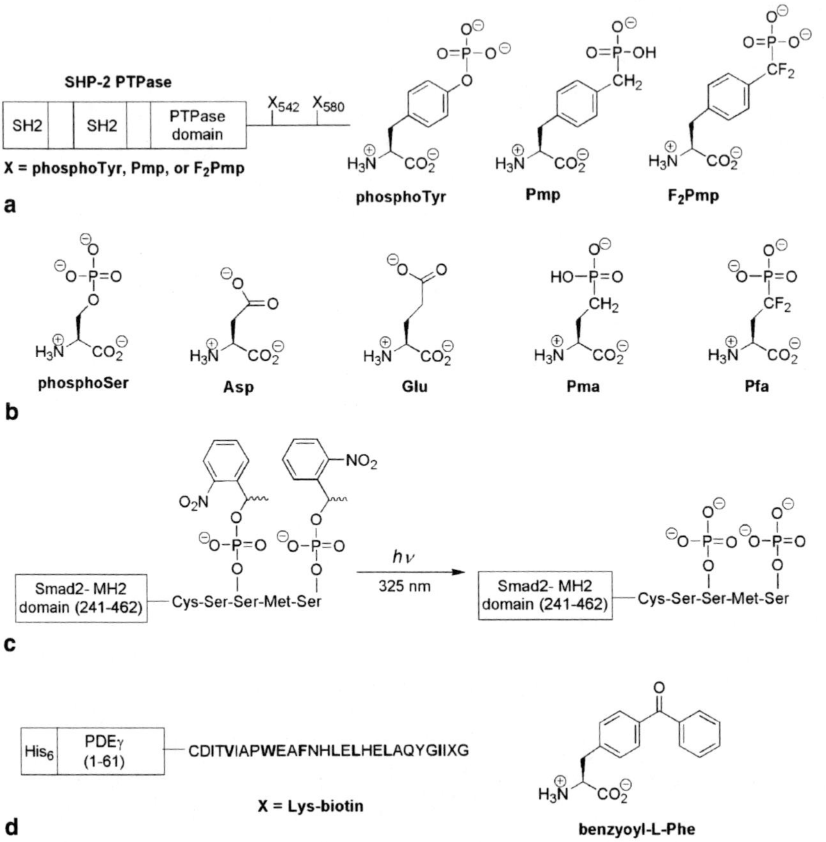

Fig. 11 Applications of unnatural amino acids introduced into proteins by NCL. **a** Replacement of tyroseine residues 542 and 580 in the protein phosphatase (*PTPase*) SHP-2 with phosphatase-resistant tyrosine mimics, phosphonomethylene phenylalanine (*Pmp*) and difluoromethylene phosphonate (*F_2Pmp*). **b** Phosphoserine and the genetically encoded phosphoserine/threonine mimics aspartic acid and glutamic acid, and the nonhydrolyzable phoshoserine/threonine mimics phosphonomethylene alanine (*Pma*) and phosphonodifluoromethylene alanine (*Pfa*). **c** Smad2-MH2 domain containing C-terminal caged phosphoserine residues. Irradiation with UV light releases the corresponding dually phosphorylated Smad2 domain. **d** Phosphodiesterase inhibitory γ-subunit (*PDEγ*) variants with the photocross-linking amino acid, benzoyl-L-phenylalanine, incorporated in place of one of seven hydrophobic residues highlighted in *bold*

6.3.2 Caged Phosphorylated Amino Acids

The caging, or photolabile protection, of biomolecules enables spatial and temporal control over the release of an activated species. The application of caged proteins holds great potential for elucidating protein function and signaling pathways but traditionally has been limited by the difficulty of preparing proteins containing masked functionalities. EPL was successfully applied to the semisynthesis of a caged domain of the signaling protein Smad2, which included two phosphoserine residues protected by the 2-nitrophenylethyl photolabile protecting group (Fig. 11c) (Hahn and Muir 2004). Previous investigations by the same laboratory used homogenous phosphoSmad2 variants to explore the effect of phosphorylation on Smad2 oligomerization (Sect. 6.1) (Ottesen et al. 2004). The authors wished to expand on that work by using a dually caged phosphoserine variant as a probe to study Smad2 pathway dynamics with spatial and temporal resolution in cellular experiments. To generate the probe, a peptide corresponding to the final five residues (463–467) of Smad2 and containing two caged phosphoserine residues was synthesized on the solid phase and ligated to a biologically expressed α-thioester of the Smad2-MH2 domain (residues 241–462). Prior to phosphorylation, Smad2 interacts with a membrane-bound complex, which includes 1:1 binding with the protein Smad anchor for receptor activation (SARA). Upon phosphorylation, Smad2 releases from the SARA complex and can form homotrimers or heterocomplexes with a co-Smad, resulting in an active complex that localizes to the nucleus for gene expression regulation. The caged phosphoSmad2 variant was shown to mimic a nonphosphorylated Smad-MH2 domain, forming a 1:1 complex with SARA. Upon irradiation, the release of the corresponding phosphodomain was detected, resulting in protein homotrimerization. In addition, a nuclear import assay was used to characterize the localization of the caged Smad2 variant before and after irradiation. As expected, prior to irradiation, caged Smad2-MH2 was excluded from the nucleus, similar to nonphosphorylated Smad2, while irradiation led to nuclear accumulation consistent with the release of phosphoSmad2. The caged phosphoSmad variant will be used in future experiments to provide quantitative information on Smad2 nuclear import and export in live cells.

EPL has also been employed for the incorporation of a caged phosphorylated residue into a full-length protein with the recent semisynthesis of a caged phosphorylated variant of the cell-migration protein paxillin (Vogel and Imperiali 2007). Paxillin is a 61-kDa, multidomain protein that contributes to the control of cellular adhesion and cell migration by acting as a dynamic scaffold for signaling and structural proteins. As a tool to investigate the influence of paxillin phosphorylation at a single site, a 2-nitrophenylethyl-caged phosphotyrosine residue was installed at position 31 of the 557-residue protein. The probe comprised the entire paxillin macromolecule, including all other binding and localization determinants, which are essential for creating a native environment to investigate phosphorylation. For the probe semisynthesis, a caged phosphotyrosine building block was used in the Fmoc-based SPPS of a hexahistidine-tagged 41-mer α-thioester, which was subsequently ligated to the biologically expressed C-terminus of paxillin (residues 38–557). Following the semisynthesis of caged phosphorylated paxillin and analogously constructed nonphosphorylated and discretely phosphorylated controls, paxillin variants were characterized biochemically and by quantitative uncaging studies. Ongoing cellular experiments using microinjected caged phosphorylated paxillin should yield important information on the effect of Tyr31 phosphorylation on cell migration.

6.3.3 Photocross-linkers

Another recent application of EPL has been the semisynthesis of protein variants incorporating a photoactivatable cross-linker, introduced as benzoyl phenylalanine, to study a central binding interaction in the phototransduction cascade involved in vision (Grant et al. 2006). The inactive form of cyclic GMP phosphodiesterase (PDE) is known to be activated by nanomolar binding of the GTP-bound form of the α-subunit of transducin ($G\alpha_t$-GTP) with the PDE inhibitory γ-subunit (PDEγ). To further characterize this interaction, a series of seven full-length PDEγ photoprobes were constructed using EPL to install the benzophenone amino acid in place of selected C-terminal hydrophobic residues (Fig. 11d). The semisynthetic domains also included a biotin affinity tag at the C-terminus for analysis. Photocross-linking experiments were performed between each of the PDEγ photoprobes and an activated form of $G\alpha_t$ ($G\alpha_t$-GTPγS). The investigators observed cross-linking with several of the photoprobes and were able to narrow down the region of photoinsertion into $G\alpha_t$, thereby revealing several previously unknown binding interactions.

6.4 Probes for Magnetic Resonance Applications: Isotopic Labeling and Spin Labels

Structural analysis of proteins by nuclear magnetic resonance (NMR) spectroscopy is facilitated by the incorporation of carbon, nitrogen, and hydrogen isotopes, which may be uniformly introduced into biologically expressed proteins. At high molecular weights, despite isotopic labeling, NMR analysis becomes complicated owing to decreased resolution caused by longer correlation times and highly complex spectra. EPL has been used to selectively introduce isotopic labels into protein domains within a full-length protein. This strategy has enabled NMR analysis of discrete protein domains or fragments within the context of large proteins, rendering a single section of the protein spectroscopically visible without disrupting domain–domain interactions. Specifically, EPL has been used to introduce a labeled domain at the terminus of a protein (Xu et al. 1999), and to demonstrate the possibility of introducing an internal isotopically labeled domain flanked by two nonlabeled domains using two sequential ligation reactions (Blaschke et al. 2000).

In a recent example, the α subunit of a G protein (Gα) was semisynthesized with ^{13}C-labeled residues within the C-terminal tail to enable NMR characterization of conformational changes upon Gα subunit activation (Anderson et al. 2005). To construct the labeled full-length subunit, a synthetic non apeptide (Gα residues 346–354) containing three ^{13}C-labeled amino acids was ligated to a recombinant thioester fragment corresponding to Gα residues 1–345. NMR characterization of the Gα subunit in its inactive (GDP-bound) form and in an active form (mimicked by the addition of AlF_4^-, a GTP γ-phosphate analog) provided evidence for an increase in the conformational order of the C-terminal tail upon Gα subunit activation.

EPL has also been employed for the incorporation of a paramagnetic amino acid into a full-length protein for electron paramagnetic resonance spectroscopy (Becker et al. 2005). A spin-labeled peptide containing a nitroxide-labeled lysine was ligated to a recombinant thioester of the Ras-binding domain (RBD) of *c*-Raf1. The ligation reaction, carried out in the presence of a thiol catalyst, resulted in reduction of the spin label, which was reoxidized prior to electron paramagnetic resonance analysis of the labeled RBD and a labeled RBD/Ras complex.

6.5 Immobilization by NCL and Use in Microarrays

The chemoselective nature of the NCL reaction has been exploited for immobilizing proteins on a solid support in strategies aimed at the development of protein microarrays. Protein microarrays have the potential to facilitate high-throughput analysis of protein interactions, including protein/protein, protein/small molecule, and protein/antibody binding. A major challenge in the construction of protein microarrays is the uniform orientation of proteins on the glass support. In several examples, NCL has been used to install a reactive group at the terminus of proteins for subsequent immobilization. For example, researchers have used EPL to C-terminally label proteins with a biotin moiety by the ligation of the expressed protein thioesters with a biotin-labeled cysteine residue (Lesaicherre et al. 2002). The biotinylated proteins were subsequently immobilized on avidin-labeled glass slides, taking advantage of the stable biotin–avidin interaction. The immobilization of an oxidoreductase enzyme, AKR1A1, for future microarray applications was also accomplished using a similar biotin-labeling strategy (Richter et al. 2004). In another immobilization strategy, researchers used EPL in the semisynthesis of protein–nucleic acid conjugates, in which recombinant thioester-containing proteins were ligated to peptides containing a C-terminal polyamide nucleic acid (Lovrinovic et al. 2003). The proteins were then incorporated into small microarrays by DNA-directed immobilization based on interaction of DNA on a glass slide to the complementary protein-conjugated polyamide nucleic acid.

Proteins have also been attached to a solid surface through the stable amide bond formed directly by NCL (Fig. 12). Expressed protein α-thioesters have been shown to selectively immobilize onto a cysteine-functionalized glass slide (Camarero et al. 2004a, b). The reverse approach has also been accomplished, in which proteins containing N-terminal cysteines were immobilized onto thioester-containing slides (Girish et al. 2005). In that case, the cysteine-containing proteins were expressed using an intein-mediated strategy, and both the subsequently purified proteins and the crude cell lysates were used effectively to immobilize the N-terminal cysteine-containing proteins onto the thioester-functionalized plates.

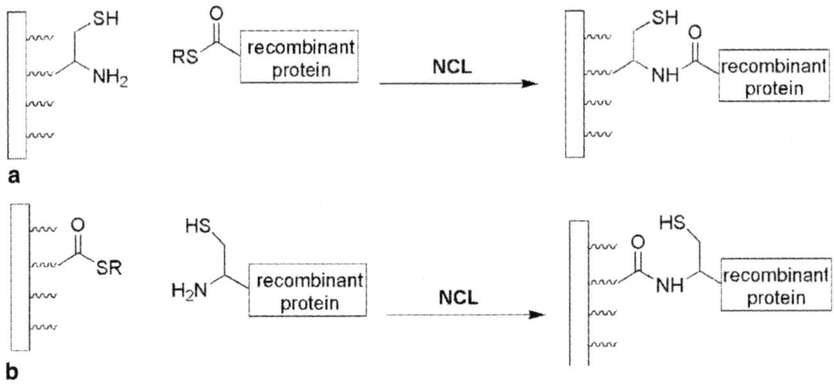

Fig. 12 Immobilization of proteins onto microarrays by **a** reaction of a thioester-containing protein with a cysteine-functionalized slide and **b** reaction of an N-terminal cysteineys-containing protein with a thioester-functionalized slide

7 Conclusion

Advances in recombinant methods for the generation of α-thioester and N-terminal cysteine-containing proteins, and in synthetic approaches for peptide α-thioester synthesis, particularly methods compatible with peptides including post-translational modifications and noncanonical amino acids, have facilitated the semisynthesis of proteins and protein domains containing unnatural amino acids and biological probes and tags in NCL applications. These advances have enabled investigators to design and construct elaborate protein biomolecules that can be used to address challenging problems not routinely accessible by traditional methods.

Acknowledgments Support from the NIH (GM64346 Cell Migration Consortium) and the Merck MIT CSBI and Charles Krakauer Graduate Fellowship programs is gratefully acknowledged.

References

Alsina J, Yokum TS, Albericio F, Barany G (1999) Backbone amide linker (BAL) strategy for Nα-9-fluorenylmethoxycarbonyl (Fmoc) solid-phase synthesis of unprotected peptide p-nitroanilides and thioesters. J Org Chem 64:8761–8769

Anderson LL, Marshall GR, Crocker E, Smith SO, Baranski TJ (2005) Motion of carboxyl terminus of Gα is restricted upon G protein activation: a solution NMR study using semisynthetic Gα subunits. J Biol Chem 280:31019–31026

Arnold U, Hinderaker MP, Nilsson BL, Huck BR, Gellman SH, Raines RT (2002) Protein prosthesis: a semisynthetic enzyme with a β-peptide reverse turn. J Am Chem Soc 124:8522–8523

Bang D, Kent SBH (2004) Protein synthesis: a one-pot total synthesis of crambin. Angew Chem Int Ed Engl 43:2534–2538

Bang D, Kent SBH (2005) His6 tag-assisted chemical protein synthesis. Proc Natl Acad Sci USA 102:5014–5019

Bang D, Makhatadze GI, Tereshko V, Kossiakoff AA, Kent SB (2005) Total chemical synthesis and X-ray crystal structure of a protein diastereomer: [D-gln 35]ubiquitin. Angew Chem Int Ed Engl 44:3852–3856

Bang D, Pentelute BL, Gates ZP, Kent SB (2006) Direct on-resin synthesis of peptide-α-thiophenylesters for use in native chemical ligation. Org Lett 8:1049–1052

Becker CFW, Hunter CL, Seidel R, Kent SBH, Goody RS, Engelhard M (2003) Total chemical synthesis of a functional interacting protein pair: the protooncogene H-Ras and the Ras-binding domain of its effector c-Raf1. Proc Natl Acad Sci USA 100:5075–5080

Becker CFW, Lausecker K, Balog M, Kalai T, Hideg K, Steinhoff H-J, Engelhard M (2005) Incorporation of spin-labelled amino acids into proteins. Magn Reson Chem 43:S34–S39

Berry SM, Ralle M, Low DW, Blackburn NJ, Lu Y (2003) Probing the role of axial methionine in the blue copper center of azurin with unnatural amino acids. J Am Chem Soc 125: 8760–8768

Blaschke UK, Cotton GJ, Muir TW (2000) Synthesis of multi-domain proteins using expressed protein ligation: strategies for segmental isotopic labeling of internal regions. Tetrahedron 56:9461–9470

Botti P, Carrasco MR, Kent SBH (2001) Native chemical ligation using removable Nα-(1-phenyl-2-mercaptoethyl) auxiliaries. Tetrahedron Lett 42:1831–1833

Brask J, Albericio F, Jensen KJ (2003) Fmoc solid-phase synthesis of peptide thioesters by masking as trithioortho esters. Org Lett 5:2951–2953

Brunsveld L, Watzke A, Durek T, Alexandrov K, Goody RS, Waldmann H (2005) Synthesis of functionalized Rab GTPases by a combination of solution- or solid-phase lipopeptide synthesis with expressed protein ligation. Chem-Eur J 11:2756–2772

Bu X, Xie G, Law CW, Guo Z (2002) An improved deblocking agent for direct Fmoc solid-phase synthesis of peptide thioesters. Tetrahedron Lett 43:2419–2422

Camarero JA, Mitchell AR (2005) Synthesis of proteins by native chemical ligation using Fmoc-based chemistry. Protein Peptide Lett 12:723–728

Camarero JA, Fushman D, Cowburn D, Muir TW (2001) Peptide chemical ligation inside living cells: In vivo generation of a circular protein domain. Bioorg Med Chem 9:2479–2484

Camarero JA, Cheung CL, Coleman MA, De Yoreo JJ (2004a) Chemoselective attachment of biologically active proteins to surfaces by native chemical ligation. Mater Res Soc Symp Proc EXS-1:237–239

Camarero JA, Hackel BJ, de Yoreo JJ, Mitchell AR (2004b) Fmoc-based synthesis of peptide alpha-thioesters using an aryl hydrazine support. J Org Chem 69:4145–4151

Canne LE, Walker SM, Kent SBH (1995) A general method for the synthesis of thioester resin linkers for use in the solid phase synthesis of peptide-α-thioacids. Tetrahedron Lett 36: 1217–1220

Canne LE, Bark SJ, Kent SBH (1996) Extending the applicability of native chemical ligation. J Am Chem Soc 118:5891–5896

Chong S, Xu MQ (2005) Harnessing inteins for protein purification and characterization. In: Gross HJ (ed) Nucleic acids and molecular biology, vol 16. Springer, Berlin, pp 273–292

Chong S, Shao Y, Paulus H, Benner J, Perler FB, Xu M-Q (1996) Protein splicing involving the Saccharomyces cerevisiae VMA intein. The steps in the splicing pathway, side reactions leading to protein cleavage, and establishment of an in vitro splicing system. J Biol Chem 271:22159–22168

Cotton GJ, Muir TW (2000) Generation of a dual-labeled fluorescence biosensor for Crk-II phosphorylation using solid-phase expressed protein ligation. Chem Biol 7:253–261

Cotton GJ, Ayers B, Xu R, Muir TW (1999) Insertion of a synthetic peptide into a recombinant protein framework: a protein biosensor. J Am Chem Soc 121:1100–1101

David R, Richter MP, Beck-Sickinger AG (2004) Expressed protein ligation. Method and applications. Eur J Biochem 271:663–677

Dawson PE, Muir TW, Clark-Lewis I, Kent SBH (1994) Synthesis of proteins by native chemical ligation. Science (Washington, DC) 266:776–779

Dawson PE, Churchill M, Ghadiri MR, Kent SBH (1997) Modulation of reactivity in native chemical ligation through the use of thiol additives. J Am Chem Soc 119:4325–4329
Durek T, Alexandrov K, Goody RS, Hildebrand A, Heinemann I, Waldmann H (2004) Synthesis of fluorescently labeled mono- and diprenylated Rab7 GTPase. J Am Chem Soc 126:16368–16378
Erlanson DA, Chytil M, Verdine GL (1996) The leucine zipper domain controls the orientation of AP-1 in the NFAT.AP-1.DNA complex. Chem Biol 3:981–991
Evans TC, Jr., Benner J, Xu M-Q (1998) Semisynthesis of cytotoxic proteins using a modified protein splicing element. Protein Sci 7:2256–2264
Flavell RR, Huse M, Goger M, Trester-Zedlitz M, Kuriyan J, Muir TW (2002) Efficient semisynthesis of a tetraphosphorylated analogue of the type I TGFbeta receptor. Org Lett 4:165–168
Futaki S, Sogawa K, Maruyama J, Asahara T, Niwa M (1997) Preparation of peptide thioesters using Fmoc-solid-phase peptide synthesis and its application to the construction of a template-assembled synthetic protein (TASP). Tetrahedron Lett 38:6237–6240
Gieselman MD, Xie L, van der Donk WA (2001) Synthesis of a selenocysteine-containing peptide by native chemical ligation. Org Lett 3:1331–1334
Girish A, Sun H, Yeo DSY, Chen GYJ, Chua T-K, Yao SQ (2005) Site-specific immobilization of proteins in a microarray using intein-mediated protein splicing. Bioorg Med Chem Lett 15:2447–2451
Gottlieb D, Grunwald C, Nowak C, Kuhlmann J, Waldmann H (2006) Intein-mediated in vitro synthesis of lipidated Ras proteins. Chem Commun 260–262
Grant JE, Guo L-W, Vestling MM, Martemyanov KA, Arshavsky VY, Ruoho AE (2006) The N terminus of GTPγS-activated transducin α-subunit interacts with the C terminus of the cGMP phosphodiesterase γ-subunit. J Biol Chem 281:6194–6202
Grogan MJ, Pratt MR, Marcaurelle LA, Bertozzi CR (2002) Homogeneous glycopeptides and glycoproteins for biological investigation. Annu Rev Biochem 71:593–634
Grogan MJ, Kaizuka Y, Conrad RM, Groves JT, Bertozzi CR (2005) Synthesis of lipidated green fluorescent protein and its incorporation in supported lipid bilayers. J Am Chem Soc 127: 14383–14387
Gross CM, Lelievre D, Woodward CK, Barany G (2005) Preparation of protected peptidyl thioester intermediates for native chemical ligation by Nα-9-fluorenylmethoxycarbonyl (Fmoc) chemistry: considerations of side-chain and backbone anchoring strategies, and compatible protection for N-terminal cysteine. J Pept Res 65:395–410
Hackenberger CPR, Friel CT, Radford SE, Imperiali B (2005) Semisynthesis of a glycosylated Im7 analogue for protein folding studies. J Am Chem Soc 127:12882–12889
Hackeng TM, Griffin JH, Dawson PE (1999) Protein synthesis by native chemical ligation: expanded scope by using straightforward methodology. Proc Natl Acad Sci USA 96:10068–10073
Hahn ME, Muir TW (2004) Bioorganic chemistry: Photocontrol of Smad2, a multiphosphorylated cell-signaling protein, through caging of activating phosphoserines. Angew Chem Int Ed Engl 43:5800–5803
He S, Bauman D, Davis Jamaine S, Loyola A, Nishioka K, Gronlund Jennifer L, Reinberg D, Meng F, Kelleher N, McCafferty Dewey G (2003) Facile synthesis of site-specifically acetylated and methylated histone proteins: reagents for evaluation of the histone code hypothesis. Proc Natl Acad Sci USA 100:12033–12038
Hojo H, Haginoya E, Matsumoto Y, Nakahara Y, Nabeshima K, Toole BP, Watanabe Y (2003) The first synthesis of peptide thioester carrying N-linked core pentasaccharide through modified Fmoc thioester preparation: synthesis of an N-glycosylated Ig domain of emmprin. Tetrahedron Lett 44:2961–2964
Huse M, Holford MN, Kuriyan J, Muir TW (2000) Semisynthesis of hyperphosphorylated type I TGFβ receptor: addressing the mechanism of kinase activation. J Am Chem Soc 122: 8337–8338
Ingenito R, Bianchi E, Fattori D, Pessi A (1999) Solid phase synthesis of peptide C-terminal thioesters by Fmoc/t-Bu chemistry. J Am Chem Soc 121:11369–11374
Jantz D, Berg JM (2003) Expanding the DNA-recognition repertoire for zinc finger proteins beyond 20 amino acids. J Am Chem Soc 125:4960–4961

Jantz D, Berg JM (2004) Reduction in DNA-binding affinity of Cys2His2 zinc finger proteins by linker phosphorylation. Proc Natl Acad Sci USA 101:7589–7593

Johnson ECB, Kent SBH (2006) Insights into the mechanism and catalysis of the native chemical ligation reaction. J Am Chem Soc 128:6640–6646

Kawakami T, Sumida M, Nakamura Ki, Vorherr T, Aimoto S (2005) Peptide thioester preparation based on an N–S acyl shift reaction mediated by a thiol ligation auxiliary. Tetrahedron Lett 46:8805–8807

Lesaicherre M-L, Lue RYP, Chen GYJ, Zhu Q, Yao SQ (2002) Intein-mediated biotinylation of proteins and its application in a protein microarray. J Am Chem Soc 124:8768–8769

Li X, Kawakami T, Aimoto S (1998) Direct preparation of peptide thioesters using an Fmoc solid-phase method. Tetrahedron Lett 39:8669–8672

Lovrinovic M, Seidel R, Wacker R, Schroeder H, Seitz O, Engelhard M, Goody RS, Niemeyer CM (2003) Synthesis of protein-nucleic acid conjugates by expressed protein ligation. Chem Commun 822–823

Lu W, Gong D, Bar-Sagi D, Cole PA (2001) Site-specific incorporation of a phosphotyrosine mimetic reveals a role for tyrosine phosphorylation of SHP-2 in cell signaling. Mol Cell 8:759–769

Lu W, Shen K, Cole PA (2003) Chemical dissection of the effects of tyrosine phosphorylation of SHP-2. Biochemistry 42:5461–5468

Maag D, Lorsch JR (2003) Communication between eukaryotic translation initiation factors 1 and 1A on the yeast small ribosomal subunit. J Mol Biol 330:917–924

Macmillan D, Anderson DW (2004) Rapid synthesis of acyl transfer auxiliaries for cysteine-free native glycopeptide ligation. Org Lett 6:4659–4662

Macmillan D, Arham L (2004) Cyanogen bromide cleavage generates fragments suitable for expressed protein and glycoprotein ligation. J Am Chem Soc 126:9530–9531

Macmillan D, Bertozzi CR (2000) New directions in glycoprotein engineering. Tetrahedron 56: 9515–9525

Macmillan D, Bertozzi CR (2004) Modular assembly of glycoproteins: towards the synthesis of GlyCAM-1 by using expressed protein ligation. Angew Chem Int Ed Engl 43:1355–1359

Marcaurelle LA, Mizoue LS, Wilken J, Oldham L, Kent SB, Handel TM, Bertozzi CR (2001) Chemical synthesis of lymphotactin: a glycosylated chemokine with a C-terminal mucin-like domain. Chemistry (Weinheim an der Bergstrasse, Germany) 7:1129–1132

Marinzi C, Offer J, Longhi R, Dawson PE (2004) An o-nitrobenzyl scaffold for peptide ligation: synthesis and applications. Bioorg Med Chem 12:2749–2757

Mezo AR, Cheng RP, Imperiali B (2001) Oligomerization of uniquely folded mini-protein motifs: development of a homotrimeric ββα peptide. J Am Chem Soc 123:3885–3891

Miller JS, Dudkin VY, Lyon GJ, Muir TW, Danishefsky SJ (2003) Toward fully synthetic N-linked glycoproteins. Angew Chem Int Ed Engl 42:431–434

Muir TW (2003) Semisynthesis of proteins by expressed protein ligation. Annu Rev Biochem 72:249–289

Muir TW, Sondhi D, Cole PA (1998) Expressed protein ligation: a general method for protein engineering. Proc Natl Acad Sci USA 95:6705–6710

Muralidharan V, Cho J, Trester-Zedlitz M, Kowalik L, Chait BT, Raleigh DP, Muir TW (2004) Domain-specific incorporation of noninvasive optical probes into recombinant proteins. J Am Chem Soc 126:14004–14012

Ottesen JJ, Huse M, Sekedat MD, Muir TW (2004) Semisynthesis of phosphovariants of Smad2 reveals substrate preference of activated TβRI kinase. Biochemistry 43:5698–5706

Paulus H (2000) Protein splicing and related forms of protein autoprocessing. Annu Rev Biochem 69:447–496

Richter MPO, Holland-Nell K, Beck-Sickinger AG (2004) Site specific biotinylation of the human aldo/keto reductase AKR1A1 for immobilization. Tetrahedron 60:7507–7513

Scheibner KA, Zhang Z, Cole PA (2003) Merging fluorescence resonance energy transfer and expressed protein ligation to analyze protein–protein interactions. Anal Biochem 317:226–232

Schwarzer D, Cole PA (2005) Protein semisynthesis and expressed protein ligation: chasing a protein's tail. Curr Opin Chem Biol 9:561–569
Schwarzer D, Zhang Z, Zheng W, Cole PA (2006) Negative regulation of a protein tyrosine phosphatase by tyrosine phosphorylation. J Am Chem Soc 128:4192–4193
Severinov K, Muir TW (1998) Expressed protein ligation, a novel method for studying protein–protein interaction in transcription. J Biol Chem 273:16205–16209
Shen K, Cole PA (2003) Conversion of a tyrosine kinase protein substrate to a high affinity ligand by ATP linkage. J Am Chem Soc 125:16172–16173
Shin Y, Winans KA, Backes BJ, Kent SBH, Ellman JA, Bertozzi CR (1999) Fmoc-based synthesis of peptide-αthioesters: application to the total chemical synthesis of a glycoprotein by native chemical ligation. J Am Chem Soc 121:11684–11689
Shogren-Knaak MA, Peterson CL (2004) Creating designer histones by native chemical ligation. Methods Enzymol 375:62–76, 61 Plate
Shogren-Knaak M, Ishii H, Sun J-M, Pazin MJ, Davie JR, Peterson CL (2006) Histone H4-K16 acetylation controls chromatin structure and protein interactions. Science (Washington, DC) 311:844–847
Tolbert TJ, Wong CH (2002) New methods for proteomic research: preparation of proteins with N-terminal cysteines for labeling and conjugation. Angew Chem Int Ed Engl 41:2171–2174
Tolbert TJ, Franke D, Wong C-H (2005) A new strategy for glycoprotein synthesis: ligation of synthetic glycopeptides with truncated proteins expressed in *E. coli* as TEV protease cleavable fusion protein. Bioorg Med Chem 13:909–915
Ueda S, Fujita M, Tamamura H, Fujii N, Otaka A (2005) Photolabile protection for one-pot sequential native chemical ligation. Chem BioChem 6:1983–1986
Villain M, Vizzavona J, Rose K (2001) Covalent capture: a new tool for the purification of synthetic and recombinant polypeptides. Chem Biol 8:673–679
Villain M, Gaertner H, Botti P (2003) Native chemical ligation with aspartic and glutamic acids as C-terminal residues: scope and limitations. Eur J Org Chem 3267–3272
Vogel EM, Imperiali B (2007) Semisynthesis of unnatural amino acid mutants of paxillin: protein probes for cell migration studies. Prot Sci 16:550–556
von Eggelkraut-Gottanka R, Klose A, Beck-Sickinger AG, Beyermann M (2003) Peptide α-thioester formation using standard Fmoc-chemistry. Tetrahedron Lett 44:3551–3554
Walsh CT (2006) Posttranslational modification of proteins: expanding nature's inventory. Roberts & Company, Englewood
Wang D, Cole PA (2001) Protein tyrosine kinase Csk-catalyzed phosphorylation of Src containing unnatural tyrosine analogues. J Am Chem Soc 123:8883–8886
Xu R, Ayers B, Cowburn D, Muir TW (1999) Chemical ligation of folded recombinant proteins: segmental isotopic labeling of domains for NMR studies. Proc Natl Acad Sci USA 96:388–393
Yan LZ, Dawson PE (2001) Synthesis of peptides and proteins without cysteine residues by native chemical ligation combined with desulfurization. J Am Chem Soc 123:526–533
Yeo DSY, Srinivasan R, Uttamchandani M, Chen GYJ, Zhu Q, Yao SQ (2003) Cell-permeable small molecule probes for site-specific labeling of proteins. Chem Commun 2870–2871
Zhang Z, Shen K, Lu W, Cole PA (2003) The role of C-terminal tyrosine phosphorylation in the regulation of SHP-1 explored via expressed protein ligation. J Biol Chem 278:4668–4674
Zheng W, Zhang Z, Ganguly S, Weller JL, Klein DC, Cole PA (2003) Cellular stabilization of the melatonin rhythm enzyme induced by nonhydrolyzable phosphonate incorporation. Nat Struct Biol 10:1054–1057
Zheng W, Schwarzer D, LeBeau A, Weller JL, Klein DC, Cole PA (2005) Cellular stability of serotonin N-acetyltransferase conferred by phosphonodifluoromethylene alanine (Pfa) substitution for Ser-205. J Biol Chem 280:10462–10467

Chemical Methods for Mimicking Post-Translational Modifications

S.I. van Kasteren, P. Garnier, and B.G. Davis(✉)

Contents

Abstract Post-translational modification (PTM) of proteins in nature is not template driven and not under tight genetic control. An a result PTM may be unpredictable and often gives rise to mixtures of modified proteins. In order to fully understand the precise effects of PTM, chemistry may offer an alternative set of methods for accessing proteins that contain PTMs. Methods available up until early 2007 are discussed in this chapter.

B.G. Davis
Department of Chemistry, University of Oxford, Chemistry Research Laboratory, 12 Mansfield Road, Oxford OX1 3TA, UK, e-mail: ben.davis@chem.ox.ac.uk

C. Köhrer and U.L. RajBhandary (eds.) *Protein Engineering.*
Nucleic Acids and Molecular Biology 22,

1 Introduction

In this postgenomic era, the realization of the emerging role of late biosynthetic events in fine-tuning of protein function is growing. Amongst such late events, the alteration of the side chains of amino acids after biosynthesis, post-translational modification (PTM), is an archetype. Present in most eukaryotic and many prokaryotic proteins, its potential to expand the structural diversity and hence fine-tune the function of proteins is vast. The lability of many of these PTMs and their often inpredictable mode of introduction, which is outside the direct control of template-driven genetics, makes this critical area of functional biology a challenging and exciting one. Advances in proteomic analysis, hand in hand with novel methods for construction of PTMs and their mimics, now allow for the first time the investigation of "top-down" analysis complemented by "bottom-up" probe design. Amongst one of the most powerful strategies is that of protein semisynthesis, which offers not only access to pure PTMs and their mimics, in contrast to the complex mixtures typically derived from nature, but also the ability to "design in" PTMs to test functional hypotheses and exploit the benefit of PTMs (Davis 2004).

This chapter will focus on the site-specific introduction of mimics of PTMs. These are modifications of the side, chains of the protein that are introduced after the expression of the protein. Often natural PTMs are not under direct genetic control, and they are often reversible. This results in the natural production of mixtures of proteins with identical peptide backbones but with different PTMs at different levels and at different sites. We shall term the components of these mixtures by the general term *modiforms*, after the nomenclature used for glycosylated PTMs *glycoforms* (Rademacher et al. 1988) first proposed by Dwek, and to distinguish from the less specific term *isoform* often applied to general protein variance. Each modiform potentially has a different function. This is best illustrated by the example of glycosylation, arguably the most widespread and complex form of PTM.

It is estimated that 50% of all human proteins are glycosylated, and approximately 2% of the open reading frames in the human genome encode for carbohydrate-modifying enzymes (Lowe and Marth 2003). The hallmark of protein glycosylation is the production of complex mixtures of glycoforms (Rademacher et al. 1988), i.e., proteins with identical peptide backbone but with unoccupied glycosylation sites and/or with variation of the attached carbohydrate structures. This phenomenon reflects the fact that glycosylation is not under direct genetic control but rather relies on a sequential enzymatic processing – trimming by glycosidases and extension by glycosyltransferases – which can prove incomplete.

A long history of protein modification techniques exists (Hermanson 1996; Lundblad 2004; Walsh 2005). Most methods have relied on the targeting of accessible (typically nucleophilic) amino acid residue side chains present on the protein surface (e.g., lysine, cysteine, aspartic/glutamic acid). Owing to the widespread presence of some of these residues (e.g., lysine) within proteins, multiple copies of modifications are often introduced with poor regio- and chemoselective control. Almost without exception, the resulting processes lead to incomplete attachments at multiple sites and at unintended sites. While these have provided a highly useful pragmatic solution for simple, qualitative labeling of proteins, dissecting the precise function and effect of the products is of low utility in the context of the goals of this chapter and this will thus largely fall outside of the scope of this review. These nonspecific processes typically bring with them more problems than they solve. The resulting chemical mixtures of near-indiscriminately modified proteins do not allow an easy understanding of either function or even identity. Instead, we will concentrate on methods that might allow successful preparation of pure or enriched samples of PTMs or mimics thereof.

While site-selectivity and/or specificity of protein modification has been recognized in several contexts ranging from tailoring of enzymatic activity using prosthetic groups (Davis 2003) to imaging (La Clair et al. 2004; Chen and Ting 2005; Miller and Cornish 2005) and proteomic labeling (Dieterich et al. 2006), it is in the hope of answering biological questions of relevance to the role of PTMs that we have constructed this review. The desire to achieve some form of qualitative chemical modification, e.g., biotinylation for affinity tagging, fluorescence labeling for visualization, has often meant that key strategic issues have been neglected. In the context of this chapter these include:

- *Conversion* levels – few chemical protein modification methods employed today have been optimized to levels that allow complete conversion. Partial conversion recreates the need for purification of modified from unmodified. We will, where known or possible, highlight conversion levels in the examples selected.
- *Selectivity* – methods designed for chemoselectivity are rarely genuinely probed in protein contexts, which often contain a multitude of functional groups as potential, competitive, reactive partners.

In this review, we will largely ignore reactions touted as potential protein modification reactions but not demonstrated either under appropriate conditions that might be of use to the protein chemist and/or on a relevant (protein) substrate, other than where these methods provide insight into future or existing chemistries. Where illustrative of the strategy, we will also include selective examples of poorly discriminating modification methods that either include elements of promise or highlight central issues.

In the past 10 years, analytical techniques, largely pioneered by the proteomic community (Aebersold and Mann 2003; Romijn et al. 2003; Haslam et al. 2006), that allow effective quantitation of both conversion (also yield or reactivity) and selectivity (chemo- and regio-) have become generally accessible. This now allows the same rigor that has long been applied to small molecule synthetic methodology development to be brought to bear in protein chemistry synthetic methodology. In this background, we highlight issues of selectivity and reactivity (as judged by conversion) through their estimation wherever possible in examples.

Broadly, two complementary strategies to PTM–protein semisynthesis exist (Fig. 1):

1. A convergent approach – the addition of a modifying group after the protein scaffold of interest has been constructed or produced.

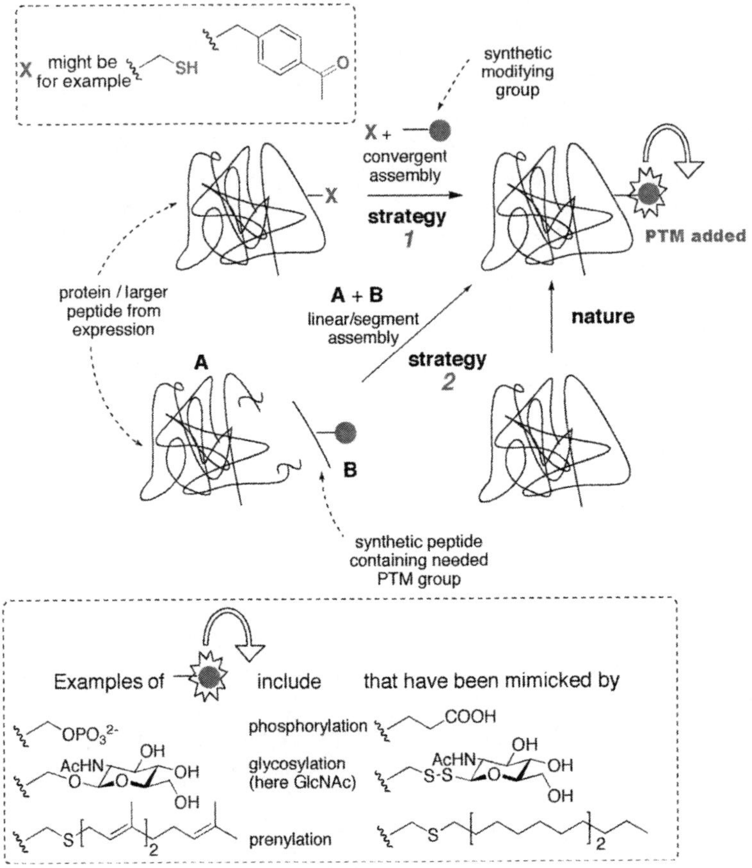

Fig. 1 A comparison of strategies for the synthesis of post-translationally modified proteins. *PTM* post-translational modification

2. A linear or segment assembly approach – the assembly of smaller peptides or amino acids that already bear the desired group into a longer protein chain.

The following striking examples illustrate what can be done. Automated synthesis machines, although far-advanced, are not capable of making peptide chains of typical protein size, and consequently bigger peptides or proteins in both strategies are typically prepared by bacterial expression. For example, in a strategy 2 manner, a truncated oncogenic protein Ras with a C-terminal cysteine at position 181 was expressed recombinantly in *Escherichia coli.* Ras is a protein that is naturally post-translationally modified with lipid at its C-terminus, which allows it to insert itself into membranes; the truncated version lacks the region that bears this lipid. The missing region could be mimicked by the chemical addition of short, chemically synthesized, replacement lipid-modified peptides. This addition onto the cysteine residue at 181 was achieved chemically and created hybrid Ras mimics that showed similar activity to natural full-length Ras in cell line differentiation assays (Bader et al. 2000). A very similar approach has recently allowed the introduction of photolabeling groups into lipids for elucidation of protein–protein interactions (Volkert et al. 2003).

Strategy 2 "linear" approaches have grown in popularity with the increasing application of so-called native chemical ligation (NCL) (Dawson et al. 1994; Muir 2003; Muralidharan and Muir 2006) that is based on the observations of thioester–amide transacylation by Wieland et al. (1953). These have allowed the incorporation of a wide range of PTMs, including glycosylation (Grogan et al. 2002; Miller et al. 2003; Macmillan and Bertozzi 2004), lipidation (Rak et al. 2003), and others. This topic has been excellently reviewed elsewhere (Dawson and Kent 2000; Khmelnitsky 2004; Schwarzer and Cole 2005) and will not be covered here in detail (see also the chapters by Merkel et al. and Imperiali and Vogel Taylor, this volume). Sections will instead highlight in passing some of the more striking examples of its application to the study of function of PTMs in proteins.

This chapter will largely focus on the flexibility offered by the convergent approach, which brings with it advantages of yield (through avoidance of compounded synthetic bottlenecks) but also functional flexibility (PTM alteration through ready retooling achieved by choice of an alternative modifying synthon, the *modifon*), concentrating on what we term the "tag-modify" or "tag-ligation" approach (Fig. 2). This convenient strategy has the potential for global site-selectivity, allowing perhaps a general technological platform for PTM mimicry and hence PTM understanding (van Kasteren et al. 2007). This two-step process involves the introduction of some form of chemical tag, followed by reaction of that tag with a chemoselective (tag-selective) reagent that introduces the desired PTM or modifon (a group that allows effective synthetic introduction of a mimic of that PTM).

Modifons are considered here in the context of desired function and hence the type of PTM to be introduced. We hope that this chapter will act as a guide to questions such as: "How can I incorporate pure complex biantennary sialyl-terminated *N*-glycan at site X of protein Y?" and "What is the *in vivo* effect of prenylation at site A in protein B?"

We would like to initiate this review by clarifying certain terms and key points associated with the strategy of protein chemistry. In this chapter, we will employ the following chemical definitions of the terms "specific" and "selective": site-*specific* modification being a method that will allow attachment at a given, existing, functional group within a protein; site-*selective* being a process that allows selection of a given site within a protein's architecture. In this way, the process of modifying a natural

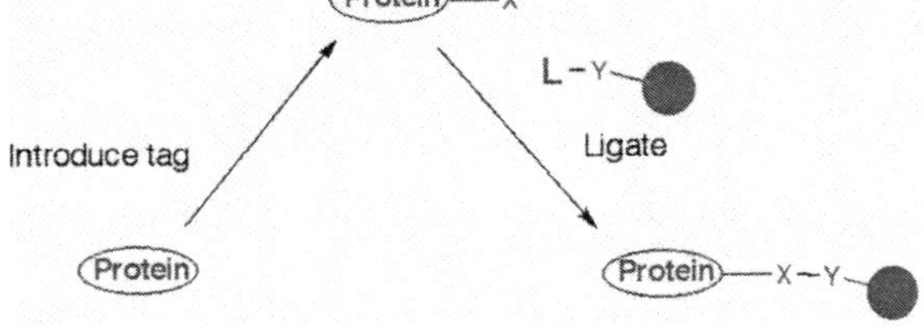

Fig. 2 The "tag-modify" approach to convergent site-selective modification

cysteine, e.g., in oxidized bovine serum albumin (BSA), is *specific* to that cysteine and to the natural site, yet an overall process that mutates out that cysteine and mutates in another cysteine to alter the position of the thiol group prior to modification provides an overall process that is *selective* for both modification and site.

Much recent research effort has gone into the modification of biomolecules with either fluorescent probes, or retrieval probes such as biotin. Again, unless these are used as functional mimics of a natural PTM, they will not be discussed in this review unless the method of introduction is of special chemical merit. The methods described will be grouped by the PTM being mimicked. The bulk of the work in the field has focused on glycosylation, as it is the most prevalent form of PTM found in nature. Lipidation has also been of interest, but chemical methods for the introduction of lipids are not manifold. Mimics of other PTMs will be discussed last, as will novel methods of introducing functionality into biomolecules, which have not yet been employed for PTM mimicking.

2 Glycosylation

As the most diverse of the PTMs, glycan introduction is an emerging area that combines the need for target synthesis of often-complex glycan reagents with strategies for their site-selective introduction. Although strikingly widespread the role of glycans in protein function is emerging only on a case-by-case basis and methods for accessing pure glycoforms or their mimics are now starting to play a critical role in unpicking precise aspects of glycan function.

2.1 Thiol Tag (X: –SH)

Owing to its low natural abundance (Fodje and Al-Karadaghi 2002) and some very selective chemistry associated with thiols, cysteine has proven an excellent source of thiol as a chemical tag and hence has been a popular target for site-specific modification. Surface-exposed cysteine residues can be readily alkylated or modified by disulfide-forming agents to yield functional mimics of glycoproteins as outlined later.

Numerous chemical methods have been developed to synthesize S-linked glycopeptides and amino acids (for a comprehensive review see Pachamuthu and Schmidt 2006), but in almost all cases their scope is to date limited to the glycosylation of peptides; the reaction conditions are often incompatible with proteins, and have therefore not been exemplified for proteins.

2.1.1 Haloacetamide

Flitsch and coworkers suggested alkylation with glycosyl iodoacetamides to selectively glycosylate amino acids (Davis and Flitsch 1991). The method was investigated in the glycosylation of three single cysteine mutants of erythropoietin (EPO) (Macmillan et al. 2001). Cysteine mutations were introduced at the three *N*-glycosylation sites (24, 38, and 83) and were treated with an *N*-acetylglucosamine (GlcNAc)–iodoacetamide reagent. No evidence of undesired glycosylation on the four cysteine residues involved in the two natural disulfide linkages present in EPO was observed. This study, however, highlighted the limitations of iodoacetamide-based glycoconjugations. Glycosylation yields of 50% were obtained (60% conversion was achieved with 500-fold excess reagents after 24 h); lectin-affinity chromatography was required to purify the glycosylated EPO and remove excess reagent. This method was also plagued by a lack of selectivity (partial selectivity), as reaction with histidine and lysine residues was observed at the basic pH required to achieve full conversion of the cysteine. At lower pHs and through the use of excess imidazole, this side reaction was partially suppressed, but incomplete conversion of the cysteine residues was observed. Glycosylation of a dehydrofolate reductase by the same method (Swanwick et al. 2005b) also suffered from partial conversion and low selectivity (Swanwick et al. 2005a); affinity purification allowed separation of glycosylated from unglycosylated material. In this model, glycosylation was shown to increase the thermal stability of the protein. The method was also suggested to be compatible with larger carbohydrates in the form of high-mannose structures (Totani et al. 2004), although in the latter case no details of characterization of the glycosylation of model protein carbonic anhydrase were given.

Bromoacetamide-mediated modification of an undecasaccharide onto a cysteine-containing peptide has also been reported. However, the maximum modification level observed was 64%, and the complex oligosaccharide required three synthetic transformations after isolation from egg yolk with 80% purity (Yamamoto et al. 2004). A yield of 77% for the final transformation (Fig. 3, steps c and d) was reported, but no yield was given for the hydrolysis and amination reactions.

2.1.2 Maleimide

Cysteines can be modified using maleimide-containing carbohydrate reagents through conjugate addition (poor selectivity). Maleimide-activated monosaccharides and disaccharides were coupled to peptides containing one reduced cysteine, and to the free cysteine residue on BSA (Shin et al. 2001). The strategy was also applied to the attachment of a high-mannose-type oligosaccharide, $Man_9GlcNAc_2Asn$ for glycosylation of a peptide in similar fashion (Ni et al. 2003). This reaction was exemplified on a 36-mer HIV-1 gp41 peptide. It should be noted that maleimides are competent electrophiles for other protein nucleophiles such as lysine and histidine.

Recently, thiol-functionality on the keyhole limpet hemocyanin (KLH) or on a tetanus peptide antigen was modified with a spacer-linked maleimide-functionalized HIV-carbohydrate epitope (Ni et al. 2006). These carbohydrates have been shown to bind the HIV-neutralizing 2G12-antibody (Scanlan et al. 2002), and conjugation of these carbohydrates was performed with the vision of raising

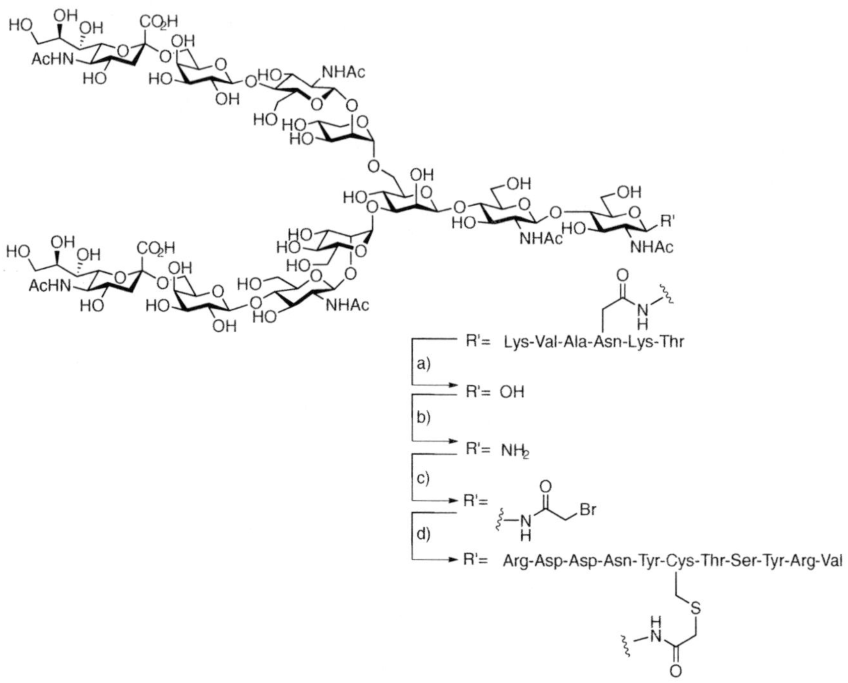

Fig. 3 Synthesis of bromoacetamide-modified oligosaccharide. **a** peptide, N-glycosidase F, 100 mM phosphate buffer; **b** saturated NH_4HCO_3, H_2O, 80% purity; **c** bromoacetic acid, *N,N′*-dicyclohexylcarbodiimide, dimethylformamide (DMF), 77% yield; **d** RDDNY*C*TSYRV, 100 mM PO_4, pH 7.0, 64% modification

antiglycan neutralizing antibodies. While it is a pragmatic approach to conjugate synthesis, this synthesis highlights how an elegant approach can be undone by less precise modification chemistry. Here, two sequential steps, thiol introduction to lysine (poor selectivity) and thiol modification with maleimide (poor selectivity), together compound the production of product mixtures. Carbohydrate incorporation levels of 15% by weight were achieved in this manner. The main additional shortcoming here is that the introduction of the thiol onto the KLH protein is determined by lysine location and so may be described as site-specific at best. This approach is a common one for the utilization of KLH as a platform for modification and has also been applied, for example, to the preparation of a potential carbohydrate-based cancer vaccine (Ragupathi et al. 2006). Another complication with maleimide-based modifications is the creation of the additional β-stereogenic centre in any resulting conjugate. In an example of modification of phosphate binding protein with a maleimide analog of the fluorophore coumarin, marked differences in fluorescence of the modified probes were observed (Hirshberg et al. 1998).

2.1.3 Disulfide Ligation – Methanethiosulfonates/Phenylthiosulfonates, Pyr-Sulfenyls, Glycoselenenylsulfides

A number of methods based on disulfide bond formation have been developed to prepare disulfide-linked proteins, which offer the most selective of the thiol-modification reactions: competing N–S or O–S products are highly labile unlike N–C or O–C counterparts and this, in part, ensures S–S product formation.

In site-selective strategies, a cysteine residue, introduced through mutagenesis to generate a protein nucleophile with a single free thiol tag, can subsequently be modified chemoselectively and quantitatively with electrophilic thiol-specific carbohydrate reagents, such as glycosyl methanethiosulfonates (GlycoMTS) (Davis et al. 1998; Davis et al. 2000b) or glycosyl phenylthiosulfonates (Gamblin et al. 2003).

Methanethiosulfonate (MTS) and phenylthiosulfonate reagents made possible the introduction of protected, spacer-free glycans, but only allowed the preparation of directly linked unprotected glycosides after enzymatic deprotection on protein, thereby potentially limiting utility (Davis et al. 2000a). The use of MTS reagents has also allowed for the site-selective modification of proteins with symmetrically branched multivalent deprotected glycans (Davis 2001). Such glycodendriproteins created from protein-degrading enzymes have been shown to reduce bacterial aggregation in *in vitro* systems (Rendle et al. 2004).

Boons and coworkers have developed a method based on nitropyridine-2-sulfenyl activated thioglycoside to modify cysteine residues, as exemplified by the modification of the single cysteine present in BSA with GlcNAc to form a disulfide-linked glycoprotein (Macindoe et al. 1998). The preparation of nitropyridine-2-sulfenyl derivatives of oligosaccharides larger than simple monosaccharides such as GlcNAc proved problematic (Watt et al. 2003) and prevented the use of this approach to glycosylate a hingeless Fc portion of human immunoglobulin G1 antibody with oligosaccharides. As an alternative, glycosyl thiols, such as thiochitobiose, were used in direct disulfide exchange (Watt et al. 2003) with the nonreduced protein and allowed yields of 60% to be obtained after purification (100% conversion as determined by mass spectrometry). Large excesses of glycosyl thiols in this type of approach are usually required, which in the case of complex carbohydrate can be a limiting factor, and aerial oxidation can require long reaction times and can be difficult to control. Moreover, methods based on disulfide exchange can potentially be problematic for proteins that contain natural disulfide bridges.

More recently, a selenenylsulfide-mediated ligation method has been reported (Gamblin et al. 2004). Two strategies can be adopted. The first is based on the activation of the cysteine-containing protein as a phenylselenenylsulfide after chemoselective reaction with phenylselenyl bromide, and subsequent reaction with glycosyl thiol to form the final mixed disulfide. In the second strategy, glycosyl thiols are initially converted to their selenenylsulfide analogs before reaction with the cysteine-containing protein. The versatility of this ligation method was evidenced by the coupling of a wide range of monosaccharides and oligosaccharides (up to heptasaccharides) to a variety of peptides, and to the single and multiple glycosylation of proteins. Furthermore, the same study showed that disulfide-linked glycoproteins can be enzymatically elongated using glycosyltransferases with full conversion, which widens the scope of the method. This method has recently been coupled with a method for direct thionation of unprotected sugars to allow direct "one-pot" glycoprotein synthesis using glycans from natural sources (Bernardes et al. 2006).

Recently, thiol tags along with the glycoselenenylsulfide and GlycoMTS methods have been used in combination with azide tags to expand the diversity of chemical protein modification (van Kasteren et al. 2007). These first examples of the use of dual tags/dual modifons allowed the creation of functional mimics of dual-PTM-modified human proteins (see Sect. 2.5).

2.2 *Amine Tag (X: –NH_2)*

Lysine is another oft-targeted residue. It is highly abundant on the surface of proteins and thus allows for the introduction of large numbers of copies of the glycan onto a protein. Thus, although it is a widely employed method for the indiscriminate modification of, for example, immunogenic carrier proteins to generate carbohydrate-based vaccines, its role in precise and selective methods is limited.

With a view to simply providing a pragmatic method for attaching small molecules to protein scaffolds to the less experienced, these methods often use amine-reactive homobifunctional coupling agents that bring together amine-bearing carbohydrates with proteinaceous amines (e.g., lysine). These have recently been reviewed and compared for their global incorporation efficiency (Izumi et al. 2003).

2.2.1 Reductive Amination Using Aldehydes

The most common method for introducing carbohydrates onto lysine side chains is reductive amination (Davis 2002). Traditionally the imine is formed between an open-chain carbohydrate and a lysine side chain, followed by reduction with sodium cyanoborohydride to give a secondary amine linkage (Gray 1974). The major disadvantages of this method are twofold. Firstly, the coupling conditions are harsh and conversion levels are generally low (four of 59 residues of BSA modified after 2 weeks, 7% conversion) (Gray 1974). Secondly, the reducing terminus generates an open-chain polyol amine linkage which alters the imparted conformation. One example of this strategy is the modification of a catalase with polysialic acid (Fernandes and Gregoriadis 1996). Aldehyde functionality was introduced in the nonreducing terminal sialic acid followed by a reductive amination to give polysialylated catalase. This modification led to improved *in vivo* lifetimes by reducing clearance, proteolytic cleavage, and inappropriate tissue uptake (Gregoriadis et al. 2005). The periodate cleavage method used for introduction of the aldehyde, however, can lead to cleavage of other vicinal diols, again generating mixtures of modiforms (Lifely et al. 1981).

Efforts have been made by using aldehyde-terminated spacer arms to combat this problem. Synthetic strategies usually carry a linker-precursor through oligosaccharide synthesis, followed by late-stage introduction of the aldehyde. Typically, this has been achieved by ozonolysis of an anomeric allyl group (Bernstein and Hall 1980). The Danishefsky group has used this strategy to conjugate complex, sialylated, tumor-associated carbohydrate antigens, such as sialylTn (Ragupathi et al. 1998) and GloboH (Ragupathi et al. 1999) to the surface of KLH. Alternative methods for the introduction of the aldehyde include the reduction of glycosyl azides and coupling to 4-pentenoic acid prior to ozonolysis (Pan et al. 2005) (Fig. 4).

2.2.2 Amidine Formation Using Imidates

One method with enhanced potential for selectivity and reactivity in creating a linkable oligosaccharide is the 2-imino-2-methoxyethyl-based modification method (Fig. 5). In this method, the active imidate form of the linker is generated from peracetylated *S*-cyanomethyl glycoside using methoxide, thereby allowing activation at the same time as deprotection (Lee et al. 1976; Stowell and Lee 1980, 1982). Recently, this method has been applied to introduce galactose onto a rhamnosidase enzyme (Robinson et al. 2004). The resulting glycosylated enzyme was successfully used as part of a bipartite drug delivery system targeted to the liver. With use of mannosyl and galactosyl 2-imino-2-methoxyethyls, high levels of glycosylation of lysine on adenoviral particles have also been achieved. The resulting modified adenoviruses are retargeted by glycosylation and hence the transfection target can be altered simply through chemistry (Pearce et al. 2005). Thus, the normal cocksackie receptor–adenovirus interaction was "switched off"; and mannose-binding-protein-mediated uptake via mannose receptor macrophages matured from human peripheral blood monocytes was "switched on." The linkage has thus usefully been proven to be of sufficient stability and compatibility *in vivo*. High modification levels can be achieved (conversion up to 70–95% of accessible surface lysines); e.g., in BSA modification, 40–45 sugar copies were incorporated per protein molecule, which corresponds to a modification of two thirds of all lysine residues (Lee et al. 1976; Stowell and Lee 1980, 1982). The success of the method hinges on the relative lability of the imidate reagents but the stability of the amidine products (formed with NH_2) – any imidates or thioimidates formed by reaction with protein *O* and *S* nucleophiles, respectively, simply hydrolyze.

2.2.3 Squarate-Mediated Amidation

Squarenes were first applied to protein glycosylation by the group of Tietze (Tietze et al. 1991a, b) and later by Hindsgaul (Kamath et al. 1996) as a way of modifying lysine residues with carbohydrates. The sugars are functionalized at the anomeric position with an amine-ended spacer arm and reacted with squaric esters. The resulting intermediates are subsequently reacted with lysine residues on proteins.

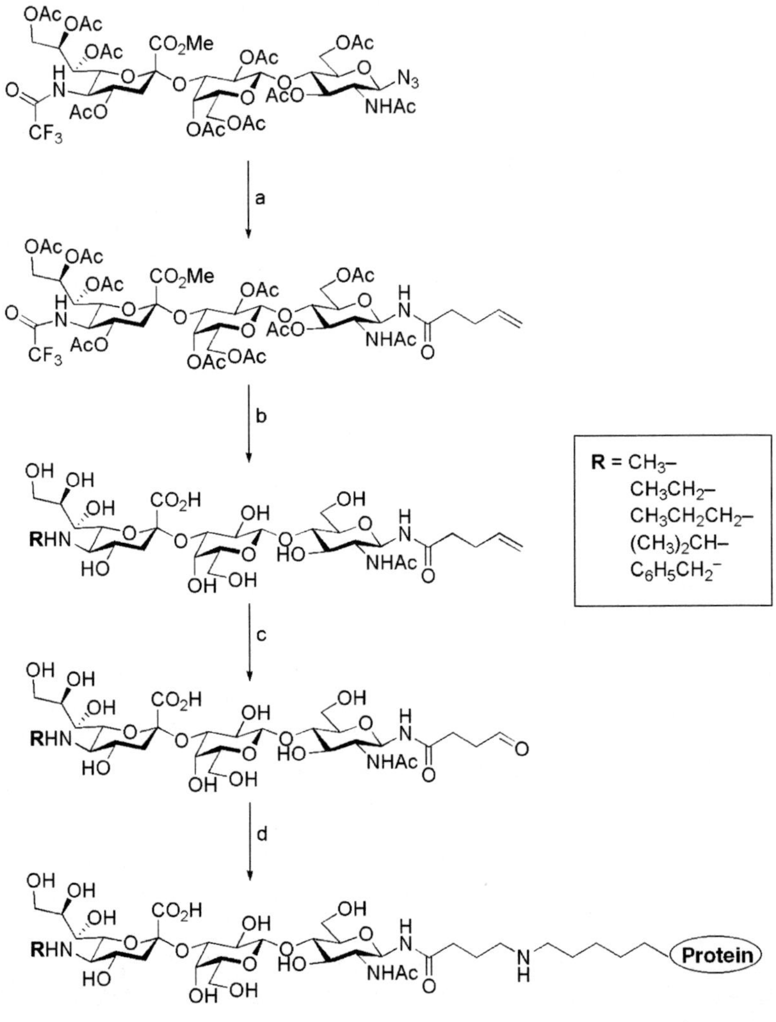

Fig. 4 Example of an aldehyde spacer in reductive amination of amine tags. **a** Lindlar's catalyst, H_2, MeOH/EtOAc (1:1), then 4-pentenoic anhydride, 90%; **b** 0.5 N NaOH then various anhydrides for introduction of the R group on neuraminic acid, 83–95% yield; **c** O_3, MeOH, −70 °C, then Me_2S, 85–90%; **d** keyhole limpet hemocyanin or human serum albumin (HSA), $NaBH_3CN$, 0.1 N, $NaHCO_3$, 3 days

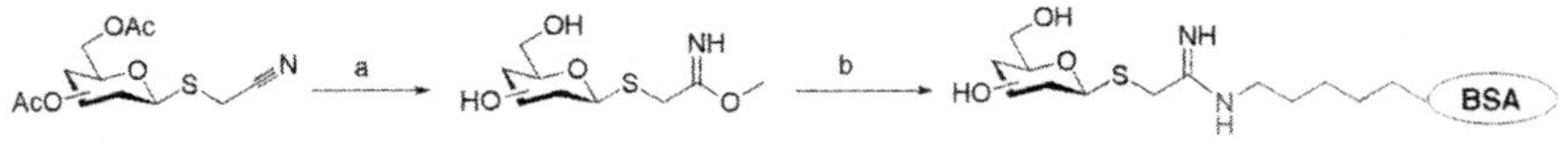

Fig. 5 Modification of a protein with a generic 2-methoxy-2-imino activated monosaccharide. **a** four equivalents of sodium methoxide; **b** pH 9.5, 2 h, 41 of 59 residues modified. *BSA* bovine serum albumin

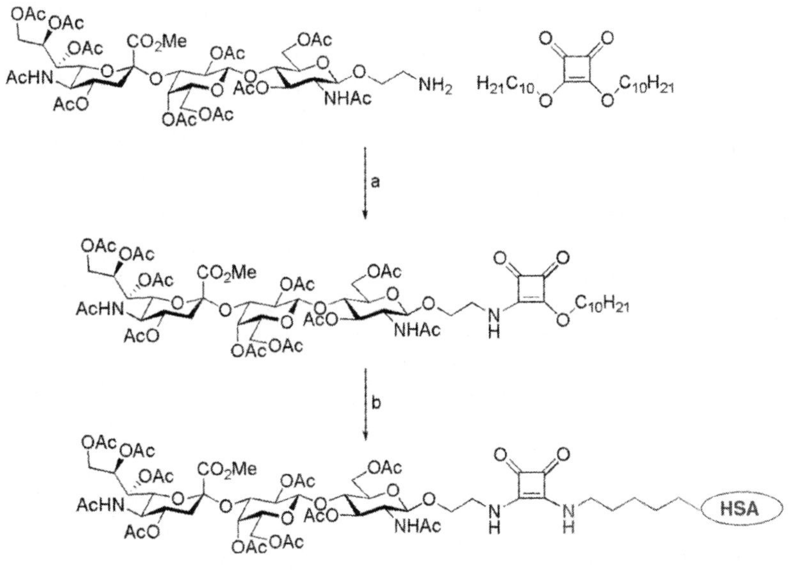

Fig. 6 Modification of HSA using squarate-mediated coupling. **a** DMF, Et_3N, 40%; **b** HSA, HCO_3 buffer, pH 9.0, 25% of lysine residues modified

This method has been shown to be compatible with oligosaccharides (Kamath et al. 1996; Saksena et al. 2005). It has been reported that approximately 25% of lysine residues of BSA could be modified by this method (Kamath et al. 1996). More recently the modification and biological application of this method was reported to create multiply modified human serum albumins carrying up to 19 sialylated *N*-acetyllactosaminyl sugars (conversion 25%) (Johansson et al. 2005). These compounds showed excellent inhibition of adhesion of adenovirus Ad37 to human corneal cells *in vitro* (Fig. 6).

The method has also been applied to the conjugation of a hexasaccharide from *Vibrio cholerae* to BSA as a potential vaccine candidate. It was shown that for one type of the lipopolysaccharide (Ogawa-type antigen), the compound was immunogenic, but no protective antibodies were generated. No yields for the carbohydrate synthesis, squarate formation and equivalents used were reported in this study (Meeks et al. 2004).

2.2.4 Isocyanates and Isothiocyanates

Isocyanates and isothiocyanates are excellent agents for the modification of amines (McBroom et al. 1972). Gabius and coworkers have successfully used this approach to couple a biantennary heptasaccharide to BSA; after acylation of the nonreducing end with a 6-aminohexanoyl spacer and activation of the free amino group as an isothiocyanate, coupling to BSA afforded neoglycoproteins with 2.4–4.6 glycan chains per carrier molecule of a total of 59 lysine residues. The versatility of this convergent approach, based on a chemoenzymatic strategy to synthesize complex oligosaccharides prior to coupling to the target protein (Andre et al. 1997, 2004), was recently used to study the effect of branching in complex-type triantennary *N*-glycans on lectin binding (Unverzagt et al. 2002), and to probe the effect of the LEC14-type branching on binding to surface lectins on cancer cells (Andre et al. 2005). One small drawback of the method is the modest yield of the introduction of the spacer arm into the complex carbohydrates late in the synthesis.

2.3 Carbonyl Tag (X: –C=O)

One of the first examples of the modification of an unnatural amino acid with a sugar was reported recently (Liu et al. 2003). By means of the amber codon suppression system (Cornish and Schultz 1994; Cornish et al. 1995), a ketone handle (Cornish et al. 1996) in the form of *p-C*-acetylphenylalanine was introduced into the 8-kDa Z-domain of staphylococcal protein A (Fig. 7). This handle was subsequently modified with an oxyamine-functionalized sugar. The resulting oxime-linked glycoprotein was then further elaborated using glycosyltransferases to yield a siaLacNAc-modified protein. The enzymatic galactosylation was reported to be 60% efficient and the sialylation proceeded with 65%. The resulting mixtures of glycoforms were characterized by liquid chromatography–mass spectrometry.

2.4 Olefin Tag (X: –=)

Despite the general utility of double bonds as functional groups in chemistry, applications in protein chemistry have largely been limited to their presence in α, β-unsaturated carbonyls, allowing modification through conjugate addition with nucleophilic reagents. Michael-type addition of 1-thio sugars on dehydroalanine-containing peptides has been demonstrated for the synthesis of *S*-linked glycopeptides, both on solid support and in solution (Zhu and van der Donk 2001). Dehydropeptides can be prepared by oxidation of *Se*-phenylselenocysteines and elimination of the corresponding selenoxide. The addition of 1-thio monosaccharides (protected or unprotected) on dehydroalanine-containing peptides proceeded with 62–74% yield in solution. A potential drawback of this method for the modification of peptide is the resulting formation of epimers in the peptide chain at the site of modification; very recently through combination with amber codon suppression methods, this approach has also been exemplified on protein systems (Wang et al. 2007) albeit with concomitant methionine oxidation.

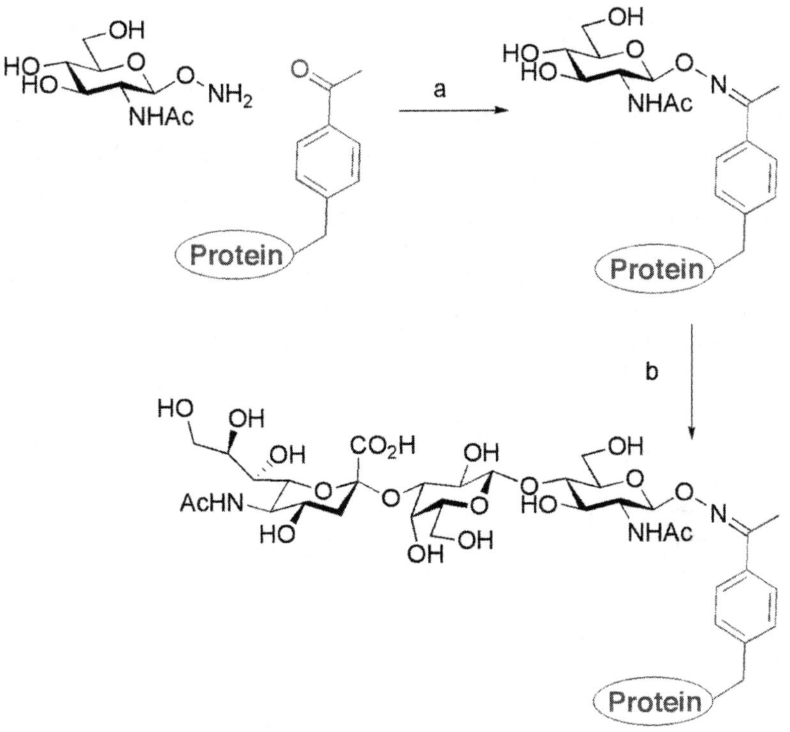

Fig. 7 Modification of ketone handle introduced in the Z-domain of staphylococcal protein A with oxyamine-functionalized *N*-acetylglucosamine. Conditions are **a** NaOAc buffer, pH 5.5 and **b** (1) UDP-Gal, β-1,4-GalT, (2) CMP-Sia, α-2,3-SiaT

2.5 *Azide Tag (X: –N_3)*

It has recently been demonstrated that a second mutually compatible chemical protein group, a second tag or modifon, can allow for dual and differential modification (van Kasteren et al. 2007). This enabled the construction of a synthetic protein probe capable of detecting mammalian brain inflammation and disease. This next-generation approach used three tags in total: thiols in natural cysteine residues, and azides and alkynes in the unnatural amino acids azidohomoalanine (Aha) and homopropargyl glycine (Hpg) (see Sect. 2.6). These chemical tags for site-selective conjugation allowed variable modification at multiple, predetermined sites in the protein backbone. The first modification was based on thiol modification using MTS reagents (Davis 2000b). The second chemoselective ligation was accomplished using copper(I)-catalyzed Huisgen–Dimroth cycloaddition (CCHC) (Rostovtsev et al. 2002; Tornoe et al. 2002). The latter reaction had been demonstrated for glycoconjugation in a variety of peptide systems, but these were the first examples of quantitative site-selective protein modification. Initial optimization studies of CCHC reactions were conducted on a fully competent enzyme as a protein substrate: SsβG-Aha43, a ten-point mutant ($(Met)_{10}(Cys)_{1} \rightarrow (Met43)_{1}(Ile)_{9}(Ser)_{1}$) of the galactosidase SsβG expressed in the methionine auxotrophic strain *E. coli* B834(DE3) in the presence of the methionine analog azidohomoalanine (Van Hest et al. 2000; Kiick et al. 2002). This protein substrate was reacted with a variety of alkyne glycosides as reagents. These studies highlighted a requirement for highly pure (99.999%) copper(I) to achieve more than 95% protein glycosylation.

The first examples of multisite differential modification were demonstrated on SsβG-Aha43-Cys439 using first a glucose MTS reagent to modify thiol, and then a galactosyl alkyne to modify azide. This protocol was also successfully applied to a Tamm–Horsfall glycoprotein fragment ($Thp_{295-306}$-Aha298-Cys303). In all cases the native enzymatic function of the protein was maintained after modification, and, in addition, all modified proteins demonstrated additional sugar-specific binding to lectins. Furthermore, this differential chemical PTM of proteins was used to mimic biological processes that are dependent upon PTMs to mediate protein–protein interfaces. Investigations into the P-selectin-glycoprotein ligand-1 (PSGL-1) have identified two critical PTMs required for binding to P-selectin in the primary rolling/adhesion phases of the inflammatory response; a sulfate attached to Tyr48 and either *O*-glycans siaLacNAc or sLe^x, attached to Thr57 (Kansas 1996; Spevak et al. 1996). As a more vigorous test of the chemical differential ligation strategy, a novel sulfotyrosine mimic (Tys) was installed onto position 439 of the lacZ-type reporter enzyme SsβG and a siaLacNAc-alkyne or sLe^x-alkyne mimic at 43, thereby reconstructing the structural parameters of the natural interaction between PSGL-1 and P-selectin.

2.6 *Alkyne Tag (X: –≡)*

The CCHC reaction (see Sect. 2.5) was also shown to be possible with alkyne-containing proteins such as SsβG-Hpg1 (in the "reverse sense") using azido-derivatized sugars as reagents.

3 PEGylation as a Mimic of Glycosylation

Covalent attachment of polymers, poly(ethylene glycol) (PEG) in particular, to therapeutic proteins is a well-established method employed to mimic some of the natural effects of glycosylation in order to improve protection of proteins towards enzymatic degradation, prolong circulation half-life and enhance solubility (Harris and Chess 2003). To date six PEGylated proteins have been approved by the Food and Drug Administration, and over a dozen are in the clinical development stage, thereby illustrating their importance in the functional engineering of therapeutic proteins (Kochendoerfer 2003, 2005; Veronese and Pasut 2005).

A variety of chemical PEGylation methods have been developed, most based on the activation of the PEG polymer with an aldehyde functionality (for ligation by reductive amination with amino groups on the protein), an activated hydroxyl group (tresyl or tosyl), succinimidyl active esters, or activated carbonates (succinimidyl carbonate, imidazoyl formate, phenyl carbonate); the merits and limitations of these methods have been reviewed elsewhere (Harris and Chess 2003; Veronese and Pasut 2005). These methods typically lack site-selectivity, as is the case for the ligation to amino groups (α-amino and ε-amino groups of lysines, N-terminus) using active carbonates and esters, aldehydes or tresylates; the indiscriminate modifications derived from these methods render batch-to-batch reproducibility more complicated. We will focus here only on recent examples of site-selective PEGylation methods.

3.1 Amine Tag (X: –NH_2)

Yamamoto et al. (2003) showed that the mutation of all lysine residues of tumor necrosis factor-α to less nucleophilic residues did not compromise bioactivity. This thereby removed competing lysine nucleophiles and allowed subsequent site-selective PEGylation of the N-terminus. It afforded higher bioactivity and an antitumor potency greater than that of nonspecifically mono-PEGylated wild-type tumor necrosis factor-α. This example of mutation for effective functional group alteration in proteins is elegant, and it will be interesting to discover how many proteins can sustain the drastic primary sequence modification necessitated by this type of approach.

3.2 Azide Tag (X: –N_3)

Cazalis et al. (2004) have reported the site-selective PEGylation of thrombomodulin at its *C*-terminus using Staudinger ligation, by reacting a methyl-PEG-triarylphosphine with a truncated thrombomodulin mutant bearing a C-terminal azidomethionine.

A PEGylation technology based on the copper-mediated Huisgen–Dimroth cycloaddition reaction has been disclosed by Schultz and coworkers (Deiters et al. 2004). The efficiency of this method was exemplified on human superoxide dismutase-1, by the introduction of a nonnatural *p*-azidophenylalanine residue into proteins in yeast, followed by formation of a stable triazole linkage with a PEG chain modified with propargylamine, which proceeded with 70–85% conversion.

3.3 NCL Assembly

Recently, Kochendoerfer and coworkers constructed a synthetic mimic of EPO (Kochendoerfer et al. 2003). They chemically coupled synthetic peptide segments with two other peptides each bearing the PEG polymer at sites corresponding to the natural glycosylation sites of EPO. The resulting protein displayed prolonged circulation *in vivo* and enhanced activity. In this example negative charges (carboxylic acid groups) attached at the end of the PEG chains were used to mimic the sialic acids featured in the natural glycosylation pattern of EPO.

3.4 Hydroxyl Tag (X: –OH)

DeFrees et al. (2006) have developed a site-specific PEGylation method combining enzymatic *N*-acetylgalactosamine (GalNAc) glycosylation at specific serine and threonine residues in proteins expressed in *E. coli*, and enzymatic transfer of PEGylated sialic acid onto the introduced GalNAc *O*-glycan using a 2,6-α-sialyltransferase. This intriguing example highlights the strong potential of enzymatic methods in determining chemo- and regioselectivity.

3.5 Thial Tag (X: –SH)

In an interesting attempt to study the effect of PEGylation on the pharmacological properties of interferon-β-1b (IFN-β-1b), Basu et al. (2006) PEGylated IFN-β-1b either randomly on the 11 lysines, or site-selectively on single cysteines (or lysine) introduced in redesigned IFN-β-1b variants by site-directed mutagenesis. Although, on the one hand, it was possible to modify lysine residues with a variety of succinimide PEG reagents, on the other hand, the alternative site-selective modification of cysteine residues with maleimide PEG polymers proved problematic owing to the lability of an internal disulfide linkage; even in acidic pH or with mild reducing conditions, complete selectivity and overall yields were shown to be insufficient. This constitutes another example of the limitation of the maleimide method of cysteine modification.

3.6 Disulfide Tag (X: –SS)

If the presence of native disulfide bonds can sometimes prove to be incompatible with the chemistries applied for the modification of reduced cysteines, they can in some cases be themselves selectively modified as shown recently by Shaunak et al. (2006). Using a cross-functionalized PEG monosulfone containing an α, β unsaturated ketone, native disulfides in human interferon α-2b and a fragment of an antibody to CD4 were site-specifically bisalkylated at the two cysteine sulfur atoms (after reduction of the disulfide bond) to form a three-carbon PEGylated bridge. This site-specific approach brings with it the advantages of leaving the tertiary structure around the native disulfide bond relatively unperturbed, and of not requiring the engineering of additional free thiols as chemical tags for modification. It will be interesting to see if other PTMs can be mimicked in this manner. However, such a method is also restricted by the limited flexibility in the positioning of native disulfide bonds (this will prevent the application of a site-selective strategy), and it might also be tricky to target a specific disulfide if more than one is present in a protein.

Although PEGylation can constitute a pragmatic means of improving the biophysical properties of therapeutic proteins, it remains clear that PEG chains are, from a chemical point of view, poor structural mimics of oligosaccharides and will be of limited use in the investigation of the precise function of the complex carbohydrate structures attached to glycoproteins.

4 Lipidation

Several types of lipidation exist, myristoylation, prenylation and palmitoylation being dominant. Surprisingly few attempts have been made to apply chemical methods to investigate their role in the modulation of protein function. Protein prenylation in nature refers to the post-translational attachment of either a farnesyl (C-15) or a geranylgeranyl (C-20) lipid chain to cysteine found in specific sequences in a protein backbone (Eastman et al. 2006; McTaggart 2006) near (and after further modification at) the C-terminus of proteins. This process is catalysed by three different enzyme types: protein farnesyltransferase and protein geranylgeranyltransferase types I and II.

Prenylation has a dramatic effect on the lipophilicity of target proteins. It aids the membrane anchoring of proteins, such as for the Ras family of small G-proteins. This family of proteins first forms a complex with its escort protein REP and is subsequently prenylated (Rak et al. 2004). Certain protein–protein interactions are also prenyl-mediated (Magee and Seabra 2003; Ramamurthy et al. 2003), which makes the rational study of prenylation an attractive target. In the case of Ras protein, which has been implicated in the insensitivity to apoptosis of certain cancer cell lines, its farnesylation and subsequent membrane targeting is crucial for its function (Cohen et al. 2000). More recently, it has been found that inhibition of the Ras-farnesyltransferase suppresses nuclear factor-κB regulated gene expression (Takada et al. 2004). Protein farnesylation and geranylgeranylation play a tantalizing role in certain human diseases, such as osteoporosis, where inhibition of geranylgeranylation prevents the formation and normal functioning of osteoclasts, the cells involved in the destruction of bone tissue (Van Beek et al. 1999; Woo et al. 2005), and in Hutchinson–Gilford progeria, the disease whereby patients age rapidly (Meta et al. 2006).

The study of these processes would benefit from the availability of a broadly applicable prenylation method, which would give access to functional mimics of farnesylated proteins. The synthesis and other applications of lipidated peptide/protein conjugates were recently comprehensively reviewed (Naider and Becker 1997; Brunsveld et al. 2006). To date, farnesylated proteins are typically synthesized by insect cell expression systems (sf9 cells), which suffer from poor yields and at least partial loss of prenyl functionality during purification (Jenkins et al. 2003).

4.1 NCL Assembly

Prenyls and other lipids have been attached to Ras proteins (which are naturally prenylated) by generating a truncated version of the protein and coupling to prenylated peptide fragments, which had been synthesized chemically (Bader et al. 2000). Despite the unnatural linkage introduced into the protein backbone, the hybrid proteins showed excellent activity by causing neurite outgrowth in a rat pheochromocytoma cell line. Fluorescent peptides carrying prenyl appendages were also introduced in this fashion (Reents et al. 2004).

4.2 Hydroxyl Tag (X: –OH)

Only one convergent chemical attachment of a farnesyl analog to a protein has been reported (Tilley and Francis 2006). This method uses palladium-mediated cross-coupling between a surface tyrosine residue on the protein and a palladium-π-allyl complex of the farnesyl lipid (Fig. 8). It allowed the modification of protein

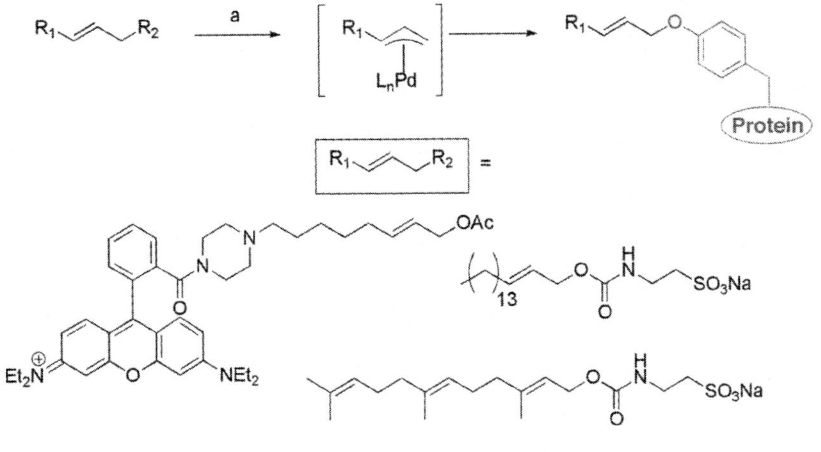

Fig. 8 Modification of a tyrosine residue with farnesyl or rhodamine by palladium-mediated π-allyl coupling of an allyl acetate or carbamate precursor. Conditions are **a** $Pd(OAc)_2$, Triphenyl phosphine trisulfonate (TPPTS), H_2O/dimethyl sulfoxide, pH 8.5–9.0

tyrosine residues in a partially selective manner in yields of up to 40% (Tilley and Francis 2006). The observed selectivity results from the lower pK_a of the phenolic hydroxyl acting as a tag.

4.3 *Thiol Tag (X: –SH)*

Although not yet applied to proteins, prenylation of a peptide has been reported by Crich et al. (2006a). This method makes use of the (2, 3)-sigmatropic rearrangement of a thioselenyl-lipid intermediate to yield a thioether-linked lipid. This method shows great promise.

4.4 *Olefins (X: –=)*

The dehydroalanine conjugate addition reaction (*vide supra*) is also amenable to the preparation of a farnesylated peptide by addition of either triisopropylsilyl-protected farnesylthiol in the presence of cesium fluoride or farnesylthioacetate with sodium methoxide (Zhu and Van Der Donk 2001).

5 Phosphorylation

Protein phosphorylation has most commonly been mimicked very simply through the use of the side chain carboxylate group of aspartic or glutamic acid as an approximation of the negative charge of the phosphate group. For example, the mutation of the penultimate serine residue at position 392 to a glutamic acid in the tumor suppressor protein p53 was shown to be an effective mimic of activated, phosphorylated p53 (Hupp and Lane 1995). The same S392E mutation of human p53 was shown to lead to an increase in nonspecific DNA affinity and thermal stability (Nichols and Matthews 2002). The mutation of serine to glutamic acid to mimic phosphorylation in the estrogen receptor is sufficient to upregulate estrogen-receptor-regulated genes (Balasenthil et al. 2004).

Considering the paramount role of protein phosphorylation in the regulation of protein function, with protein kinases being involved in most signal transduction events, it is surprising that – so far – relatively little effort has been put into the development of an efficient, reliable, chemical method of mimicking phosphorylation.

6 Sulfation

The use of a thiol tag combined with an aryl sulfonate MTS reagent has allowed the site-selective incorporation of a sulfotyrosine mimic into a protein scaffold (van Kasteren et al. 2007). This allowed the creation of an effective mimic of human protein PSGL-1, which was both site-selectively sulfated and glycosylated (see Sect. 2.5)

7 Conclusion

The key to understanding PTMs is by considering their role as one of fine-tuning protein function through diversification. Their only very loose connection to template-driven biosynthesis highlights the consequent need for methods that will generate more precise protein probes to allow the elucidation of their function in a flexible and empirical manner. This may be achieved, as we have outlined, through protein semisynthesis in an approach that incorporates the desired PTM or mimics of that PTM. If we wish to make sense of this diversity we need to break it into its individual elements – its *modiforms*. These will reveal structure–activity relationships in a realm that has been almost untouched by such precise investigation. This challenge may be met by the protein chemist through the invention of genuinely protein-friendly coupling reactions. If they are applied in a convergent manner, both yield and flexibility will be enhanced. The development of protein-compatible conditions, coupled with modification methods that may be evaluated using state-of-the-art analytical proteomics tools, will address key issues of conversion and selectivity. There are already promising examples of novel chemical modification methods that may ultimately fit these criteria of utility, but these have yet to show their suitability for the investigation of PTMs. Recently, an iridium-mediated reductive amination was reported, but only modification with simple aromatic aldehydes was achieved (McFarland and Francis 2005).

As well as novel reagents, novel tags are now emerging as possible sites for ligations. Tryptophan modification using carbenoids (Antos and Francis 2004) has been applied to myoglobin (Fig. 9), although with moderate yield and regioselectivity. Conversion levels of more than 50% were observed for myoglobin at pH 3.5 with a 7-h reaction time, although modification of the single tryptophan residue in subtilisin Carlsberg under the same conditions failed, and recent reports highlight that cysteine and methionine and their derivatives are also reactive partners to carbenoids (Crich et al. 2006b).

Such new reactions for natural amino acids as tags are being complemented by methods for the introduction of non-natural amino acids as tags. Select examples have already emerged (Liu et al. 2003; van Kasteren et al. 2007). As technologies for their incorporation develop, further opportunities will emerge.

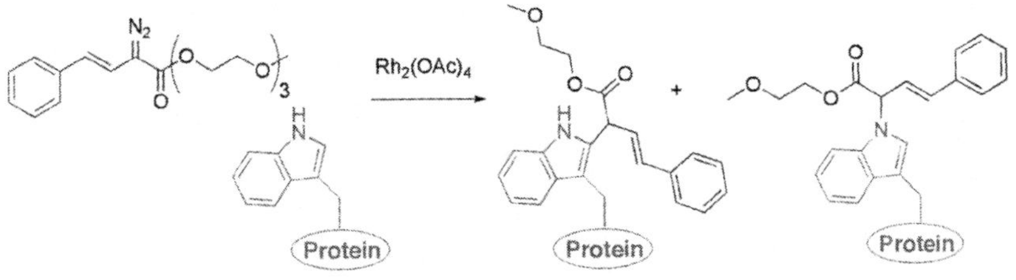

Fig. 9 Carbenoid reactions with tryptophan as an aromatic tag. Although currently nonselective, these reactions highlight a useful new reaction type for protein modification

Other competitive techniques will develop for PTM incorporation. Although incorporation and production levels are still low, the use of cotranslational methods that utilize transfer RNA intermediates acylated with amino acid esters already bearing requisite PTMs may start to become competitive. Early exciting examples of this use of the ribosome as the synthetic biocatalytic machinery for PTM introduction include the incorporation of a GlcNAc-modified serine (Zhang et al. 2004) or threonine (Xu et al. 2004) into a myoglobin via TAG-codon suppression. The incorporation level of the GlcNAc-threonine was up to 40%.

Key challenges remain. Multiple site-selective chemical modifications have only recently been disclosed (van Kasteren et al. 2007), an important addition given the many protein systems known to be functionally dependent on more than one PTM at precise sites. In addition, many PTMs have been unusually neglected. These include the small amount of effort applied to prenylation and phosphorylation. Others, such as sulfation, have been almost entirely neglected (Liu and Schultz 2006).

Moreover, considering the growing importance of protein-based biopharmaceuticals and the difficulties encountered when expressing these therapeutic proteins, for example production of mixtures of glycoforms of glycoproteins (Doores et al. 2006; Walsh and Jefferis 2006), there is an impetus for the development of either bioengineered expression systems that can reliably avoid the production of *modiforms*, or semisynthetic mimicking methods based on highly selective chemistries to generate homogeneous therapeutic compounds.

Acknowledgments The authors would like to thank all members of the Davis group past and present and the following organizations for financial support of work in the Davis group: EPSRC, BBSRC, MRC, IAVI, The European Union FPG Program, The Royal Society, The Leverhulme Trust, The Rhodes Trust, The Mitzutani Foundation, The Samsung Corporation, Genencor, GlycoForm, Syngenta, AstraZeneca, GlaxoSmithKline, High Force Research, and UCB Celltech.

References

Aebersold R, Mann M (2003) Mass spectrometry-based proteomics. Nature 422:198–207

Andre S, Kojima S, Prahl I, Lensch M, Unverzagt C, Gabius HJ (2005) Introduction of extended LEC14-type branching into core-fucosylated biantennary N-glycan: glycoengineering for enhanced cell binding and serum clearance of the neoglycoprotein. FEBS J 272:1986–1998

Andre S, Unverzagt C, Kojima S, Dong X, Fink C, Kayser K, Gabius HJ (1997) Neoglycoproteins with the synthetic complex biantennary nonasaccharide or its α-2,3- and α-2,6-sialylated derivatives: their preparation, assessment of their ligand properties for purified lectins, for tumor cells in vitro, and in tissue sections, and their biodistribution in tumor-bearing mice. Bioconjugate Chem 8:845–855

Andre S, Unverzagt C, Kojima S, Frank M, Seifert J, Fink C, Kayser K, Von Der Lieth CW, Gabius HJ (2004) Determination of modulation of ligand properties of synthetic complex-type biantennary N-glycans by introduction of bisecting GlcNAc in silico, in vitro and in vivo. Eur J Biochem 271:118–134

Antos JM, Francis MB (2004) Selective tryptophan modification with rhodium carbenoids in aqueous solution. J Am Chem Soc 126:10256–10257

Bader B, Kuhn K, Owen DJ, Waldmann H, Wittinghofer A, Kuhlmann J (2000) Bioorganic synthesis of lipid-modified proteins for the study of signal transduction. Nature 403:223–226

Balasenthil S, Barnes CJ, Rayala SK, Kumar R (2004) Estrogen receptor activation at serine 305 is sufficient to upregulate cyclin D1 in breast cancer cells. FEBS Lett 567:243–247

Basu A, Yang K, Wang M, Liu S, Chintala R, Palm T, Zhao H, Peng P, Wu D, Zhang Z, Hua J, Hsieh MC, Zhou J, Petti G, Li X, Janjua A, Mendez M, Liu J, Longley C, Zhang Z, Mehlig M, Borowski V, Viswanathan M, Filpula D (2006) Structure–function engineering of interferon-β-1b for improving stability, solubility, potency, immunogenicity, and pharmacokinetic properties by site-selective mono-PEGylation. Bioconjugate Chem 17:618–630

Bernardes GJL, Gamblin DP, Davis BG (2006) The direct formation of glycosyl thiols from reducing sugars allows one-pot protein glycoconjugation. Angew Chem Int Ed 45:4007–4011

Bernstein MA, Hall LD (1980) A general synthesis of model glycoproteins: coupling of alkenyl glycosides to proteins, using reductive ozonolysis followed by reductive amination with sodium cyanoborohydride. Carbohydr Res 78:C1–C3

Brunsveld L, Kuhlmann J, Alexandrov K, Wittinghofer A, Goody RS, Waldmann H (2006) Lipidated Ras and Rab peptides and proteins – synthesis, structure, and function. Angew Chem Int Ed 45:6622–6646

Cazalis CS, Haller CA, Sease-Cargo L, Chaikof EL (2004) C-terminal site-specific PEGylation of a truncated thrombomodulin mutant with retention of full bioactivity. Bioconjugate Chem 15:1005–1009

Chen I, Ting AY (2005) Site-specific labeling of proteins with small molecules in live cells. Curr Opin Biotechnol 16:35–40

Cohen LH, Pieterman E, van Leeuwen REW, Overhand M, Burm BEA, van Der Marel GA, van Boom JH (2000) Inhibitors of prenylation of Ras and other G-proteins and their application as therapeutics. Biochem Pharmacol 60:1061–1068

Cornish VW, Hahn KM, Schultz PG (1996) Site-specific protein modification using a ketone handle. J Am Chem Soc 118:8150–8151

Cornish VW, Mendel D, Schultz PG (1995) Probing protein-structure and function with an expanded genetic-code. Angew Chem Int Ed 34:621–633

Cornish VW, Schultz PG (1994) A new tool for studying protein-structure and function. Curr Opin Struct Biol 4:601–607

Crich D, Krishnamurthy V, Hutton TK (2006a) Allylic selenosulfide rearrangement: a method for chemical ligation to cysteine and other thiols. J Am Chem Soc 128:2544–2545

Crich D, Zou Y, Brebion F (2006b) Sigmatropic rearrangements as tools for amino acid and peptide modification: application of the allylic sulfur ylide rearrangement to the preparation of neoglycoconjugates and other conjugates. J Org Chem 71:9172–9177

Davis BG (2001) The controlled glycosylation of a protein with a bivalent glycan: towards a new class of glycoconjugates, glycodendriproteins. Chem Commun 351–352

Davis BG (2002) Synthesis of glycoproteins. Chem Rev 102:579–601

Davis BG (2003) Chemical modification of biocatalysts. Curr Opin Biotechnol 14:379–386

Davis BG (2004) Mimicking posttranslational modifications of proteins. Science 303:480–482

Davis BG, Lloyd RC, Jones JB (1998) Controlled site-selective glycosylation of proteins by a combined site-directed mutagenesis and chemical modification approach. J Org Chem 63:9614–9615

Davis BG, Lloyd RC, Jones JB (2000a) Controlled site-selective protein glycosylation for precise glycan structure-catalytic activity relationships. Bioorg Med Chem 8:1527–1535

Davis BG, Maughan MAT, Green MP, Ullman A, Jones JB (2000b) Glycomethanethiosulfonates: Powerful reagents for protein glycosylation. Tetrahedron: Asymmetry 11:245–262

Davis NJ, Flitsch SL (1991) A novel method for the specific glycosylation of proteins. Tetrahedron Lett 32:6793–6796

Dawson PE, Kent SBH (2000) Synthesis of native proteins by chemical ligation. Ann Rev Biochem 69:923–960

Dawson PE, Muir TW, Clark-Lewis I, Kent SBH (1994) Synthesis of proteins by native chemical ligation. Science 266:776–779

DeFrees S, Wang ZG, Xing R, Scott AE, Wang J, Zopf D, Gouty DL, Sjoberg ER, Panneerselvam K, Brinkman-Van der Linden ECM, Bayer RJ, Tarp MA, Clausen H (2006) GlycoPEGylation of recombinant therapeutic proteins produced in *Escherichia coli*. Glycobiology 16:833–843

Deiters A, Cropp TA, Summerer D, Mukherji M, Schultz PG (2004) Site-specific PEGylation of proteins containing unnatural amino acids. Bioorg Med Chem Lett 14:5743–5745

Dieterich DC, Link AJ, Graumann J, Tirrell DA, Schuman EM (2006) Selective identification of newly synthesized proteins in mammalian cells using bioorthogonal noncanonical amino acid tagging (BONCAT). Proc Natl Acad Sci USA 103:9482–9487

Doores KJ, Gamblin DP, Davis BG (2006) Exploring and exploiting the therapeutic potential of glycoconjugates. Chem Eur J 12:656–665

Eastman RT, Buckner FS, Yokoyama K, Gelb MH, Van Voorhis WC (2006) Fighting parasitic disease by blocking protein farnesylation. J Lipid Res 47:233–240

Fernandes AI, Gregoriadis G (1996) Synthesis, characterization and properties of sialylated catalase. Biochim Biophys Acta 1293:90–96

Fodje MN, Al-Karadaghi S (2002) Occurrence, conformational features and amino acid propensities for the α-helix. Protein Eng 15:353–358

Gamblin DP, Garnier P, Van Kasteren SI, Oldham NJ, Fairbanks AJ, Davis BG (2004) Glyco-SeS: Selenenylsulfide-mediated protein glycoconjugation – a new strategy in post-translational modification. Angew Chem Int Ed 43:828–833

Gamblin DP, Garnier P, Ward SJ, Oldham NJ, Fairbanks AJ, Davis BG (2003) Glycosyl phenylthiosulfonates (Glyco-PTS): novel reagents for glycoprotein synthesis. Org Biomol Chem 1: 3642–3644

Gray GR (1974) The direct coupling of oligosaccharides to proteins and derivatized gels. Arch Biochem Biophys 163:426–428

Gregoriadis G, Jain S, Papaioannou I, Laing P (2005) Improving the therapeutic efficacy of peptides and proteins: a role for polysialic acids. Int J Pharm 300:125–130

Grogan MJ, Pratt MR, Marcaurelle LA, Bertozzi CR (2002) Homogeneous glycopeptides and glycoproteins for biological investigation. Ann Rev Biochem 71:593–634

Harris JM, Chess RB (2003) Effect of PEGylation on pharmaceuticals. Nat Rev Drug Disc 2:214–221

Haslam SM, North SJ, Dell A (2006) Mass spectrometric analysis of N- and O-glycosylation of tissues and cells. Curr Opin Struct Biol 16:584–591

Hermanson GT (1996) Bioconjugate techniques, 1st edn. Academic Press, San Diego

Hirshberg M, Henrick K, Haire LL, Vasisht N, Brune M, Corrie JET, Webb MR (1998) Crystal structure of phosphate binding protein labeled with a coumarin fluorophore, a probe for inorganic phosphate. Biochemistry 37:10381–10385

Hupp TR, Lane DP (1995) Two distinct signaling pathways activate the latent DNA binding function of p53 in a casein kinase II-independent manner. J Biol Chem 270:18165–18174

Izumi M, Okumura S, Yuasa H, Hashimoto H (2003) Mannose-BSA conjugates: comparison between commercially available linkers in reactivity and bioactivity. J Carbohydr Chem 22:317–329

Jenkins CM, Han X, Yang J, Mancuso DJ, Sims HF, Muslin AJ, Gross RW (2003) Purification of recombinant human cPLA2γ and identification of C-terminal famesylation, proteolytic processing, and carboxymethylation by MALDI-TOF-TOF analysis. Biochemistry 42: 11798–11807

Johansson SMC, Arnberg N, Elofsson M, Wadell G, Kihlberg J (2005) Multivalent HSA conjugates of 3′-sialyllactose are potent inhibitors of adenoviral cell attachment and infection. ChemBioChem 6:358–364

Kamath VP, Diedrich P, Hindsgaul O (1996) Use of diethyl squarate for the coupling of oligosaccharide amines to carrier proteins and characterization of the resulting neoglycoproteins by MALDI-TOF mass spectrometry. Glycoconj J 13:315–319

Kansas GS (1996) Selectins and their ligands: current concepts and controversies. Blood 88:3259–3287

Khmelnitsky YL (2004) Current strategies for in vitro protein glycosylation. J Mol Catal B 31:73–81

Kiick KL, Saxon E, Tirrell DA, Bertozzi CR (2002) Incorporation of azides into recombinant proteins for chemoselective modification by the Staudinger ligation. Proc Natl Acd Sci USA 99:19–24
Kochendoerfer G (2003) Chemical and biological properties of polymer-modified proteins. Exp Opin Biol Ther 3:1253–1261
Kochendoerfer GG (2005) Site-specific polymer modification of therapeutic proteins. Curr Opin Chem Biol 9:555–560
Kochendoerfer GG, Chen SY, Mao F, Cressman S, Traviglia S, Shao H, Hunter CL et al (2003) Design and chemical synthesis of a homogeneous polymer-modified erythropoiesis protein. Science 299:884–887
La Clair JJ, Foley TL, Schegg TR, Regan CM, Burkart MD (2004) Manipulation of carrier proteins in antibiotic biosynthesis. Chem Biol 11:195–201
Lee YC, Stowell CP, Krantz MJ (1976) 2-Imino-2-methoxyethyl-1-thioglycosides: New reagents for attaching sugars to proteins. Biochemistry 15:3956–3963
Lifely MR, Gilbert AS, Moreno C (1981) Sialic acid polysaccharide antigens of Neisseria meningitidis and *Escherichia coli*: Esterification between adjacent residues. Carbohydr Res 94:193–203
Liu CC, Schultz PG (2006) Recombinant expression of selectively sulfated proteins in *Escherichia coli*. Nat Biotechnol 24:1436–1440
Liu H, Wang L, Brock A, Wong CH, Schultz PG (2003) A method for the generation of glycoprotein mimetics. J Am Chem Soc 125:1702–1703
Lowe JB, Marth JD (2003) A genetic approach to mammalian glycan function. Ann Rev Biochem 72:643–691
Lundblad RL (2004) Chemical reagents for protein modification – 3rd edn. CRC Press, Boca Raton
Macindoe WM, Van Oijen AH, Boons G-J (1998) A unique highly facile method for synthesising disulfide linked neoglycoconjugates: a new approach for remodelling of peptides and proteins. Chem Commun:847–848
Macmillan D, Bertozzi CR (2004) Modular assembly of glycoproteins: towards the synthesis of GlyCAM-1 by using expressed protein ligation. Angew Chem Int Ed 43:1355–1359
Macmillan D, Bill RM, Sage KA, Fern D, Flitsch SL (2001) Selective in vitro glycosylation of recombinant proteins: semi-synthesis of novel homogeneous glycoforms of human erythropoietin. Chem Biol 8:133–145
Magee AI, Seabra MC (2003) Are prenyl groups on proteins sticky fingers or greasy handles? Biochem J 376:e3–e4
McBroom CR, Samanen CH, Goldstein IJ (1972) Carbohydrate antigens: coupling of carbohydrates to proteins by diazonium and phneylisothiocyanate reactions. Methods Enzymol 28:212–219
McFarland JM, Francis MB (2005) Reductive alkylation of proteins using iridium catalyzed transfer hydrogenation. J Am Chem Soc 127:13490–13491
McTaggart SJ (2006) Isoprenylated proteins. Cell Mol Life Sci 63:255–267
Meeks MD, Saksena R, Ma X, Wade TK, Taylor RK, Kovac P, Wade WF (2004) Synthetic fragments of *Vibrio cholerae* O1 Inaba O-specific polysaccharide bound to a protein carrier are immunogenic in mice but do not induce protective antibodies. Infect Immun 72:4090–4101
Meta M, Yang SH, Bergo MO, Fong LG, Young SG (2006) Protein farnesyltransferase inhibitors and progeria. Trends Mol Med 12:480–487
Miller JS, Dudkin VY, Lyon GJ, Muir TW, Danishefsky SJ (2003) Toward fully synthetic N-linked glycoproteins. Angew Chem Int Ed 42:431–434
Miller LW, Cornish VW (2005) Selective chemical labeling of proteins in living cells. Curr Opin Chem Biol 9:56–61
Muir TW (2003) Semisynthesis of proteins by expressed protein ligation. Ann Rev Biochem 72:249–289
Muralidharan V, Muir TW (2006) Protein ligation: An enabling technology for the biophysical analysis of proteins. Nat Methods 3:429–438
Naider FR, Becker JM (1997) Synthesis of prenylated peptides and peptide esters. Biopolymers 43:3–14

Ni J, Singh S, Wang LX (2003) Synthesis of maleimide-activated carbohydrates as chemoselective tags for site-specific glycosylation of peptides and proteins. Bioconjugate Chem 14:232–238

Ni J, Song H, Wang Y, Stamatos NM, Wang LX (2006) Toward a carbohydrate-based HIV-1 vaccine: Synthesis and immunological studies of oligomannose-containing glycoconjugates. Bioconjugate Chem 17:493–500

Nichols NM, Matthews KS (2002) Human p53 phosphorylation mimic, S392E, increases nonspecific DNA affinity and thermal stability. Biochemistry 41:170–178

Pachamuthu K, Schmidt RR (2006) Synthetic routes to thiooligosaccharides and thioglycopeptides. Chem Rev 106:160–187

Pan Y, Chefalo P, Nagy N, Harding C, Guo Z (2005) Synthesis and immunological properties of N-modified GM3 antigens as therapeutic cancer vaccines. J Med Chem 48:875–883

Pearce OMT, Fisher KD, Humphries J, Seymour LW, Smith A, Davis BG (2005) Glycoviruses: chemical glycosylation retargets adenoviral gene transfer. Angew Chem Int Ed 44:1057–1061

Rademacher TW, Parekh RB, Dwek RA (1988) Glycobiology. Ann Rev Biochem 57:785–838

Ragupathi G, Koganty RR, Qiu D, Lloyd KO, Livingston PO (1998) A novel and efficient method for synthetic carbohydrate conjugate vaccine preparation: synthesis of sialyl Tn-KLH conjugate using a 4-(4-N-maleimidomethyl) cyclohexane-1-carboxyl hydrazide (MMCCH) linker arm. Glycoconjugate J 15:217–221

Ragupathi G, Koide F, Livingston PO, Cho YS, Endo A, Wan Q, Spassova MK, Keding SJ, Allen J, Ouerfelli O, Wilson RM, Danishefsky SJ (2006) Preparation and evaluation of unimolecular pentavalent and hexavalent antigenic constructs targeting prostate and breast cancer: a synthetic route to anticancer vaccine candidates. J Am Chem Soc 128:2715–2725

Ragupathi G, Slovin SF, Adluri S, Sames D, Kim IJ, Kim HM, Spassova M, Bornmann WG, Lloyd KO, Scher HI, Livingston PO, Danishefsky SJ (1999) A fully synthetic globo H carbohydrate vaccine induces a focused humoral response in prostate cancer patients: a proof of principle. Angew Chem Int Ed Engl 38:563–566

Rak A, Pylypenko O, Durek T, Watzke A, Kushnir S, Brunsveld L, Waldmann H, Goody RS, Alexandrov K (2003) Structure of Rab GDP-dissociation inhibitor in complex with prenylated YPT1 GTPase. Science 302:646–650

Rak A, Pylypenko O, Niculae A, Pyatkov K, Goody RS, Alexandrov K (2004) Structure of the Rab7: REP-1 complex: insights into the mechanism of Rab prenylation and choroideremia disease. Cell 117:749–760

Ramamurthy V, Roberts M, Van den Akker F, Niemi G, Reh TA, Hurley JB (2003) AIPL1, a protein implicated in Leber's congenital amaurosis, interacts with and aids in processing of farnesylated proteins. Proc Natl Acad Sci USA 100:12630–12635

Reents R, Wagner M, Kuhlmann J, Waldmann H (2004) Synthesis and application of fluorescence-labeled Ras-proteins for live-cell imaging. Angew Chem Int Ed 43:2711–2714

Rendle PM, Seger A, Rodrigues J, Oldham NJ, Bott RR, Jones JB, Cowan MM, Davis BG (2004) Glycodendriproteins: a synthetic glycoprotein mimic enzyme with branched sugar-display potently inhibits bacterial aggregation. J Am Chem Soc 126:4750–4751

Robinson MA, Charlton ST, Garnier P, Wang XT, Davis SS, Perkins AC, Frier M, Duncan R, Savage TJ, Wyatt DA, Watson SA, Davis BG (2004) LEAPT: Lectin-directed enzyme-activated prodrug therapy. Proc Natl Acad Sci USA 101:14527–14532

Romijn EP, Krijgsveld J, Heck AJR (2003) Recent liquid chromatographic-(tandem) mass spectrometric applications in proteomics. J Chromatography A 1000:589–608

Rostovtsev VV, Green LG, Fokin VV, Sharpless KB (2002) A stepwise Huisgen cycloaddition process: copper(I)-catalyzed regioselective "ligation" of azides and terminal alkynes. Angew Chem Int Ed 41:2596–2599

Saksena R, Zhang J, Kovac P (2005) Immunogens from a synthetic hexasaccharide fragment of the O-SP of *Vibrio cholerae* O:1, serotype Ogawa. Tetrahedron: Asymmetry 16:187–197

Scanlan CN, Pantophlet R, Wormald MR, Ollmann Saphire E, Stanfield R, Wilson IA, Katinger H, Dwek RA, Rudd PM, Burton DR (2002) The broadly neutralizing anti-human immunodeficiency virus type 1 antibody 2G12 recognizes a cluster of α 1$\rightarrow$2 mannose residues on the outer face of gp120. J Virol 76:7306–7321

Schwarzer D, Cole PA (2005) Protein semisynthesis and expressed protein ligation: chasing a protein's tail. Curr Opin Chem Biol 9:561–569

Shaunak S, Godwin A, Choi JW, Balan S, Pedone E, Vijayarangam D, Heidelberger S, Teo I, Zloh M, Brocchini S (2006) Site-specific PEGylation of native disulfide bonds in therapeutic proteins. Nat Chem Biol 2:312–313

Shin I, Jung HJ, Lee MR (2001) Chemoselective ligation of maleimidosugars to peptides/protein for the preparation of neoglycopeptides/neoglycoprotein. Tetrahedron Lett 42:1325–1328

Spevak W, Foxall C, Charych DH, Dasgupta F, Nagy JO (1996) Carbohydrates in an acidic multivalent assembly: nanomolar P-selectin inhibitors. J Med Chem 39:1018–1020

Stowell CP, Lee VC (1980) Neoglycoproteins: the preparation and application of synthetic glycoproteins. Adv Carbohydr Chem 37:225–281

Stowell CP, Lee YC (1982) Preparation of neoglycoproteins using 2-imino-2-methoxyethyl 1-thioglycosides. Methods Enzymol 83:278–288

Swanwick RS, Daines AM, Flitsch SL, Allemann RK (2005a) Synthesis of homogenous site-selectively glycosylated proteins. Org Biomol Chem 3:572–574

Swanwick RS, Daines AM, Tey LH, Flitsch SL, Allemann RK (2005b) Increased thermal stability of site-selectively glycosylated dihydrofolate reductase. ChemBioChem 6:1338–1340

Takada Y, Khuri FR, Aggarwal BB (2004) Protein farnesyltransferase inhibitor (SCH 66336) abolishes NF-κB activation induced by various carcinogens and inflammatory stimuli leading to suppression of NF-κB-regulated gene expression and up-regulation of apoptosis. J Biol Chem 279:26287–26299

Tietze L, Arlt M, Beller M, Glusenkamp KH, Jahde E, Rajewsky MF (1991a) Squaric acid diethyl ester: a new coupling reagent for the formation of drug biopolymer conjugates. Synthesis of squaric acid ester amides and diamides. Chem Ber 124:1215–1219

Tietze LF, Schroter C, Gabius S, Brinck U, Goerlach-Graw A, Gabius HJ (1991b) Conjugation of p-aminophenyl glycosides with squaric acid diester to a carrier protein and the use of neoglycoprotein in the histochemical detection of lectins. Bioconjugate Chem 2:148–153

Tilley SD, Francis MB (2006) Tyrosine-selective protein alkylation using π-allylpalladium complexes. J Am Chem Soc 128:1080–1081

Tornoe CW, Christensen C, Meldal M (2002) Peptidotriazoles on solid phase: [1,2,3]-triazoles by regiospecific copper(I)-catalyzed 1,3-dipolar cycloadditions of terminal alkynes to azides. J Org Chem 67:3057–3064

Totani K, Matsuo I, Takatani M, Arai MA, Hagihara S, Ito Y (2004) Synthesis of glycoprotein molecular probes for the analyses of protein quality control system. Glycoconjugate J 21:69–74

Unverzagt C, Andre S, Seifert J, Kojima S, Fink C, Srikrishna G, Freeze H, Kayser K, Gabius HJ (2002) Structure–activity profiles of complex biantennary glycans with core fucosylation and with/without additional α–2,3/α–2,6 sialylation: synthesis of neoglycoproteins and their properties in lectin assays, cell binding, and organ uptake. J Med Chem 45:478–491

Van Beek E, Lowik C, Van Der Pluijm G, Papapoulos S (1999) The role of geranylgeranylation in bone resorption and its suppression by bisphosphonates in fetal bone explants in vitro: a clue to the mechanism of action of nitrogen-containing bisphosphonates. J Bone Miner Res 14:722–729

Van Hest JCM, Kiick KL, Tirrell DA (2000) Efficient incorporation of unsaturated methionine analogues into proteins in vivo. J Am Chem Soc 122:1282–1288

Van Kasteren SI, Kramer HB, Jensen HH, Campbell SJ, Kirkpatrick J, Oldham NJ, Anthony DC, Davis BG (2007) Expanding the diversity of chemical protein modification allows post-translational mimicry. Nature 446:1105–1109

Veronese FM, Pasut G (2005) PEGylation, successful approach to drug delivery. Drug Discovery Today 10:1451–1458

Volkert M, Uwai K, Tebbe A, Popkirova B, Wagner M, Kuhlmann J, Waldmann H (2003) Synthesis and biological activity of photoactivatable N-Ras peptides and proteins. J Am Chem Soc 125:12749–12758

Walsh CT (2005) Posttranslational modification of proteins: expanding nature's inventory. Roberts and Co. Englewood, Colorado, U.S.A.

Walsh G, Jefferis R (2006) Post-translational modifications in the context of therapeutic proteins. Nat Biotechnol 24:1241–1252

Wang J, Schiller SM, Schultz PG (2007) A biosynthetic route to dehydroalanine-containing proteins. Angew Chem Intl Ed 46:6849–6851

Watt GM, Lund J, Levens M, Kolli VSK, Jefferis R, Boons GJ (2003) Site-specific glycosylation of an aglycosylated human IgG1-Fc antibody protein generates neoglycoproteins with enhanced function. Chem Biol 10:807–814

Wieland T, Bokelmann E, Bauer L, Lang HU, Lau H (1953) Liebigs Ann Chem 583:129

Woo JT, Nakagawa H, Krecic AM, Nagai K, Hamilton AD, Sebti SM, Stern PH (2005) Inhibitory effects of mevastatin and a geranylgeranyl transferase I inhibitor (GGTI-2166) on mononuclear osteoclast formation induced by receptor activator of NFκB ligand (RANKL) or tumor necrosis factor-α (TNF-α). Biochemical Pharmacol 69:87–95

Xu R, Hanson SR, Zhang Z, Yang YY, Schultz PG, Wong CH (2004) Site-specific incorporation of the mucin-type N-acetylgalactosamine-β-O-threonine into protein in *Escherichia coli*. J Am Chem Soc 126:15654–15655

Yamamoto N, Sakakibara T, Kajihara Y (2004) Convenient synthesis of a glycopeptide analogue having a complex type disialyl-undecasaccharide. Tetrahedron Lett 45:3287–3290

Yamamoto Y, Tsutsumi Y, Yoshioka Y, Nishibata T, Kobayashi K, Okamoto T, Mukai Y, Shimizu T, Nakagawa S, Nagata S, Mayumi T (2003) Site-specific PEGylation of a lysine-deficient TNF-α with full bioactivity. Nat Biotechnol 21:546–552

Zhang Z, Gildersleeve J, Yang YY, Xu R, Loo JA, Uryu S, Wong CH, Schultz PG (2004) A new strategy for the synthesis of glycoproteins. Science 303:371–373

Zhu Y, Van Der Donk WA (2001) Convergent synthesis of peptide conjugates using dehydroalanines for chemoselective ligations. Org Lett 3:1189–1192

Noncanonical Amino Acids in Protein Science and Engineering

K.E. Beatty and D.A. Tirrell(✉)

Contents

D.A. Tirrell
Division of Chemistry and Chemical Engineering, California Institute of Technology, Pasadena, CA 91125, USA, e-mail: tirrell@caltech.edu

C. Köhrer and U.L. RajBhandary (eds.) *Protein Engineering.*
Nucleic Acids and Molecular Biology 22,

Abstract Whether guided by computation, intuition, or evolution, recombinant DNA methods have enabled the preparation of an astonishing array of new protein variants of fundamental and practical importance. In recent years, it has become apparent that the power of protein engineering can be extended further through the use of an expanded set of amino acid constituents (Budisa 2006). The use of "noncanonical" amino acids creates new possibilities for protein design, and can be integrated in straightforward fashion into either "rational" or evolutionary design strategies. This chapter describes some of these new possibilities, with emphasis on the use of noncanonical amino acids to interrogate or change the reactivity, stability, or spectral properties of engineered proteins. Examples include the use of "bio-orthogonal" ligation reactions to enable selective dye-labeling and affinity-tagging, the use of aromatic amino acid analogs to alter the emission properties of luminescent proteins, the use of photosensitive amino acids to effect controlled protein cross-linking, and the use of fluorinated amino acids to control protein stability and protein–protein interactions. While many important experiments in this field have utilized in vitro translation, this chapter will focus primarily on cellular synthesis of proteins that contain noncanonical amino acids.

1 Incorporation of Noncanonical Amino Acids into Engineered Proteins

There are two generic (and complementary) strategies for metabolic incorporation of noncanonical amino acids into proteins – the so-called *residue-specific* and *site-specific* methods (Fig. 1) (Budisa 2004; Hendrickson et al. 2004; Hohsaka and Sisido 2002; Link et al. 2003; Wang and Schultz 2005). The residue-specific approach involves replacement of all (or a fraction) of one of the natural amino acid residues. This method has its origins in the work of Cohen and coworkers, who showed in the 1950s that near-quantitative replacement of methionine by selenomethionine could be accomplished in bacterial cells (Cohen and Cowie 1957; Cowie and Cohen 1957; Cowie et al. 1959; Munier and Cohen 1956). This observation has had revolutionary consequences for protein science and engineering, in that it provides the basis of the multiwavelength anomalous diffraction method for crystallographic structure determination (Hendrickson et al. 1990).

The site-specific approach allows replacement of a single amino acid residue by a noncanonical analog. In this approach, a heterologous transfer RNA(tRNA)/aminoacyl-tRNA synthetase pair is used to deliver the analog in response to a nonsense or four-base codon. In 1996, Drabkin and coworkers used an *Escherichia coli* tRNA/glutaminyl-tRNA synthetase pair for amber codon suppression in mammalian cells (Drabkin et al. 1996), and showed that the suppressor tRNA was not charged by any of the mammalian aminoacyl-tRNA synthetases. Shortly thereafter, Furter (1998) introduced a yeast tRNA/phenylalanyl-tRNA synthetase (PheRS)

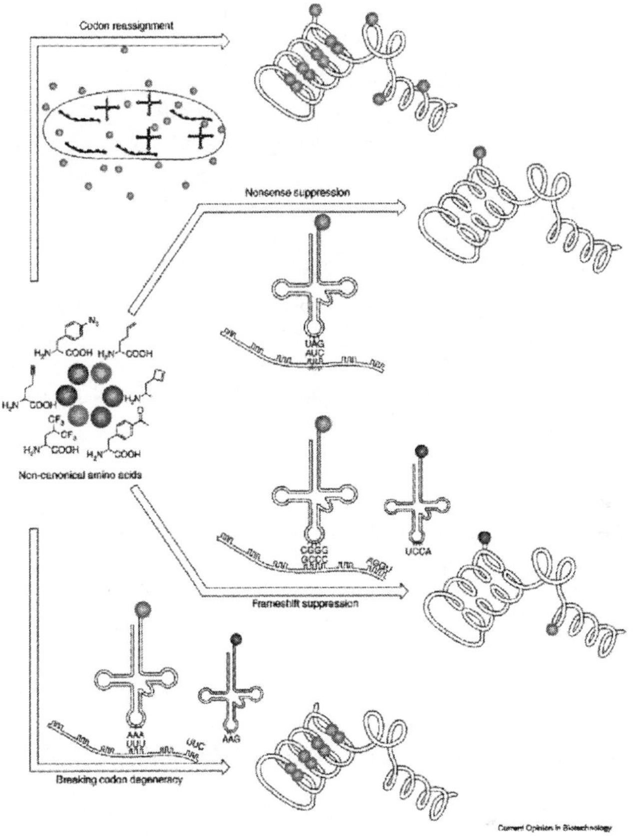

Fig. 1 Methods for incorporation of noncanonical amino acids. Residue-specific incorporation by sense codon reassignment enables replacement of all, or a fraction, of the corresponding canonical residues. Nonsense suppression, frameshift suppression, and breaking codon degeneracy can all be used to place noncanonical amino acids at specific sites. (Reprinted from Link et al. 2003; with permission from Elsevier)

pair into *E. coli* for site-specific incorporation of the noncanonical amino acid *p*-fluorophenylalanine. Since then, amber codon suppression has become the most common method for site-specific incorporation of noncanonical amino acids in vivo (Chin et al. 2003a; Kowal et al. 2001; Liu and Schultz 1999; Sakamoto et al. 2002; Wang et al. 2001). Schultz and coworkers have been especially successful in producing orthogonal suppressor tRNA/aminoacyl-tRNA synthetase pairs for incorporation of chemically, structurally, and spectroscopically diverse amino acid analogs (Wang and Schultz 2005; Xie and Schultz 2005). Site-specific incorporation has also been accomplished in *Xenopus* oocytes using microinjected messenger RNAs and chemically misacylated amber suppressor tRNAs (Dougherty 2000; see also the chapter by Dougherty, this volume). In a recent example, Dougherty and coworkers have reported the incorporation of two noncanonical amino acids into an ion channel protein by using both nonsense and four-base codon suppression (Rodriguez et al. 2006).

Sisido and coworkers have pioneered the use of four-base codons (frameshift suppression) for site-specific introduction of noncanonical amino acids into proteins (Hohsaka et al. 2001b; Hohsaka and Sisido 2002), and have employed this strategy to label streptavidin with fluorophores for fluorescence resonance energy transfer (FRET) experiments (Taki et al. 2002). Much of the work reported to date with four-base codons involves in vitro translation, but design of appropriate orthogonal tRNA/aminoacyl-tRNA synthetase pairs enables use of the method in bacterial cells. Anderson and coworkers have reported orthogonal tRNA/leucyl-tRNA synthetase (LeuRS) pairs for four-base, amber, and opal suppression (Anderson and Schultz 2003). Anderson et al. (2004) have reported use of a four-base codon with an amber codon for incorporation of two noncanonical amino acids into a recombinant protein using two orthogonal sets. An analogous five-base codon strategy has also been described (Anderson et al. 2002; Hohsaka et al. 2001a).

Reassignment of sense codons can also be used for site-specific incorporation of noncanonical amino acids, although the fidelity of the method is lower than that of nonsense or frameshift suppression (Fig. 2) (Kwon et al. 2003). Because the 20 canonical amino acids are encoded by 61 sense codons, the genetic code is highly degenerate. For example, phenylalanine is coded by two codons, UUC and UUU. In *E. coli*, both codons are read by a single tRNA, which decodes UUC via Watson–Crick base-pairing and UUU through a "wobble" interaction. Reassignment of the UUU codon was achieved by introducing into an *E. coli* expression host a mutant yeast PheRS capable of charging 2-naphthylalanine, and a mutant yeast

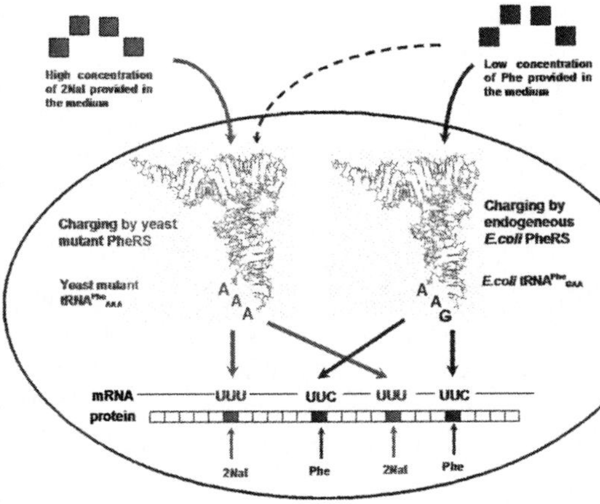

Fig. 2 Breaking the degeneracy of phenylalanine codons in *Escherichia coli*. The endogenous *E. coli* phenylalanyl-tRNA synthetase (*PheRS*) charges Phe to tRNA$^{Phe}_{GAA}$. The plasmid-borne yeast PheRS charges 2-naphthylalanine (*2Nal*) to yeast tRNA$^{Phe}_{AAA}$. UUC codons are decoded predominantly as Phe, while UUU codons are decoded predominantly as 2-naphthylalanine. *mRNA* messenger RNA, *tRNA* transfer RNA. (Reprinted with permission from Kwon et al. 2003. Copyright 2003 American Chemical Society)

$tRNA^{Phe}$ equipped with an AAA anticodon. Expression of dehydrofolate reductase led to preferential incorporation of phenylalanine at UUC codons and of 2-naphthylalanine at UUU codons. The generality and quantitative specificity of this method have not yet been established.

2 Translational Fidelity: Aminoacyl-tRNA Synthetases

Translational fidelity is controlled in large measure by the aminoacyl-tRNA synthetases (Ibba and Soll 2000), which match the 20 canonical amino acids with their cognate tRNAs. The remarkable capacity of the synthetases to discriminate among the natural amino acids might lead one to expect noncanonical substrates to be excluded by the translational apparatus (for more details see the chapter by Mascarenhas et al., this volume). In fact, many noncanonical amino acids are activated by the wild-type synthetases at rates that support efficient protein synthesis in bacterial cells. For analogs that are activated more slowly, addition of plasmid-encoded copies of the cognate synthetase can restore the rate of protein synthesis to levels characteristic of overexpressed recombinant proteins (Kiick et al. 2000; Tang and Tirrell 2001), and synthetase engineering has enabled further expansion of the set of useful amino acids (Datta et al. 2002; Hamano-Takaku et al. 2000; Ibba et al. 1994; Ibba and Hennecke 1995; Kirshenbaum et al. 2002). Szostak and coworkers have described a screen for identifying noncanonical amino acid substrates that are susceptible to enzymatic aminoacylation (Hartman et al. 2006). Using the screen, they identified 59 previously unknown amino acid substrates.

3 Reactive Amino Acids

Post-translational modification of proteins is essential for many aspects of protein function (Walsh 2006). Chemical methods for protein modification are equally important, in that they provide powerful approaches to protein engineering and analysis (Prescher and Bertozzi 2005). For example, "PEGylation" – the attachment of poly(ethylene glycol) – is widely used to enhance the performance of therapeutic proteins, and labeling with fluorescent dyes has enabled both visualization of proteins in cells and instructive in vitro studies of protein structure and dynamics (Griffin et al. 1998). The most common methods of protein modification exploit the high reactivity of cysteine and lysine residues (Hermanson 1996). For some purposes, however, the ubiquity of these amino acids is problematic, in that it may preclude discrimination among proteins or among sites on a single protein. In this section, we illustrate the use of noncanonical amino acids to effect selective protein modification through the chemistry of alkynes, azides, ketones, aryl halides, and alkenes.

3.1 Azide–Alkyne Ligation

Sharpless and coworkers and Tornoe and coworkers reported independently in 2002 that the [3+2] cycloaddition of azides and alkynes can be accelerated substantially by copper catalysis (Fig. 3) (Lewis et al. 2002; Rostovtsev et al. 2002; Tornoe et al. 2002). Since the initial reports, there has been a steady increase in the use of azide–alkyne ligations of biomolecules. The appeal of this reaction is due to its selectivity and the ease with which it can be implemented, even in complex biomolecular or cellular milieux. Labeling of proteins via the azide–alkyne ligation becomes especially straightforward if metabolic incorporation of azide- or alkyne-functionalized amino acids can be accomplished. In recent years, such reactive analogs have been incorporated into bacterial proteins by both residue-specific (Kiick et al. 2001, 2002; Kirshenbaum et al. 2002; Link et al. 2004; van Hest et al. 2000) and site-specific (Chin et al. 2002b; Deiters and Schultz 2005) methods. Azide- and alkyne-functional analogs have also been incorporated site-specifically in phage-displayed peptides (Feng et al. 2004) and in proteins expressed in *Saccharomyces cerevisiae* (Chin et al. 2003b; Deiters et al. 2003, 2004). After incorporation of the reactive analog, the protein can be modified by ligation of dyes, polymers, or affinity tags. The ligation rate is usually enhanced by addition of a copper catalyst (Rostovtsev et al. 2002; Tornoe et al. 2002), but for systems that are sensitive to copper, strain-promoted azide–alkyne ligations may be a suitable alternative (Agard et al. 2004; Link et al. 2006).

Residue-specific incorporation of reactive amino acids into *E. coli* outer membrane protein C can be detected following covalent modification (Link and Tirrell 2003), and has been used to develop a simple screening method for the discovery of new aminoacyl-tRNA synthetase activities (Link et al. 2006). Link and coworkers demonstrated that ligation of azide-bearing amino acids in outer membrane protein C with alkyne–biotin (Link and Tirrell 2003; Link et al. 2004) or cyclooctyne–biotin reagents (Agard et al. 2004) followed by cell staining with fluorescent avidin enables analysis of the extent of cell-surface labeling by flow cytometry. Using this method, Link and coworkers screened a library of methionyl-tRNA synthetase variants to identify methionyl-tRNA synthetase mutants that enable efficient incorporation of azidonorleucine into bacterial proteins (Link et al. 2006). The screen yielded three variants, all of which shared an L13G mutation. The single-site L13G mutant proved to be more active than

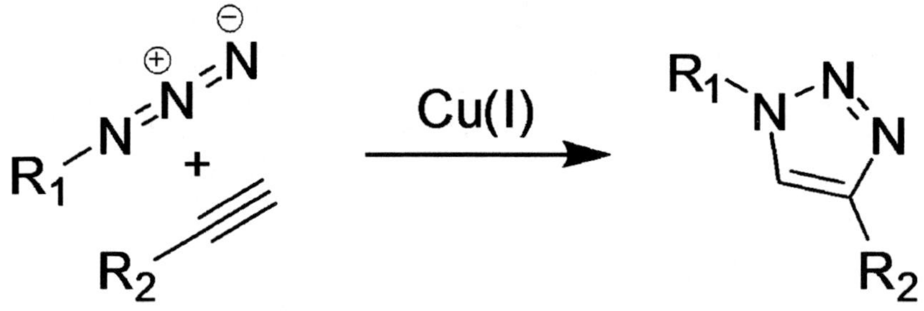

Fig. 3 The Cu(I)-catalyzed azide–alkyne ligation

any of the three variants that emerged from the screen, and enabled replacement of methionine by azidonorleucine with near-quantitative efficiency. The use of cell-surface display for the discovery of new synthetase activities should be quite general.

Similar chemistry has been used for direct ligation of fluorescent dyes for the purpose of imaging newly synthesized proteins in cells (Fig. 4). Fluorogenic azide (Sivakumar et al. 2004) and alkyne (Zhou and Fahrni 2004) dyes have been especially useful in such experiments (Beatty et al. 2005). For example, cotranslational introduction of an alkynyl amino acid (either homopropargylglycine or ethynylphenylalanine) into bacterial proteins was followed by selective Cu(I)-catalyzed ligation of the alkynyl side chain to a fluorogenic azidocoumarin. Extension of the method to mammalian cells has allowed visualization of a "temporal slice" of the proteome in a variety of cells (Beatty et al. 2006), and should allow imaging of localized protein translation. An advantage of the residue-specific method for proteomic imaging is that no prior knowledge of the location, function, or identity of the target proteins is needed for effective labeling. For modification of purified proteins of predetermined sequence, site-specific Cu(I)-catalyzed labeling has been reported (Deiters et al. 2003, 2004; Deiters and Schultz 2005).

The azide–alkyne ligation has been used to develop a new method (bio-orthogonal non-canonical amino acid tagging, BONCAT) for the study of proteome dynamics (Fig. 4) (Dieterich et al. 2006). In the first report on the method, azidohomoalanine was used as a methionine surrogate to effect pulse-labeling of newly synthesized

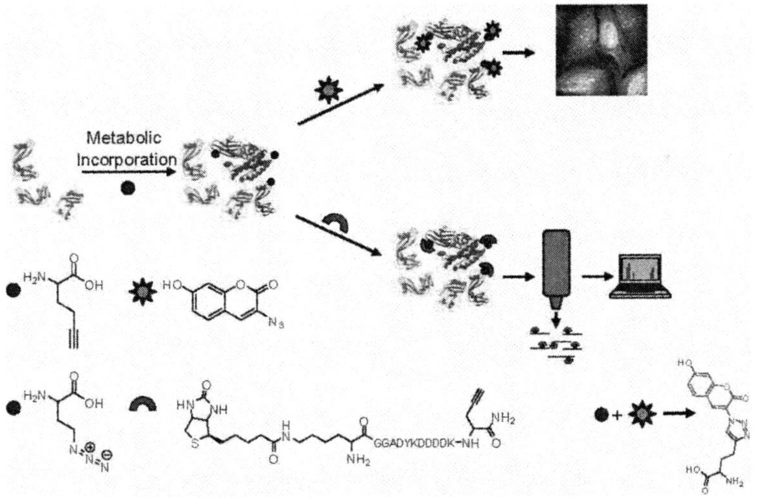

Fig. 4 Metabolic incorporation of reactive analogs enables visualization (*top*) or identification (*bottom*) of spatially and/or temporally defined subsets of the proteome. Newly synthesized proteins can be distinguished from all pre-existing proteins by the incorporation of a reactive amino acid analog. To visualize newly synthesized proteins, the analog is labeled with a fluorescent probe via the azide–alkyne ligation, as illustrated for the reaction of homopropargylglycine with 3-azido-7-hydroxycoumarin. Proteins are visualized in situ using fluorescence microscopy. To identify proteins translated during the pulse, an affinity purification tag is attached to the reactive amino acid. After enrichment, the proteins are cleaved from the affinity column and the peptide fragments are identified by mass spectrometry

proteins in human embryonic kidney (HEK 293) cells in culture. Labeled cells were lysed, and proteins bearing azidohomoalanine were ligated to a biotin tag for affinity purification. Resin-bound proteins were digested with trypsin, and the resulting fragments (and the proteins from which they were derived) were identified by mass spectrometry. The BONCAT method should be generally useful for probing dynamic changes in microbial and mammalian proteomes.

3.2 *Staudinger Ligation*

The Staudinger ligation allows metabolic incorporation of azides to be detected selectively by reaction with phosphine probes (Fig. 5) (Kiick et al. 2002; Kohn and Breinbauer 2004; Prescher et al. 2004; Saxon and Bertozzi 2000). Kiick et al. (2002) demonstrated the modification of a model protein, dehydrofolate reductase, containing azidohomoalanine. In purified protein samples and in complex cellular lysates, the azide reacted selectively with an antigenic FLAG-tag peptide bearing a reactive phosphine moiety. The purified protein was also labeled successfully with

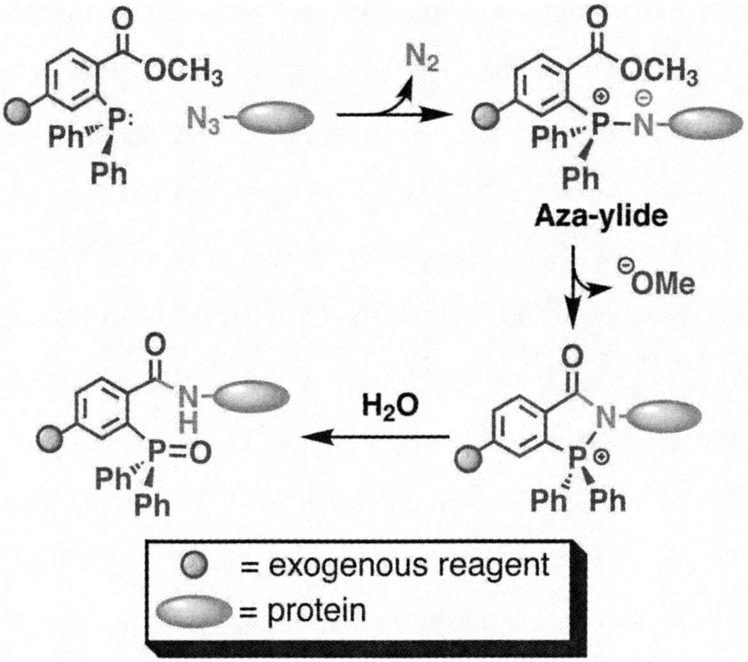

Fig. 5 Staudinger ligation of a phosphine to an azide. Metabolic incorporation of azide-bearing analogs can be detected or visualized selectively following reaction with a phosphine probe. (Reprinted with permission from Kiick et al. 2002. Copyright 2002 National Academy of Sciences, USA)

a fluorogenic phosphine dye (Lemieux et al. 2003). In complementary studies, Tsao et al. (2005) have shown that *p*-azidophenylalanine can be incorporated site-specifically into proteins or phage-displayed peptides for in vitro labeling with fluorescent phosphines. A detailed comparison of the Staudinger and azide–alkyne ligations has been reported (Agard et al. 2006).

3.3 Ketone Addition

Ketones undergo chemoselective ligation with hydrazides, hydroxylamines, and thiosemicarbazides under physiological conditions (Lemieux and Bertozzi 1998; Rodriguez et al. 1998). Acetylphenylalanine provides a convenient source of ketone functionality for use in protein modification, and has been incorporated into expressed proteins in both residue-specific and site-specific fashion. Residue-specific incorporation has been accomplished through development of a computationally designed PheRS containing an expanded amino acid binding pocket, and has enabled selective protein modification with biotin hydrazide (Datta et al. 2002). Site-specific incorporation of acetylphenylalanine into bacterial proteins has also been demonstrated (Wang et al. 2003a; Zhang et al. 2003). Reaction of *m*-acetylphenylalanine has allowed fluorescence imaging of dye-labeled proteins on the surface of *E. coli* (Zhang et al. 2003). Purified proteins displaying *p*-acetylphenylalanine were successfully modified by treatment with aminooxy-derivatized sugars (Liu et al. 2003), with fluorescein hydrazide, or with biotin hydrazide (Wang et al. 2003a). Site-specific incorporation of a β-diketone has been used to ligate biotin reagents and fluorophores to model protein (Zeng et al. 2006). Selective labeling with sugars provides a useful model for protein glycosylation, an important post-translational modification with implications for protein folding, stability, and activity, and for protein–protein interactions. Direct incorporation of glycosylated amino acids has also been reported (Zhang et al. 2004b).

3.4 Palladium-Catalyzed Cross-Coupling

Palladium-catalyzed cross-coupling provides efficient means of ligating aryl halides – most commonly bromides and iodides – to terminal alkenes or alkynes. Metabolic incorporation of bromophenylalanine or iodophenylalanine into recombinant proteins has been demonstrated (Ibba et al. 1994; Ibba and Hennecke 1995; Kast and Hennecke 1991; Kirshenbaum et al. 2002; Kwon et al. 2006), and recombinant barstar containing iodophenylalanine has been labeled with an alkyne–rhodamine dye under mild aqueous conditions via palladium-catalyzed coupling (Fig. 6) (Carrico 2003). The accessibility of the iodophenylalanine residues affected the conjugation efficiency, with labeling limited to a surface-exposed iodophenylalanine. Site-specific incorporation of iodophenylalanine in mammalian cells has

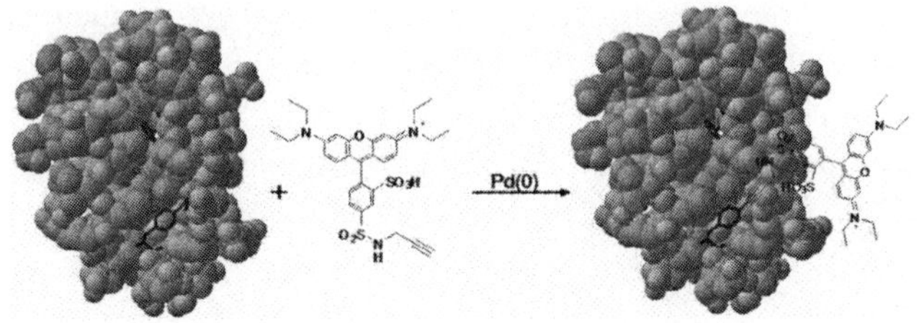

Fig. 6 Selective dye-labeling of recombinant barstar. Palladium-catalyzed cross-coupling was used to ligate a fluorophore (lissamine rhodamine propargyl sulfonamide) to a surface-accessible iodophenylalanine residue in barstar (Carrico 2003; Ratnaparkhi et al. 1998). Swiss-PdbViewer was used to model barstar

also been reported (Chin et al. 2003b; Kiga et al. 2002; Sakamoto et al. 2002). The Sonagashira coupling has been used to ligate iodophenylalanine in synthetic peptides to a trialkyne scaffold (Bong and Ghadiri 2001), and to biotin–alkyne (Dibowski and Schmidtchen 1998) and ferrocene–alkyne reagents (Hoffmanns and Metzler-Nolte 2006). Suzuki coupling (which uses an aryl boronate in place of the aryl halide) of a coumarin dye to a synthetic protein has been used to prepare a fluorescent biosensor for phosphorylated peptides (Ojida et al. 2005). Finally, a Mizoroki–Heck reaction was used to biotinylate a modified Ras protein made in a cell-free translation system, although the yield of biotinylated product was low and the reaction was complicated by dehalogenation and by competitive reaction at cysteine sites in the protein (Kodama et al. 2006a, b). Routine use of palladium-catalyzed coupling reactions on engineered proteins may have to await the development of reliable methods of suppressing interference by cysteine.

3.5 *Olefin Metathesis*

Alkenyl amino acids offer protein engineers additional options for protein modification via olefin metathesis (Blackwell 1998; Blackwell et al. 2001; Grubbs and Chang 1998). Tirrell and coworkers have demonstrated that homoallylglycine and 2-amino-3-methyl-4-pentenoic acid can be incorporated into recombinant proteins in place of methionine and isoleucine, respectively (Mock et al. 2006; van Hest et al. 2000). Site-specific incorporation of *O*-allyl-L-tyrosine might enable modification at a preselected site in a protein (Zhang et al. 2002). Recombinant proteins containing these amino acids should be subject to selective metathesis coupling reactions, but the chemistry reported to date has been restricted to synthetic peptides. Clark and Ghadiri (1995) have used olefin metathesis to stabilize cyclic dimers containing homoallylglycine. Similarly, Schafmeister et al. (2000) reported that the helicity and proteolytic stability of peptides were improved by judicious intrahelical coupling of alkene side chains.

4 Fluorescent Amino Acids

Proteins containing noncanonical amino acids have been labeled with extrinsic fluorescent probes via the azide–alkyne, Staudinger, and ketone-mediated ligations, as described earlier. Incorporation of intrinsic fluorescent probes is also possible through translational introduction of fluorescent amino acids. Fluorescent analogs have been introduced by both residue-specific and site-specific methods to yield proteins with altered spectral properties.

4.1 Tryptophan Analogs

In proteins containing only the 20 canonical amino acids, the most intense fluorescence is from the indole ring of tryptophan (Lakowicz 1999). Tryptophan absorption is maximal at 280nm; the maximum in the emission spectrum lies near 340nm. Modification of the indole ring can shift both the excitation and the emission spectra. Residue-specific replacement of tryptophan by azatryptophan analogs in bacteria was first reported more than 50 years ago (Brawerman and Ycas 1957; Pardee et al. 1956). More recently, residue-specific incorporation of 5-hydroxytryptophan was reported by Hogue et al. (1992) (see the review in Correa and Farah 2005). The excitation maximum of this analog is distinct from that of tryptophan, enabling 5-hydroxytryptophan to be used for monitoring protein–protein interactions, even in mixtures of proteins containing tryptophan. Hoque et al. (1992) prepared a variant of the Ca^{2+}-binding protein oncomodulin containing a single encoded tryptophan residue (Y57W), which was replaced by 5-hydroxytryptophan. Changes in the fluorescence anisotropy of the modified protein were observed upon treatment with antioncomodulin antibodies. More recently, residue-specific incorporation of 5-hydroxytryptophan into phosphoglycerate kinase has enabled fluorescence visualization of the protein in yeast cells (Botchway et al. 2005).

Site-specific incorporation of 5-hydroxytryptophan in mammalian cells has also been reported (Zhang et al. 2004a). Zhang and coworkers generated an orthogonal *Bacillus subtilis* tRNA/tryptophanyl-tRNA synthetase pair that selectively charges 5-hydroxytryptophan. A variant of the foldon protein containing a single 5-hydroxytryptophan residue exhibited fluorescence 11-fold more intense than that of the wild-type protein.

Li et al. (2003) reported the use of 5-hydroxytryptophan and 7-azatryptophan in studies of antibody binding to *Staphylococcus* protein G. Fluorescence detection of other protein–protein interactions has also been reported, including the use of 5-hydroxytryptophan in *Pseudomonas aeruginosa* exotoxin A to detect interactions with elongation factor-2 (Mohammadi et al. 2001). The authors reported a 260-fold decrease in k_{cat} for toxin containing 5-hydroxytryptophan compared with toxin containing the 20 natural amino acids. The loss in activity might be recoverable through evolution (Montclare and Tirrell 2006).

Muralidharan et al. (2004) have described a method for incorporating 7-azatryptophan into a single domain of a multidomain protein. Using residue-specific incorporation, they expressed an SH3 domain containing 7-azatryptophan. Expressed protein ligation to the SH2 domain gave the final protein c-Crk-I, with a domain-specific label. This method should be generally useful for the synthesis of multidomain proteins containing a single genetically encoded fluorophore (for more details see the chapters by Merkel et al. and Imperiali and Vogel Taylor, this volume).

Aminotryptophans have been incorporated into proteins to make fluorescent pH sensors (Budisa et al. 2002). Replacement of three tryptophan residues in barstar with aminotryptophan resulted in pH-dependent fluorescence. Titration from pH 4 to 9 showed decreasing fluorescence with increasing pH.

4.2 Altering the Spectra of Fluorescent Proteins

4-Aminotryptophan has also been used to alter the spectra of fluorescent proteins (Bae et al. 2003). The fluorescence of the *Aequoria victoria* green fluorescent protein (GFP) is dependent on an aromatic amino acid at position 66, which is part of a triad of amino acids that forms the chromophore (Tsien 1998). Incorporation of noncanonical amino acids at position 66 enables engineering of the fluorescence properties of GFP and related proteins. Using an auxotrophic bacterial strain, Bae et al. (2003) replaced tryptophan with 4-aminotryptophan in enhanced cyan fluorescent protein (ECFP). Addition of the electron-donating amino group yielded a red-shifted, thermostable, monomeric "gold" fluorescent protein (Fig. 7). Similarly,

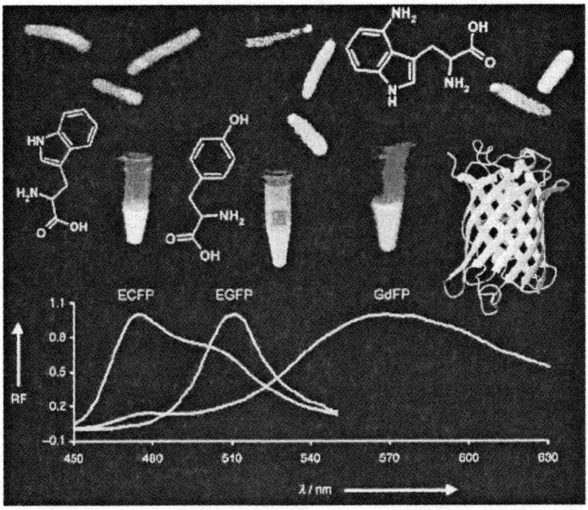

Fig. 7 Altering the amino acid at position 66 in the chromophore of fluorescent proteins derived from *Aequoria victoria* can shift the fluorescence from cyan (tryptophan) to green (tyrosine) to gold (4-aminotryptophan). The noncanonical amino acid 4-aminotryptophan forms a chromophore that converts the protein to a redshifted, thermostable, monomeric fluorescent protein. (From Budisa 2004. Reproduced by permission of Wiley)

replacement of tryptophan has been used to examine chromophore and protein structure, pH-dependent behavior, and fluorescence properties of enhanced GFP and ECFP containing fluorine-, chalcogen-, and methyl-substituted analogs (Budisa et al. 2004b). Incorporation of 4-methyltryptophan resulted in a 4-nm redshift in emission, while incorporation of fluorinated tryptophan analogs caused a small blueshift. Incorporation of chalcogen-substituted analogs in enhanced GFP yielded fluorescent variants, while replacement in ECFP resulted in nonfluorescent proteins. Incorporation of fluorinated tyrosine analogs in enhanced GFP and in yellow fluorescent protein produced proteins with altered spectral features, including blueshifted emission for proteins containing 2-fluorotyrosine and redshifted emission for proteins containing 3-fluorotyrosine (Pal et al. 2005).

Schultz and coworkers have reported that site-specific incorporation of *para*-substituted phenylalanine analogs at position 66 resulted in fluorescent proteins with emission maxima ranging from 428 to 498 nm (Wang et al. 2003b). Use of a cell-free translation system by Sisido and coworkers demonstrated that large aromatic amino acids at position 66 resulted in nonfluorescent variants of GFP (Kajihara 2005). Incorporation of aminophenylalanine and *o*-methyltyrosine resulted in blueshifted, weak fluorescence; however, the fluorescence could be enhanced by random mutagenesis. Randomization of position 145 allowed the intensity of fluorescence to be increased 1.5- to 4-fold.

4.3 Other Fluorescent Analogs

Site-specific incorporation of other fluorescent amino acids has been described by Schultz and coworkers (Summerer et al. 2006; Wang et al. 2006). A coumarin amino acid was incorporated into recombinant proteins using an evolved *Methanococcus jannaschii* tRNA/tyrosyl-tRNA synthetase (TyrRS) pair (Wang et al. 2006). Fluorescence monitoring of myoglobin denaturation was used to demonstrate the utility of the amino acid as a site-specific probe of changes in protein conformation. Similarly, incorporation of dansylalanine enabled fluorescence detection of local conformation changes (Summerer et al. 2006). Dansylalanine was incorporated site-specifically in *S. cerevisiae* by using an engineered *E. coli* LeuRS. The suppression efficiency was optimized by using an evolved LeuRS with a redesigned editing domain and by increasing the in vivo levels of suppressor tRNA. Proteins containing other dansylated amino acids as fluorescent electrostatic probes have been reported previously (Cohen et al. 2002; Hohsaka et al. 2004; Nitz et al. 2002).

The large size of fluorescent amino acids makes them challenging targets for cotranslational incorporation into proteins in cells (Hohsaka et al. 1999; Nakata et al. 2006). In an alternative approach, Sisido and coworkers have shown the incorporation of many useful fluorophores at four-base codons by using cell-free translation systems (Hamada et al. 2005; Hohsaka et al. 2004; Murakami et al. 2000; Ohtsuki et al. 2005; Taki et al. 2001, 2002). These systems employ chemical ligation or ribozyme-based aminoacylation methods to attach the analog to the tRNA.

It is likely that newly evolved aminoacyl-tRNA synthetases, such as those reported by Schultz and coworkers, will enable the in vivo incorporation of an expanded set of fluorescent amino acids (Summerer et al. 2006; Wang et al. 2006).

4.4 Generating FRET Pairs

Incorporation of FRET pairs into engineered proteins enables the study of a wide variety of issues in protein structure, dynamics, and function. Many in vitro examples of FRET labeling have been described by Sisido and coworkers, and in vivo incorporation of fluorescent probes will further expand the power of the method (Kajihara et al. 2006; Ohtsuki et al. 2005; Taki et al. 2002). Taki et al. (2002) have described the incorporation of a fluorophore–quencher pair into streptavidin. An *E. coli* in vitro translation system was used to place the fluorescent amino acid (β-anthraniloyl-L-α,β-diaminopropionic acid) and the quencher (*p*-nitrophenylalanine) at defined positions. Ohtsuki et al. (2005) have also reported that a combination of two fluorophores and a quencher (a total of three analogs) has been incorporated into a single streptavidin chain by using four-base suppression. In vivo incorporation of donor–acceptor pairs consisting of two noncanonical amino acids has yet to be demonstrated; however, placement of a fluorescence donor–quencher pair is possible using tryptophan paired with the quencher *p*-nitrophenylalanine. *p*-Nitrophenylalanine quenches the intrinsic fluorescence of tryptophan, and was utilized by Tsao et al. (2006) to demonstrate distance-dependent fluorescence quenching in a model leucine zipper peptide.

5 Photosensitive Amino Acids

Incorporation of photosensitive noncanonical amino acids into proteins has enabled immobilization of proteins on surfaces, isolation of protein–protein complexes, and time-resolved activation of proteins.

5.1 Aryl Azides

The residue-specific replacement of phenylalanine by *p*-azidophenylalanine was described in 2002 by Kirshenbaum et al. (2002). They suggested that the aryl azide would be a valuable tool for photoactivated cross-linking and photoaffinity labeling. More recently, Zhang et al. (Zhang et al. 2005) have reported selective immobilization of proteins on surfaces coated with photocross-linked artificial proteins containing leucine-zipper domains (Fig. 8). Extension of this approach to the preparation of protein microarrays and gradient biomaterials should be straightforward.

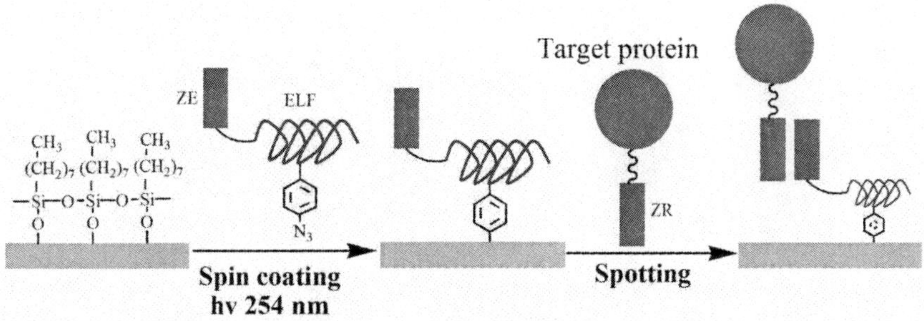

Fig. 8 Azidophenylalanine in protein immobilization. The surface was prepared by irradiation of spin-coated zipper (*ZE*)–elastin fusion proteins (*ELF*) containing *p*-azidophenylalanine. Target proteins were immobilized via high-affinity interactions between their basic zipper (*ZR*) domains and the acidic zipper (*ZE*) on the surface. (Reprinted with permission from Zhang et al. 2005. Copyright 2005 American Chemical Society)

Chin et al. (2002b) have described the evolution of a *M. jannaschii* tRNA/TyrRS pair for site-specific incorporation of *p*-azidophenylalanine into bacterial proteins. Irradiation of dimeric glutathione *S*-transferase containing *p*-azidophenylalanine resulted in covalently cross-linked dimers. Cross-linking was observed only for proteins containing the aryl azide at the interhelical dimer interface.

5.2 Benzophenones

Site-specific incorporation of *p*-benzoylphenylalanine has been demonstrated by using an evolved *M. jannaschii* tRNA/TyrRS pair (Chin et al. 2002a; Chin and Schultz 2002). Incorporation of the benzophenone moiety into dimeric glutathione *S*-transferase was followed by excitation at 365 nm for 1–5 min. Cross-linked protein dimers could be observed on denaturing gels. Chin et al. (2003b) have also reported site-specific incorporation of this analog into proteins expressed in *S. cerevisiae*, by using an *E. coli* tRNA/TyrRS pair. The same report detailed the incorporation of *p*-acetylphenylalanine, *p*-azidophenylalanine, *o*-methyltyrosine, and *p*-iodophenylalanine. Detailed instructions for cross-linking of proteins containing *p*-benzoylphenylalanine have been reported by Farrell et al. (2005). Incorporation of this analog has been used to examine the bacterial protein translocation machinery through cross-linking of SecY to SecA (Mori and Ito 2006).

Using the previously reported *E. coli* tRNA/TyrRS pair, Hino and coworkers have reported site-specific incorporation of *p*-benzoylphenylalanine into mammalian proteins (Chin et al. 2003b; Hino et al. 2005). Cross-linking of Grb-2 containing *p*-benzoylphenylalanine to epidermal growth factor receptor as well as to endogenous proteins was described, demonstrating the utility of this amino acid for identifying protein–protein interactions in mammalian cells (Fig. 9).

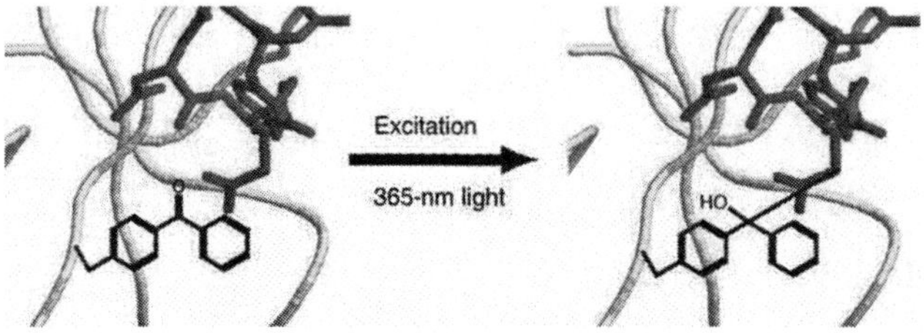

Fig. 9 Incorporation of benzoylphenylalanine into Grb-2 for identification of protein–protein interactions inside mammalian cells. Excitation with 365 nm light produces a reactive triplet state in the amino acid analog, which can then react with a CH bond on an adjacent protein. (Reprinted with permission from Macmillan Publishers Ltd: *Nature Methods*; Hino et al. 2005, Copyright 2005)

5.3 Benzofurans

Incorporation of the photoreactive phenylalanine analog benzofuranylalanine into dihydrofolate reductase has been reported by Nielsen and coworkers, who suggest that this analog might be useful for cross-linking under long-wavelength (UVA) irradiation (Bentin et al. 2004). No cross-linking studies have been reported to date.

5.4 Diazirines

Leucine and methionine analogs containing photoactivatible diazirine rings have been reported by Suchanek et al. (2005). These analogs were incorporated in residue-specific fashion into mammalian proteins without prior modification of the cellular translational machinery. Replacement of fraction of the encoded methionine residues with the diazirine analog was sufficient to allow identification of covalently cross-linked protein complexes. Several systems were examined, including the interaction of the membrane protein PGRMC1 with Insig-1, a regulator of cholesterol homeostasis. A recently reported synthesis should foster the use of L-4-[3-trifluoromethyl)-3*H*-diazirin-3-yl]phenylalanine for in vivo photocross-linking (Nakashima et al. 2006). A review provides additional information on the use of diazirines, benzophenones, and aryl azides as photochemical cross-linking agents (Sadakane and Hatanaka 2006).

5.5 Photocaged Amino Acids

Photocaged amino acids have enabled time-resolved analysis of the functional roles of individual side chains in ion channels, enzymes, and phosphorylated proteins

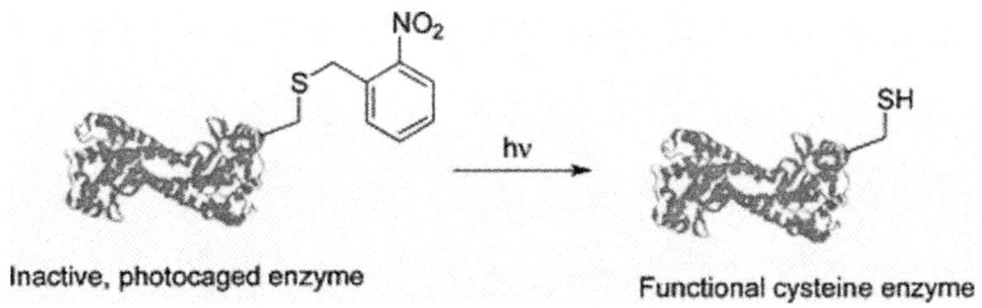

Fig. 10 Irradiation promotes decaging to form the natural amino acid. Here, photocaged cysteine masks the activity of caspase until the active-site cysteine is revealed. (Reprinted with permission from Wu et al. 2004. Copyright 2004 American Chemical Society)

(see the reviews in Beene et al. 2003; Lawrence 2005). *o*-Nitrobenzyl tyrosine has been incorporated into specific sites in ion channel proteins in *Xenopus* oocytes and into recombinant enzymes in *E. coli*. Incorporation in oocytes has enabled detailed study of individual residues in ion channel proteins (Miller et al. 1998; Philipson et al. 2001; Tong et al. 2001). For example, decaging of *o*-nitrobenzyl tyrosine leads to endocytosis of channel proteins under conditions favoring tyrosine phosphorylation (Tong et al. 2001). Incorporation of photocaged tyrosine into recombinant *E. coli* proteins utilized an evolved *M. jannaschii* tRNA/TyrRS pair (Deiters et al. 2006). Bacteria with the photocaged amino acid inserted into the active site of β-galactosidase were irradiated for 30 min, and the enzyme activity was restored to approximately 67% of that characteristic of cells expressing conventional β-galactosidase. Photocaged cysteine has also been incorporated into cellular proteins via nonsense suppression in *Xenopus* oocytes (Philipson et al. 2001) and yeast (Wu et al. 2004) (Fig. 10). Incorporation into a yeast caspase utilized an evolved *E. coli* tRNA/LeuRS pair. Dougherty and Imperiali have reported the incorporation of caged phosphoamino acids into full-length proteins via in vitro translation (Rothman et al. 2005; see also the chapters by Dougherty and by Imperiali and Vogel Taylor, this volume).

5.6 Photoisomerizable Amino Acids

Schultz and coworkers have described the site-specific incorporation of the photoisomerizable amino acid phenylalanine-4′-azobenzene into the *E. coli* catabolite activator protein, a homodimeric transcriptional activator (Bose et al. 2006). The DNA binding affinity of the modified catabolite activator protein was reduced approximately fourfold upon photoisomerization of the side chain from the all-*trans* form to the photostationary state containing roughly 50% of the *cis* form of the chromophore. The authors suggest that photoisomerization might be used to trigger a variety of biological processes in vitro and in live cells.

6 Fluorinated Amino Acids

Fluorinated amino acids can be used to alter the properties of proteins or to create structural probes (Kukhar and Soloshonok 1995). Incorporation of fluorinated analogs should cause minimal perturbation of protein structure because the van der Waals radii of fluorine (1.35 Å) and hydrogen (1.2 Å) are similar. On the other hand, the solubility properties of hydrocarbons and fluorocarbons are strikingly different. Residue-specific replacement of leucine by trifluoroleucine was first reported in 1963 (Rennert and Anker 1963). Fluorinated proteins continue to be designed and evolved today (see the review in Yoder and Kumar 2002).

6.1 Protein Stabilization

Incorporation of fluorinated amino acids has been used to stabilize engineered proteins. Tang and coworkers used trifluoroleucine to stabilize the leucine zipper protein A1 (Petka et al. 1998; Tang et al. 2001a). The fluorinated protein, in which 92% of the leucine residues were replaced by trifluoroleucine, retained its secondary structure and showed enhanced thermal and chemical stability compared with the leucine form of the protein. Recombinant protein was expressed in a leucine auxotrophic strain, and the extent of substitution was controlled by altering the amount of leucine added to the expression medium. Unfolding in urea demonstrated that increasing the extent of fluorination resulted in an increase in stability. Similarly, Tang et al. (2001b) have described the incorporation of trifluoroleucine into a leucine zipper DNA-binding peptide. The fluorinated peptides demonstrated increased stability while retaining their affinity and selectivity for DNA binding. Son et al. (2006) have reported the residue-specific incorporation of 5,5,5-trifluoroisoleucine and (2*S*,3*R*)-4,4,4-trifluorovaline in place of isoleucine in basic leucine zipper peptides. The fluorinated peptides showed enhanced stability and retained their DNA-binding specificity. Incorporation of trifluoroleucine produced melittin peptides with improved membrane binding (Niemz and Tirrell 2001). Similarly, Budisa et al. (2004a) have described the residue-specific incorporation of trifluoroleucine or trifluoromethionine into recombinant proteins. Although they reported difficulty obtaining good substitution for proteins larger than 10 kDa, we have not encountered this problem (Montclare and Tirrell 2006).

Elevating the LeuRS activity of the host strain enables the incorporation of hexafluoroleucine (Tang and Tirrell 2001). Incorporation of this analog resulted in a zipper protein with defined secondary structure and enhanced thermal and chemical stability. Compared with trifluoroleucine, hexafluoroleucine demonstrated an additional enhancement in stability. Hexafluoroleucine has also been used to examine the stability and flexibility of designed four-helix bundles (Lee et al. 2006). By systematic replacement of leucine with hexafluoroleucine in the hydrophobic core, the stability of the bundles could be modulated. It was found that increasing the extent of fluorination resulted in a near-linear increase in stability, as well as an increase in structural rigidity.

The analog (2*S*,3*R*)-trifluorovaline can serve as a surrogate for either valine or isoleucine. By elevating the cellular activity of either isoleucyl-tRNA synthetase or valyl-tRNA synthetase, Wang et al. (2004) were able to decode a single messenger RNA as two different proteins. For *E. coli* with elevated isoleucyl-tRNA synthetase, the analog served as an isoleucine surrogate, whereas in *E. coli* with elevated valyl-tRNA synthetase, the analog served as a valine surrogate. The corresponding natural amino acid was removed from the culture medium in each case.

6.2 *Evolution of Fluorinated Enzymes*

Panchenko et al. (2006) have examined the effect of global fluorination on the activity and stability of chloramphenicol acetyltransferase. Complete replacement resulted in fluorinated enzymes with good activity but reduced stability. A method to compensate for losses in activity and stability has been described by Montclare and Tirrell (2006). In this work, the residue-specific replacement of leucine by trifluoroleucine resulted in a marked decrease (27-fold) in the half-life of thermal inactivation of chloramphenicol acetyltransferase (Fig. 11). However, the thermostability of the fluorinated enzyme was restored to that characteristic of the leucine form through two rounds of random mutagenesis and screening. The evolved enzyme contained three amino acid substitutions, none of which removed any of the 13 leucine residues.

6.3 *Peptide Self-Sorting*

Kumar and coworkers have reported intriguing self-sorting behavior in mixtures of conventional and fluorinated coiled-coil peptides (Fig. 12) (Bilgicer et al. 2001;

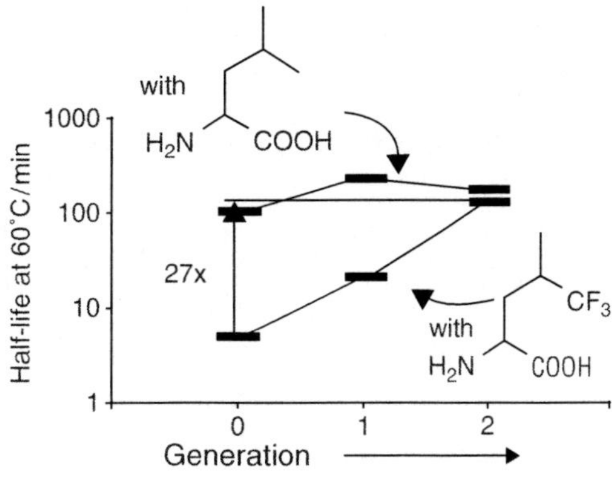

Fig. 11 Fluorination of chloramphenicol acetyltransferase with trifluoroleucine resulted in a marked decrease in the stability of the enzyme. However, the thermostability of the fluorinated enzyme was restored to that characteristic of the wild-type enzyme through two rounds of random mutagenesis and screening. (From Montclare and Tirrell 2006. Reproduced with permission of Wiley)

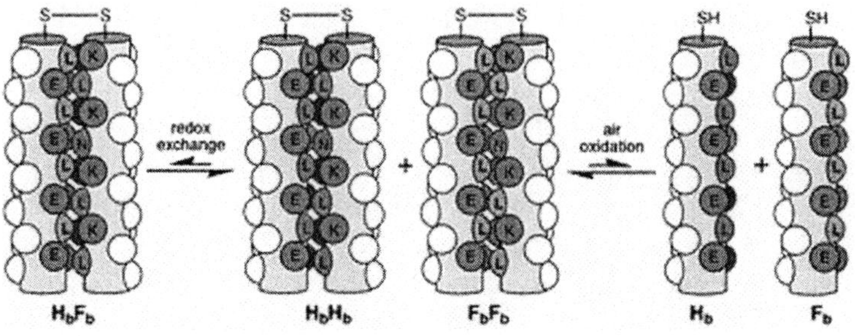

Fig. 12 Fluorinated (F_b) and hydrogenated (H_b) coiled-coil peptides exhibit self-sorting behavior. The homodimers, H_bH_b and F_bF_b, represent more than 90% of the dimers present at equilibrium. (From Yoder and Kumar 2002. Reproduced with permission of the Royal Society of Chemistry)

Bilgicer and Kumar 2002). Otherwise-identical peptides were prepared, with either fluorinated or hydrogenated leucine side chains. A striking preference for self-sorting into hexafluoroleucine homodimers and leucine homodimers was observed (Bilgicer et al. 2001); at equilibrium, only trace amounts of the heterodimer remain.

7 Conclusion

Although it is not a new field, the use of noncanonical amino acids in protein science and engineering is moving in important new directions. Methods for introducing amino acid analogs into proteins, whether in residue-specific or in site-specific fashion, are now firmly established, and dozens of active analogs have been developed to impart useful new properties to natural and artificial proteins. The residue-specific replacement of methionine by selenomethionine has had genuinely revolutionary consequences for protein science, and newly developed strategies for protein labeling, stabilization, evolution, and analysis offer substantial promise for the future. The coming years should witness a shift from methods development toward the use of noncanonical amino acids to address important chemical and biological problems in protein science and engineering.

Acknowledgments Research at Caltech on noncanonical amino acids has been supported by NIH grant GM62523. K.E.B. is a Fellow of the Fannie and John Hertz Foundation.

References

Agard NJ, Prescher JA, Bertozzi CR (2004) A strain-promoted [3+2] azide–alkyne cycloaddition for covalent modification of biomolecules in living systems. J Am Chem Soc 126:15046–15047

Agard NJ, Baskin JM, Prescher JA, Lo A, Bertozzi CR (2006) A comparative study of bioorthogonal reactions with azides. ACS Chem Biol 1:644–648

Anderson JC, Schultz, PG (2003) Adaptation of an orthogonal archaeal leucyl-tRNA and synthetase pair for four-base, amber, and opal suppression. Biochemistry 42:9598–9608

Anderson JC, Magliery TJ, Schultz PG (2002) Exploring the limits of codon and anticodon size. Chem Biol 9:237–244

Anderson JC, Wu N, Santoro SW, Lakshman V, King DS, Schultz PG (2004) An expanded genetic code with a functional quadruplet codon. Proc Natl Acad Sci USA 101:7566–7571

Bae JH, Rubini M, Jung G, Wiegand G, Seifert MHJ, Azim MK, Kim JS, Zumbusch A, Holak TA, Moroder L, Huber R, Budisa N (2003) Expansion of the genetic code enables design of a novel "gold" class of green fluorescent proteins. J Mol Biol 328:1071–1081

Beatty KE, Xie F, Wang Q, Tirrell DA (2005) Selective dye-labeling of newly synthesized proteins in bacterial cells. J Am Chem Soc 127:14150–14151

Beatty KE, Liu JC, Xie F, Dieterich DC, Schuman EM, Wang Q, Tirrell DA (2006) Fluorescence visualization of newly synthesized proteins in mammalian cells. Angew Chem Int Edn Engl 45:7364–7367

Beene DL, Dougherty DA, Lester HA (2003) Unnatural amino acid mutagenesis in mapping ion channel function. Curr Opin Neurobiol 13:264–270

Bentin T, Hamzavi R, Salomonsson J, Roy H, Ibba M, Nielsen PE (2004) Photoreactive bicyclic amino acids as substrates for mutant *Escherichia coli* phenylalanyl-tRNA synthetases. J Biol Chem 279:19839–19845

Bilgicer B, Kumar K (2002) Synthesis and thermodynamic characterization of self-sorting coiled coils. Tetrahedron 58:4105–4112

Bilgicer B, Xing X, Kumar K (2001) Programmed self-sorting of coiled coils with leucine and hexafluoroleucine cores. J Am Chem Soc 123:11815–11816

Blackwell HE (1998) Highly efficient synthesis of covalently cross-linked peptide helices by ring-closing metathesis. Angew Chem Int Edn Engl 37:3281–3284

Blackwell HE, Sadowsky JD, Howard RJ, Sampson JN, Chao JA, Steinmetz WE, O'Leary DJ, Grubbs RH (2001) Ring-closing metathesis of olefinic peptides: design, synthesis, and structural characterization of macrocyclic helical peptides. J Org Chem 66:5291–5302

Bong DT, Ghadiri MR (2001) Chemoselective Pd(0)-catalyzed peptide coupling in water. Org Lett 3:2509–2511

Bose M, Groff D, Xie JM, Brustad E, Schultz PG (2006) The incorporation of a photoisomerizable amino acid into proteins in *E. coli*. J Am Chem Soc 128:388–389

Botchway SW, Barba I, Jordan R, Harmston R, Haggie PM, William SP, Fulton AM, Parker AW, Brindle KM (2005) A novel method for observing proteins in vivo using a small fluorescent label and multiphoton imaging. Biochem J 390:787–790

Brawerman G, Ycas M (1957) Incorporation of the amino acid analog tryptazan into the protein of *Escherichia coli*. Arch Biochem Biophys 68:112–117

Budisa N (2004) Prolegomena to future experimental efforts on genetic code engineering by expanding its amino acid repertoire. Angew Chem Int Edn Engl 43:6426–6263

Budisa N (2006) Engineering the genetic code: Expanding the amino acid repertoire for the design of novel proteins. Wiley-VCH, New York

Budisa N, Pipitone O, Siwanowicz I, Rubini M, Pal PP, Holak TA, Gelmi ML (2004a) Efforts towards the design of "Teflon" proteins: In vivo translation with trifluorinated leucine and methionine analogues. Chem Biodiversity 1:1465–1475

Budisa N, Rubini M, Bae JH, Weyher E, Wenger W, Golbik R, Huber R, Moroder L (2002) Global replacement of tryptophan with aminotryptophans generates non-invasive protein-based optical pH sensors. Angew Chem Int Edn Engl 41:4066–4069

Budisa N, Pal PP, Alefelder S, Birle P, Krywcun T, Rubini M, Wenger W, Bae JH, Steiner T (2004b) Probing the role of tryptophans in *Aequorea victoria* green fluorescent proteins with an expanded genetic code. Biol Chem 385:191–202

Carrico IS (2003) Protein engineering through in vivo incorporation of phenylalanine analogs. Ph.D. Dissertation. California Institute of Technology

Chin JW, Schultz PG (2002) In vivo photocrosslinking with unnatural amino acid mutagenesis. ChemBioChem 3:1135–1137

Chin JW, Martin AB, King DS, Wang L, Schultz PG (2002a) Addition of a photocrosslinking amino acid to the genetic code of *Escherichia coli*. Proc Natl Acad Sci USA 99: 11020–11024

Chin JW, Cropp TA, Chu S, Meggers E, Schultz PG (2003a) Progress toward an expanded eukaryotic genetic code. Chem Biol 10:511–519

Chin JW, Santoro SW, Martin AB, King DS, Wang L, Schultz PG (2002b) Addition of p-azido-L-phenylalanine to the genetic code of *Escherichia coli*. J Am Chem Soc 124:9026–9027

Chin JW, Cropp TA, Anderson JC, Mukherji M, Zhang ZW, Schultz PG (2003b) An expanded eukaryotic genetic code. Science 301:964–967

Clark TD, Ghadiri MR (1995) Supramolecular design by covalent capture – design of a peptide cylinder via hydrogen-bond-promoted intermolecular olefin metathesis. J Am Chem Soc 117: 12364–12365

Cohen BE, McAnaney TB, Park ES, Jan YN, Boxer SG, Jan LY (2002) Probing protein electrostatics with a synthetic fluorescent amino acid. Science 296:1700–1703

Cohen GN, Cowie DB (1957) Total replacement of methionine by selenomethionine in the proteins of *Escherichia coli*. Comptes Rendus 244:680–683

Correa F, Farah CS (2005) Using 5-hydroxytryptophan as a probe to follow protein–protein interactions and protein folding transitions. Protein Pept Lett 12:241–244

Cowie DB, Cohen GN (1957) Biosynthesis by *Escherichia coli* of active altered proteins containing selenium instead of sulfur. Biochim Biophys Acta 26:252–261

Cowie DB, Cohen GN, Bolton ET, Derobichonszulmajster H (1959) Amino acid analog incorporation into bacterial proteins. Biochim Biophys Acta 34:39–46

Datta D, Wang P, Carrico IS, Mayo SL, Tirrell DA (2002) A designed phenylalanyl-tRNA synthetase variant allows efficient in vivo incorporation of aryl ketone functionality into proteins. J Am Chem Soc 124:5652–5653

Deiters A, Schultz PG (2005) In vivo incorporation of an alkyne into proteins in *Escherichia coli*. Bioorg Med Chem Lett 15:1521–1524

Deiters A, Cropp TA, Summerer D, Mukherji M, Schultz PG (2004) Site-specific PEGylation of proteins containing unnatural amino acids. Bioorg Med Chem Lett 14:5743–5745

Deiters A, Groff D, Ryu YH, Xie JM, Schultz PG (2006) A genetically encoded photocaged tyrosine. Angew Chem Int Edn Engl 45:2728–2731

Deiters A, Cropp TA, Mukherji M, Chin JW, Anderson JC, Schultz PG (2003) Adding amino acids with novel reactivity to the genetic code of *Saccharomyces cerevisiae*. J Am Chem Soc 125:11782–11783

Dibowski H, Schmidtchen FP (1998) Bioconjugation of peptides by palladium-catalyzed C–C cross-coupling in water. Angew Chem Int Edn Engl 37:476–478

Dieterich DC, Link AJ, Graumann J, Tirrell DA, Schuman EM (2006) Selective identification of newly synthesized proteins in mammalian cells using bioorthogonal noncanonical amino acid tagging (BONCAT). Proc Natl Acad Sci USA 103:9482–9487

Dougherty DA (2000) Unnatural amino acids as probes of protein structure and function. Curr Opin Chem Biol 4:645–652

Drabkin HJ, Park HJ, RajBhandary UL (1996) Amber suppression in mammalian cells dependent upon expression of an *Escherichia coli* aminoacyl-tRNA synthetase gene. Mol Cell Biol 16:907–913

Farrell IS, Toroney R, Hazen JL, Mehl RA, Chin JW (2005) Photo-cross-linking interacting proteins with a genetically encoded benzophenone. Nat Methods 2:377–384

Feng T, Tsao ML, Schultz PG (2004) A phage display system with unnatural amino acids. J Am Chem Soc 126:15962–15963

Furter R (1998) Expansion of the genetic code: Site-directed p-fluoro-phenylalanine incorporation in *Escherichia coli*. Protein Sci 7:419–426

Griffin BA, Adams SR, Tsien RY (1998) Specific covalent labeling of recombinant protein molecules inside live cells. Science 281:269–272

Grubbs RH, Chang S (1998) Recent advances in olefin metathesis and its application in organic synthesis. Tetrahedron 54:4413–4450

Hamada H, Kameshima N, Szymanska A, Wegner K, Lankiewicz L, Shinohara H, Taki M, Sisido M (2005) Position-specific incorporation of a highly photodurable and blue-laser excitable fluorescent amino acid into proteins for fluorescence sensing. Bioorg Med Chem 13:3379–3384

Hamano-Takaku F, Iwama T, Saito-Yano S, Takaku K, Monden Y, Kitabatake M, Soll D, Nishimura S (2000) A mutant *Escherichia coli* tyrosyl-tRNA synthetase utilizes the unnatural amino acid azatyrosine more efficiently than tyrosine. J Biol Chem 275:40324–40328

Hartman MCT, Josephson K, Szostak JW (2006) Enzymatic aminoacylation of tRNA with unnatural amino acids. Proc Natl Acad Sci USA 103:4356–4361

Hendrickson TL, de Crecy-Lagard V, Schimmel P (2004) Incorporation of nonnatural amino acids into proteins. Annu Rev Biochem 73:147–176

Hendrickson WA, Horton JR, Lemaster DM (1990) Selenomethionyl proteins produced for analysis by multiwavelength anomalous diffraction (MAD) – a vehicle for direct determination of 3-dimensional structure. EMBO J 9:1665–1672

Hermanson GT (1996) Bioconjugate techniques. Academic Press, London

Hino N, Okazaki Y, Kobayashi T, Hayashi A, Sakamoto K, Yokoyama S (2005) Protein photo-cross-linking in mammalian cells by site-specific incorporation of a photoreactive amino acid. Nat Methods 2:201–206

Hoffmanns U, Metzler-Nolte N (2006) Use of the Sonogashira coupling reaction for the "two-step" labeling of phenylalanine peptide side chains with organometallic compounds. Bioconjugate Chem 17:204–213

Hogue CWV, Rasquinha I, Szabo AG, Macmanus JP (1992) A new intrinsic fluorescent-probe for proteins – biosynthetic incorporation of 5-hydroxytryptophan into oncomodulin. FEBS Lett 310:269–272

Hohsaka T, Sisido M (2002) Incorporation of non-natural amino acids into proteins. Curr Opin Chem Biol 6:809–815

Hohsaka T, Ashizuka Y, Murakami H, Sisido M (2001a) Five-base codons for incorporation of nonnatural amino acids into proteins. Nucleic Acids Res 29:3646–3651

Hohsaka T, Kajihara D, Ashizuka, Y, Murakami H, Sisido M (1999) Efficient incorporation of nonnatural amino acids with large aromatic groups into streptavidin in in vitro protein synthesizing systems. J Am Chem Soc 121:34–40

Hohsaka T, Ashizuka Y, Taira H, Murakami H, Sisido M (2001b) Incorporation of nonnatural amino acids into proteins by using various four-base codons in an *Escherichia coli* in vitro translation system. Biochemistry 40:11060–11064

Hohsaka T, Muranaka N, Komiyama C, Matsui K, Takaura S, Abe R, Murakami H, Sisido M (2004) Position-specific incorporation of dansylated non-natural amino acids into streptavidin by using a four-base codon. FEBS Lett 560:173–177

Ibba M, Hennecke H (1995) Relaxing the substrate specificity of an aminoacyl-tRNA synthetase allows in vitro and in vivo synthesis of proteins containing unnatural amino acids. FEBS Lett 364:272–275

Ibba M, Soll D (2000) Aminoacyl-tRNA synthesis. Annu Rev Biochem 69:617–650

Ibba M, Kast P, Hennecke H (1994) Substrate specificity is determined by amino acid binding pocket size in *Escherichia coli* phenylalanyl-tRNA synthetase. Biochemistry 33:7107–7112

Kajihara D (2005) Synthesis and sequence optimization of GFP mutants containing aromatic non-natural amino acids at the Tyr66 position. Protein Eng Des Sel 18:273–278

Kajihara D, Abe R, Iijima I, Komiyama C, Sisido, M, Hohsaka T (2006) FRET analysis of protein conformational change through position-specific incorporation of fluorescent amino acids. Nat Methods 3:923–929

Kast P, Hennecke H (1991) Amino acid substrate specificity of *Escherichia coli* phenylalanyl-tRNA synthetase altered by distinct mutations. J Mol Biol 222:99–124

Kiga D, Sakamoto K, Kodama K, Kigawa T, Matsuda T, Yabuki T, Shirouzu M, Harada Y, Nakayama H, Takio K, Hasegawa Y, Endo Y, Hirao I, Yokoyama S (2002) An engineered *Escherichia coli* tyrosyl-tRNA synthetase for site-specific incorporation of an unnatural amino acid into proteins in eukaryotic translation and its application in a wheat germ cell-free system. Proc Natl Acad Sci USA 99:9715–9720

Kiick KL, van Hest JCM, Tirrell DA (2000) Expanding the scope of protein biosynthesis by altering the methionyl-tRNA synthetase activity of a bacterial expression host. Angew Chem Int Edn Engl 39:2148–2152

Kiick KL, Weberskirch R, Tirrell DA (2001) Identification of an expanded set of translationally active methionine analogues in *Escherichia coli.* FEBS Lett 502:25–30

Kiick KL, Saxon E, Tirrell DA, Bertozzi CR (2002) Incorporation of azides into recombinant proteins for chemoselective modification by the Staudinger ligation. Proc Natl Acad Sci USA 99:19–24

Kirshenbaum K, Carrico IS, Tirrell DA (2002) Biosynthesis of proteins incorporating a versatile set of phenylalanine analogues. ChemBioChem 3:235–237

Kodama K, Fukuzawa S, Sakamoto K, Nakayama H, Kigawa T, Yabuki T, Matsuda N, Shirouzu M, Takio K, Tachibana K, Yokoyama S (2006a) A new protein engineering approach combining chemistry and biology, part 1; site-specific incorporation of 4-iodo-L-phenylalanine in vitro by using misacylated suppressor tRNA(Phe). ChemBioChem 7: 1577–1581

Kodama K, Fukuzawa S, Nakayama H, Kigawa T, Sakamoto K, Yabuki T, Matsuda N, Shirouzu M, Takio K, Tachibana K, Yokoyama S (2006b) Regioselective carbon–carbon bond formation in proteins with palladium catalysis; new protein chemistry by organometallic chemistry. ChemBioChem 7:134–139

Kohn M, Breinbauer R (2004) The Staudinger ligation – a gift to chemical biology. Angew Chem Int Edn Engl 43:3106–3116

Kowal AK, Koehrer C, RajBhandary UL (2001) Twenty-first aminoacyl-tRNA synthetase-suppressor tRNA pairs for possible use in site-specific incorporation of amino acid analogues into proteins in eukaryotes and in eubacteria. Proc Natl Acad Sci USA 98:2268–2273

Kukhar VP, Soloshonok VA (1995) Fluorine-containing amino acids: Synthesis and properties. Wiley, New York

Kwon I, Kirshenbaum K, Tirrell DA (2003) Breaking the degeneracy of the genetic code. J Am Chem Soc 125:7512–7513

Kwon I, Wang P, Tirrell DA (2006) Design of a bacterial host for site-specific incorporation of p-bromophenylalanine into recombinant proteins. J Am Chem Soc 128:11778–11783

Lakowicz JR (1999) Principles of fluorescence spectroscopy, 2nd edn. Kluwer Academic/Plenum Pubishers, New York

Lawrence DS (2005) The preparation and in vivo applications of caged peptides and proteins. Curr Opin Chem Biol 9:570–575

Lee HY, Lee KH, Al-Hashimi HM, Marsh ENG (2006) Modulating protein structure with fluorous amino acids: increased stability and native-like structure conferred on a 4-helix bundle protein by hexafluoroleucine. J Am Chem Soc 128:337–343

Lemieux GA, Bertozzi CR (1998) Chemoselective ligation reactions with proteins, oligosaccharides and cells. Trends Biotechnol 16:506–513

Lemieux GA, de Graffenried CL, Bertozzi CR (2003) A fluorogenic dye activated by the Staudinger ligation. J Am Chem Soc 125:4708–4709

Lewis WG, Green LG, Grynszpan F, Radic Z, Carlier PR, Taylor P, Finn MG, Sharpless KB (2002) Click chemistry in situ: acetylcholinesterase as a reaction vessel for the selective assembly of a femtomolar inhibitor from an array of building blocks. Angew Chem Int Edn Engl 41:1053–1057

Li Q, Du HN, Hu HY (2003) Study of protein–protein interactions by fluorescence of tryptophan analogs: application to immunoglobulin G binding domain of streptococcal protein G. Biopolymers 72:116–122

Link AJ, Tirrell DA (2003) Cell surface labeling of *Escherichia coli* via copper(I)-catalyzed [3+2] cycloaddition. J Am Chem Soc 125:11164–11165

Link AJ, Mock ML, Tirrell DA (2003) Non-canonical amino acids in protein engineering. Curr Opin Biotechnol 14:603–609

Link AJ, Vink MKS, Tirrell DA (2004) Presentation and detection of azide functionality in bacterial cell surface proteins. J Am Chem Soc 126:10598–10602

Link AJ, Vink MKS, Agard NJ, Prescher JA, Bertozzi CR, Tirrell DA (2006) Discovery of aminoacyl-tRNA synthetase activity through cell-surface display of noncanonical amino acids. Proc Natl Acad Sci USA 103:10180–10185

Liu DR, Schultz PG (1999) Progress toward the evolution of an organism with an expanded genetic code. Proc Natl Acad Sci USA 96:4780–4785

Liu HT, Wang L, Brock A, Wong CH, Schultz PG (2003) A method for the generation of glycoprotein mimetics. J Am Chem Soc 125:1702–1703

Miller JC, Silverman SK, England PM, Dougherty DA, Lester HA (1998) Flash decaging of tyrosine sidechains in an ion channel. Neuron 20:619–624

Mock ML, Michon T, van Hest JCM, Tirrell DA (2006) Stereoselective incorporation of an unsaturated isoleucine analogue into a protein expressed in *E. coli*. ChemBioChem 7:83–87

Mohammadi F, Prentice GA, Merrill AR (2001) Protein–protein interaction using tryptophan analogues: novel spectroscopic probes for toxin-elongation factor-2 interactions. Biochemistry 40:10273–10283

Montclare JK, Tirrell DA (2006) Evolving proteins of novel composition. Angew Chem Int Edn Engl 45:4518–4521

Mori H, Ito K (2006) Different modes of SecY–SecA interactions revealed by site-directed in vivo photo-cross-linking. Proc Natl Acad Sci USA 103:16159–16164

Munier R, Cohen GN (1956) Incorporation of structural analogues of amino acids in bacterial proteins. Biochim Biophys Acta 21:592–593

Murakami H, Hohsaka T, Ashizuka Y, Hashimoto K, Sisido M (2000) Site-directed incorporation of fluorescent nonnatural amino acids into streptavidin for highly sensitive detection of biotin. Biomacromolecules 1:118–125

Muralidharan V, Cho JH, Trester-Zedlitz M, Kowalik L, Chait BT, Raleigh DP, Muir TW (2004) Domain-specific incorporation of noninvasive optical probes into recombinant proteins. J Am Chem Soc 126:14004–14012

Nakashima H, Hashimoto M, Sadakane Y, Tomohiro T, Hatanaka Y (2006) Simple and versatile method for tagging phenyldiazirine photophores. J Am Chem Soc 128:15092–15093

Nakata H, Ohtsuki T, Abe R, Hohsaka T, Sisido M (2006) Binding efficiency of elongation factor Tu to tRNAs charged with nonnatural fluorescent amino acids. Anal Biochem 348:321–323

Niemz A, Tirrell DA (2001) Self-association and membrane-binding behavior of melittins containing trifluoroleucine. J Am Chem Soc 123:7407–7413

Nitz M, Mezo AR, Ali MH, Imperiali B (2002) Enantioselective synthesis and application of the highly fluorescent and environment-sensitive amino acid 6-(2-dimethylaminonaphthoyl) alanine (DANA). Chem Commun 17:1912–1913

Ohtsuki T, Manabe T, Sisido M (2005) Multiple incorporation of non-natural amino acids into a single protein using tRNAs with non-standard structures. FEBS Lett 579:6769–6774

Ojida A, Tsutsumi H, Kasagi N, Hamachi I (2005) Suzuki coupling for protein modification. Tetrahedron Lett 46:3301–3305

Pal PP, Bae JH, Azim MK, Hess P, Friedrich R, Huber R, Moroder L, Budisa N (2005) Structural and spectral response of *Aequorea victoria* green fluorescent proteins to chromophore fluorination. Biochemistry 44:3663–3672

Panchenko T, Zhu WW, Montclare JK (2006) Influence of global fluorination on chloramphenicol acetyltransferase activity and stability. Biotechnol Bioeng 94:921–930

Pardee AB, Shore VG, Prestidge LS (1956) Incorporation of azatryptophan into proteins of bacteria and bacteriophage. Biochim Biophys Acta 21 (2):406–407

Petka WA, Harden JL, McGrath KP, Wirtz D, Tirrell DA (1998) Reversible hydrogels from self-assembling artificial proteins. Science 281:389–392

Philipson KD, Gallivan JP, Brandt GS, Dougherty DA, Lester HA (2001) Incorporation of caged cysteine and caged tyrosine into a transmembrane segment of the nicotinic ACh receptor. Am J Physiol Cell Physiol 281:C195–C206

Prescher JA, Bertozzi CR (2005) Chemistry in living systems. Nat Chem Biol 1:13–21

Prescher JA, Dube DH, Bertozzi CR (2004) Chemical remodelling of cell surfaces in living animals. Nature 430:873–877

Ratnaparkhi GS, Ramachandran S, Udgaonkar JB, Varadarajan R (1998) Discrepancies between the NMR and X-ray structures of uncomplexed barstar: analysis suggests that packing densities of protein structures determined by NMR are unreliable. Biochemistry 37:6958–6966

Rennert OM, Anker HS (1963) On incorporation of 5′,5′,5′-trifluoroleucine into proteins of *E. coli*. Biochemistry 2:471–476

Rodriguez EA, Lester HA, Dougherty DA (2006) In vivo incorporation of multiple unnatural amino acids through nonsense and frameshift suppression. Proc Natl Acad Sci USA 103: 8650–8655

Rodriguez EC, Marcaurelle LA, Bertozzi CR (1998) Aminooxy-, hydrazide-, and thiosemicarbazide-functionalized saccharides: versatile reagents for glycoconjugate synthesis. J Org Chem 63: 7134–7135

Rostovtsev VV, Green LG, Fokin VV, Sharpless KB (2002) A stepwise Huisgen cycloaddition process: copper(I)-catalyzed regioselective "ligation" of azides and terminal alkynes. Angew Chem Int Edn Engl 41:2596–2599

Rothman DM, Petersson EJ, Vazquez ME, Brandt GS, Dougherty DA, Imperiali B (2005) Caged phosphoproteins. J Am Chem Soc 127:846–847

Sadakane Y, Hatanaka Y (2006) Photochemical fishing approaches for identifying target proteins and elucidating the structure of a ligand-binding region using carbene-generating photoreactive probes. Anal Sci 22:209–218

Sakamoto K, Hayashi A, Sakamoto A, Kiga D, Nakayama H, Soma A, Kobayashi T, Kitabatake M, Takio K, Saito K, Shirouzu M, Hirao I, Yokoyama S (2002) Site-specific incorporation of an unnatural amino acid into proteins in mammalian cells. Nucleic Acids Res 30: 4692–4699

Saxon E, Bertozzi CR (2000) Cell surface engineering by a modified Staudinger reaction. Science 287:2007–2010

Schafmeister CE, Po J, Verdine GL (2000) An all-hydrocarbon cross-linking system for enhancing the helicity and metabolic stability of peptides. J Am Chem Soc 122:5891–5892

Sivakumar K, Xie F, Cash BM, Long S, Barnhill HN, Wang Q (2004) A fluorogenic 1,3-dipolar cycloaddition reaction of 3-azidocoumarins and acetylenes. Org Lett 6:4603–4606

Son S, Tanrikulu IC, Tirrell DA (2006) Stabilization of bzip peptides through incorporation of fluorinated aliphatic residues. ChemBioChem 7:1251–1257

Suchanek M, Radzikowska A, Thiele C (2005) Photo-leucine and photo-methionine allow identification of protein–protein interactions in living cells. Nat Methods 2:261–267

Summerer D, Chen S, Wu N, Deiters A, Chin JW, Schultz PG (2006) A genetically encoded fluorescent amino acid. Proc Natl Acad Sci USA 103:9785–9789

Taki M, Hohsaka T, Murakami H, Taira K, Sisido M (2001) A non-natural amino acid for efficient incorporation into proteins as a sensitive fluorescent probe. FEBS Lett 507:35–38

Taki M, Hohsaka T, Murakami H, Taira K, Sisido M (2002) Position-specific incorporation of a fluorophore-quencher pair into a single streptavidin through orthogonal four-base codon/anticodon pairs. J Am Chem Soc 124:14586–14590

Tang Y, Tirrell DA (2001) Biosynthesis of a highly stable coiled-coil protein containing hexafluoroleucine in an engineered bacterial host. J Am Chem Soc 123:11089–11090

Tang Y, Ghirlanda G, Petka WA, Nakajima T, DeGrado WF, Tirrell DA (2001a) Fluorinated coiled-coil proteins prepared in vivo display enhanced thermal and chemical stability. Angew Chem Int Edn Engl 40:1494–1496

Tang Y, Ghirlanda G, Vaidehi N, Kua J, Mainz DT, Goddard WA, DeGrado WF, Tirrell DA (2001b) Stabilization of coiled-coil peptide domains by introduction of trifluoroleucine. Biochemistry 40:2790–2796

Tong YH, Brandt GS, Li M, Shapovalov G, Slimko E, Karschin A, Dougherty DA, Lester HA (2001) Tyrosine decaging leads to substantial membrane trafficking during modulation of an inward rectifier potassium channel. J Gen Physiol 117:103–118

Tornoe CW, Christensen C, Meldal M (2002) Peptidotriazoles on solid phase: 1,2,3-triazoles by regiospecific copper(I)-catalyzed 1,3-dipolar cycloadditions of terminal alkynes to azides. J Org Chem 67:3057–3064

Tsao ML, Tian F, Schultz PG (2005) Selective Staudinger modification of proteins containing p-azidophenylalanine. ChemBioChem 6:2147–2149

Tsao ML, Summerer D, Ryu YH, Schultz PG (2006) The genetic incorporation of a distance probe into proteins in *Escherichia coli.* J Am Chem Soc 128:4572–4573

Tsien RY (1998) The green fluorescent protein. Annu Rev Biochem 67:509–44

van Hest JCM, Kiick KL, Tirrell DA (2000) Efficient incorporation of unsaturated methionine analogues into proteins in vivo. J Am Chem Soc 122:1282–1288

Walsh CT (2006) Posttranslational modification of proteins: Expanding nature's inventory. Roberts and Company Publishers, Englewood

Wang J, Xie J, Schultz PG (2006) A genetically encoded fluorescent amino acid. J Am Chem Soc 128:8738–8739

Wang L, Schultz PG (2005) Expanding the genetic code. Angew Chem Int Edn Engl 44:34–66

Wang L, Brock A, Herberich B, Schultz PG (2001) Expanding the genetic code of *Escherichia coli.* Science 292:498–500

Wang L, Zhang ZW, Brock A, Schultz PG (2003a) Addition of the keto functional group to the genetic code of *Escherichia coli.* Proc Natl Acad Sci USA 100:56–61

Wang L, Xie JM, Deniz AA, Schultz PG (2003b) Unnatural amino acid mutagenesis of green fluorescent protein. J Org Chem 68:174–176

Wang P, Fichera A, Kumar K, Tirrell DA (2004) Alternative translations of a single RNA message: an identity switch of (2S,3R)-4,4,4-trifluorovaline between valine and isoleucine codons. Angew Chem Int Edn Engl 43:3664–3666

Wu N, Deiters A, Cropp TA, King D, Schultz PG (2004) A genetically encoded photocaged amino acid. J Am Chem Soc 126:14306–14307

Xie JM, Schultz PG (2005) Adding amino acids to the genetic repertoire. Curr Opin Chem Biol 9:548–554

Yoder NC, Kumar K (2002) Fluorinated amino acids in protein design and engineering. Chem Soc Rev 31:335–341

Zeng H, Xie J, Schultz PG (2006) Genetic introduction of a diketone-containing amino acid into protiens. Bioorg Med Chem Lett 16:5356–5359

Zhang KC, Diehl MR, Tirrell DA (2005) Artificial polypeptide scaffold for protein immobilization. J Am Chem Soc 127:10136–10137

Zhang ZW, Wang L, Brock A, Schultz PG (2002) The selective incorporation of alkenes into proteins in *Escherichia coli.* Angew Chem Int Edn Engl 41:2840–2842

Zhang ZW, Smith BAC, Wang L, Brock A, Cho C, Schultz, PG (2003) A new strategy for the site-specific modification of proteins in vivo. Biochemistry 42:6735–6746

Zhang ZW, Alfonta L, Tian F, Bursulaya B, Uryu S, King DS, Schultz PG (2004a) Selective incorporation of 5-hydroxytryptophan into proteins in mammalian cells. Proc Natl Acad Sci USA 101:8882–8887

Zhang ZW, Gildersleeve J, Yang YY, Xu R, Loo JA, Uryu S, Wong CH, Schultz PG (2004b) A new strategy for the synthesis of glycoproteins. Science 303:371–373

Zhou Z, Fahrni CJ (2004) A fluorogenic probe for the copper(I)-catalyzed azide–alkyne ligation reaction: modulation of the fluorescence emission via (3)(n,pi*)-(1)(pi,pi*) inversion. J Am Chem Soc 126:8862–8863

Fidelity Mechanisms of the Aminoacyl-tRNA Synthetases

A.P. Mascarenhas, S. An, A.E. Rosen, S.A. Martinis(✉), and K. Musier-Forsyth(✉)

Contents

1 Abstract

The central dogma and its accurate interpretation by a large array of biomolecules remains a fascinating process. Understanding its mechanisms of decoding holds tremendous potential for the future of protein engineering. The family of aminoacyl

---

S.A. Martinis
Department of Biochemistry, University of Illinois at Urbana-Champaign, Urbana, IL 61801, USA, e-mail: martinis@life.uiuc.edu

K. Musier-Forsyth
Departments of Chemistry and Biochemistry, The Ohio State University, Columbus, OH 43210, USA, e-mail: musier@chemistry.ohio-state.edu

C. Köhrer and U.L. RajBhandary (eds.) *Protein Engineering.*
Nucleic Acids and Molecular Biology 22,

transfer RNA (tRNA) synthetases (AARSs) is at the forefront of this field. While aminoacylation is the primary role of these enzymes, they possess various additional functions that are important to cell survival. An editing activity, which clears incorrectly attached amino acids, minimizes errors in protein synthesis. If this error correction mechanism is disabled, the incorporation of novel amino acids into proteins offers an exciting approach to expand the genetic code (Döring et al. 2001; Nangle et al. 2006). This chapter focuses on the AARSs that have amino acid editing functions with an emphasis on their continuing dynamic role in the field of protein engineering.

2 Aminoacylation and Specificity

Translation employs a large number of diverse proteins, RNAs, and small molecules to decode information contained in DNA, via RNA intermediates, into functional proteins. At each step, accurate interactions are crucial to correctly interpret the genetic code. Poor or inaccurate translation of the genetic code could threaten the viability or function of living cells (Karkhanis et al. 2006; Lee et al. 2006).

In the initial step of protein synthesis, a specific amino acid is attached to its cognate tRNA isoacceptor. Aminoacylation of the tRNA is conferred by the AARSs, which are an ancient family of enzymes (Ibba et al. 2000). These AARSs have the formidable task of selecting their respective amino acids and tRNAs from large pools of similar substrates within the cell. Each cell typically has one AARS that is responsible for activating one of the 20 amino acids. Organelle protein synthesis relies on a completely separate set of synthetases that are also encoded in the nucleus.

AARSs aminoacylate or "charge" tRNA with amino acids via a two-step aminoacylation reaction (Fig. 1). The enzyme binds adenosine triphosphate (ATP) and amino acid to catalyze the formation of an aminoacyl adenylate with concomitant release of pyrophosphate. The activated amino acid is then transferred specifically to the 2′ or the 3′ ribose hydroxyl of the tRNA's terminal adenosine to form an aminoacyl-tRNA (Freist 1989; Ibba and Söll 2000; First 2005). The charged tRNA

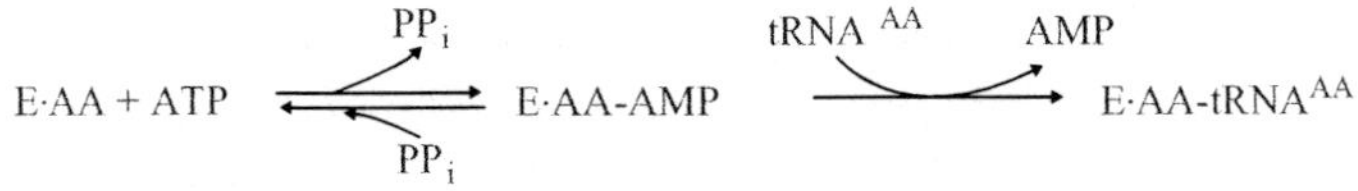

Fig. 1 Two-step aminoacylation reaction catalyzed by the aminoacyl transfer RNA (tRNA) synthetases (AARSs). The enzyme (*E*) hydrolyzes ATP to activate amino acid (*AA*) and form an aminoacyl adenylate intermediate complex (*E·AA-AMP*). Inorganic pyrophosphate (*PP$_i$*) is released as a by-product. The activated amino acid is transferred to its cognate transfer RNA (*tRNA^{AA}*), producing aminoacylated tRNA (*AA-tRNA^{AA}*)

is transferred to an elongation factor (EF-Tu in bacteria), which delivers the charged tRNA to the ribosome, where the attached amino acid is incorporated into the growing polypeptide chain.

Selection of cognate tRNA isoacceptors is largely dependent on diverse sets of identity determinants that facilitate interaction with the cognate AARSs (Giegé et al. 1998). Positive determinants are specific nucleotides and/or functional groups that confer tRNA recognition for specific aminoacylation. In contrast, negative determinants prevent noncognate tRNA molecules from misaminoacylation. Most identity elements are located in the acceptor stem and the anticodon region, although they can be found throughout the entire tRNA molecule. Specific structural features in the core of the tRNA can also promote recognition by the synthetase. Interestingly, many AARSs can tolerate significant variations in their tRNA substrates, including deletions of entire domains or the absence of base modifications, as long as their respective identity elements remain intact.

Complete discrimination of amino acids by AARSs is achieved via different strategies. Certain amino acids have unique side chains that are easily distinguished at the molecular level (Fersht and Dingwall 1979b). For example, cysteinyl-tRNA synthetase (CysRS) utilizes an active site with highly conserved residues, as well as a zinc ion that interacts specifically with the negatively charged thiolate of cysteine (Newberry et al. 2002). This mechanism allows selection against even the structurally similar serine (Fig. 2), which has a much lower zinc binding affinity. Likewise, the active site of tyrosyl-tRNA synthetase (TyrRS) is highly evolved to preferentially bind tyrosine rather than the isosteric phenylalanine (Fersht et al. 1980). In contrast, isosteric sets of amino acids that differ by a single methyl group (Fig. 2; i.e., isoleucine and valine or alanine and glycine) cannot be easily distinguished by the AARSs, and therefore require an editing mechanism that clears mistakes to maintain thresholds of fidelity that are sufficient for cell viability (Hendrickson and Schimmel 2003; Hendrickson et al. 2004).

AARSs are divided into two classes based on the folding topologies of their ATP-binding domains (Webster et al. 1984; Ludmerer and Schimmel 1987; Cusack et al. 1990; Eriani et al. 1990). The class I proteins possess a Rossmann fold characterized by conserved HIGH (Ludmerer and Schimmel 1987) and KMSKS (Hountondji et al. 1986) signature sequences. The ten class II members have an antiparallel β-sheet arrangement and three degenerate sequence motifs called motifs 1, 2, and 3 (Cusack et al. 1990; Eriani et al. 1990, 1995). Each class is further divided into three subclasses (a, b, and c) depending on related primary sequences, organization of class-defining structural features, anticodon-binding domain similarity, and mechanistic distinctions (Moras 1992; Cusack 1995; Landès et al. 1995; Ribas de Pouplana and Schimmel 2001). However, for both classes, the canonical core containing the aminoacylation active site is considered the ancient, historical AARS (Schimmel et al. 1993). Over time, the ancient synthetase core incorporated insertions, additional motifs, and appendages, including the editing domains described here, to evolve into the modern family of synthetases (Burbaum and Schimmel 1991). Some of these additions even introduced secondary functions for the AARSs in the cell (Martinis et al. 1999a, b).

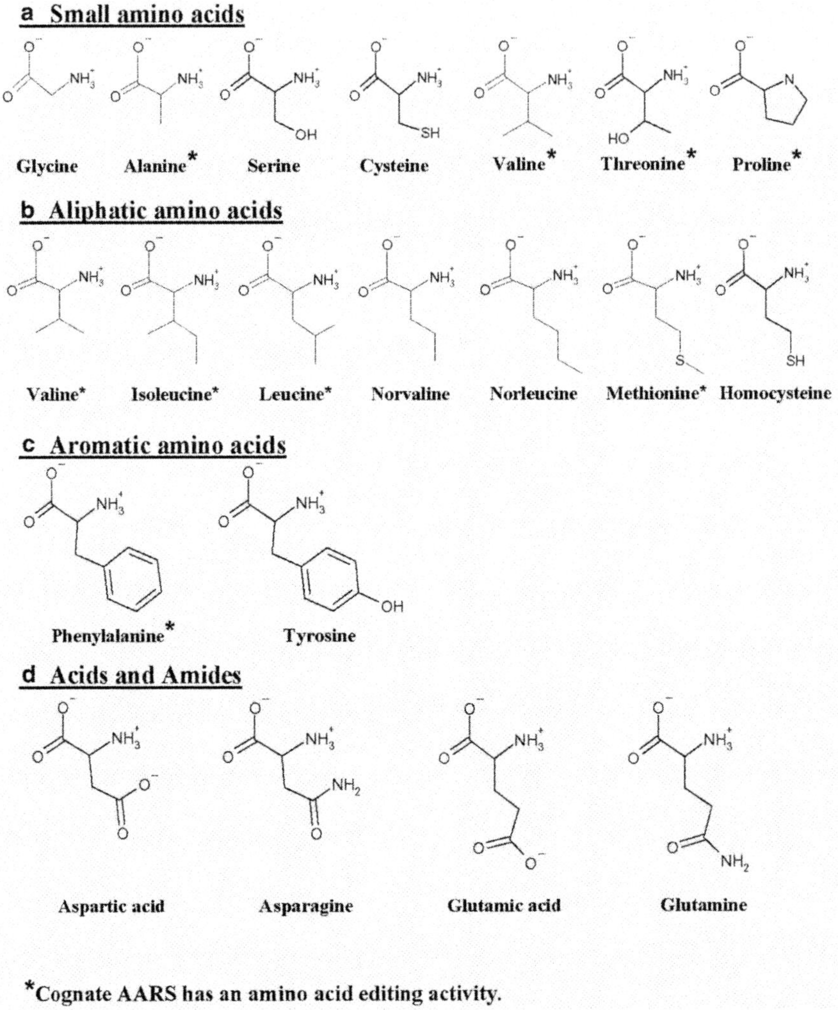

Fig. 2 Structurally similar amino acids. Many amino acid side chains share common shapes, functional groups and/or sizes. **a** Small amino acids with similar shape and size. **b** Aliphatic amino acids. **c** Aromatic amino acids that differ by a hydroxyl group. **d** Acids and amides. The *asterisks* indicate those amino acids that are activated by cognate AARSs that have amino acid editing activities

2.1 Historical Background on Amino Acid Editing

The fidelity of protein synthesis relies on the efficient production of correctly charged tRNAs and thus is completely dependent on the specificity of the AARSs. In 1957, Linus Pauling calculated the theoretical hurdles for protein discrimination of isosteric amino acid substrates that differed by one methyl group (Pauling 1958).

He predicted an error rate, for example, of a valine substitution for isoleucine in an isoleucine binding pocket, of about 1 in 20. Subsequent in vivo measurements by Loftfield estimated the rate of erroneous incorporations of valine for isoleucine into ovalbumin and globin proteins at less than 1 in 3,000 (Loftfield 1963; Loftfield and Vanderjagt 1972). This increase in fidelity of 2 orders of magnitude suggested that amino acid specificity is achieved by molecular recognition mechanisms that extend beyond the highly complementary substrate binding pocket of the AARS.

Baldwin and Berg (1966) showed that *Escherichia coli* isoleucyl-tRNA synthetase (IleRS) catalyzed the formation of both isoleucyl adenylate and valyl adenylate (Val-AMP), but only isoleucine was transferred to tRNA to form Ile-$tRNA^{Ile}$. Significantly, the addition of $tRNA^{Ile}$ to the IleRS·Val-AMP complex induced the hydrolysis of the adenylate intermediate rather than formation of Val-$tRNA^{Ile}$. In 1972, the Schimmel laboratory determined that *E. coli* IleRS rapidly hydrolyzed Val-$tRNA^{Ile}$ (Schreier and Schimmel 1972). It was later reported for *E. coli* IleRS that the misactivated Val-AMP is hydrolyzed before valine is transferred to $tRNA^{Ile}$ (Fersht and Dingwall 1979c). Thus, the concept of amino acid "editing" mechanisms was introduced for the AARSs. Hydrolysis of misactivated adenylate intermediates and of mischarged tRNAs was termed "pre-transfer editing" and "post-transfer editing," respectively.

To explain how AARSs can confer sufficiently high levels of amino acid discrimination to fully distinguish between structurally similar amino acids (Fig. 2), Fersht proposed the "double sieve" mechanism (Fig. 3a) (Fersht 1977a). Using IleRS as an example, the model predicted that, whereas one active site could not completely discriminate isoleucine and valine, two separate active sites with distinct strategies for recognition could significantly enhance fidelity (Fersht 1977a). The aminoacylation active site of the AARS would act as a "coarse" sieve for adenylate synthesis, activating the cognate amino acid but also allowing, to a lesser extent, activation of isosteric or smaller amino acids that could fit into the amino acid binding pocket. The second "fine" sieve would selectively bind misactivated amino acids for editing, while excluding the original cognate amino acid. Thus, substrates synthesized in the first sieve would be further screened by the second sieve to enhance fidelity by as much as tenfold.

The aminoacylation site or coarse sieve has been well defined by X-ray crystallography, biochemical, and mutagenesis investigations (Carter 1993; Martinis and Schimmel 1999). A separate editing active site was first identified by the Schimmel group, when they recombinantly expressed a discrete domain, called connective polypeptide 1 (CP1) (Starzyk et al. 1987), from IleRS and valyl-tRNA synthetase (ValRS) and showed that it could specifically hydrolyze mischarged Val-$tRNA^{Ile}$ and Thr-$tRNA^{Val}$, respectively (Lin et al. 1996). CP1 is highly conserved through evolution and a homologous domain is present in all class I editing enzymes that clear misactivated standard amino acids (Table 1).

Over the past two decades, numerous editing domains have been characterized via X-ray crystallography. In 1998, the Yokoyama laboratory solved the cocrystal structure of *Thermus thermophilus* IleRS bound to valine (Nureki et al. 1998), which revealed that the fine sieve for editing is about 35 Å from the aminoacylation active

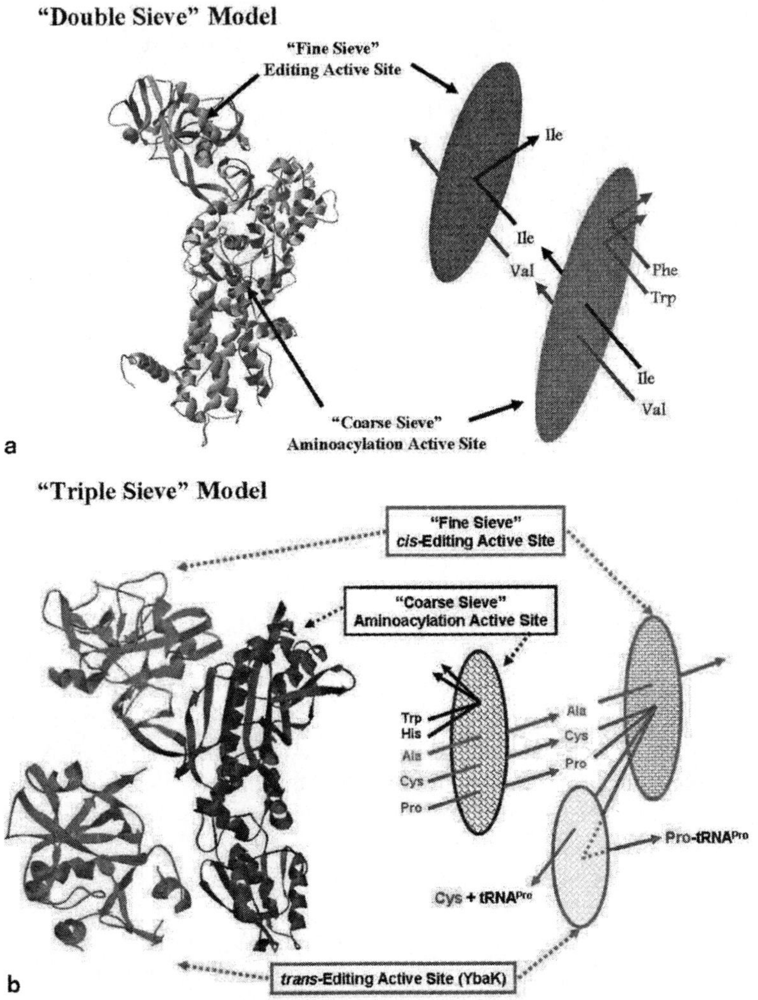

Fig. 3 Models for AARS fidelity. **a** "Double sieve" model for isoleucyl-tRNA synthetase (IleRS). The crystal structure of IleRS (Protein Data Bank code 1ILE) is depicted with the aminoacylation active site (*light gray*), which acts as the "coarse sieve" to block structurally dissimilar amino acids, such as phenylalanine and tryptophan. The editing active site or "fine sieve" (*dark gray*) binds valine for hydrolysis. **b** Proposed "triple sieve" model for prolyl-tRNA synthetase (ProRS). The crystal structure of ProRS (Protein Data Bank code 2J3M) is depicted with the aminoacylation domain or "coarse sieve" which blocks large amino acids such as histidine and tryptophan. The editing domain or "fine sieve" hydrolyzes noncognate alanine but blocks cysteine and proline. A freestanding YbaK protein (Protein Data Bank code 1DBU) acts in *trans* to hydrolyze mischarged Cys-tRNA^{Pro}, but not correctly formed Pro-tRNA^{Pro}

site of the AARS. Subsequently, the Steitz group solved the *Staphylococcus aureus* IleRS structure with tRNA^{Ile} bound in an "editing" conformation (Silvian et al. 1999). In this structure, the acceptor stem was positioned such that the CCA-3′ end

Table 1 Amino acid editing targets

AARS	Editing domain	Misactivated amino acids		References
		Standard	Nonstandard	
Class I				
LeuRS	CP1	Isoleucine Methionine Asparagine Aspartic acid Valine	Norvaline, homocysteine Norleucine, homoserine Homocysteine α-Aminobutyrate γ-Hydroxyleucine γ-Hydroxyisoleucine δ-Hydroxyleucine δ-Hydroxyisoleucine γ,δ-Dihydroxyisoleucine	Englisch et al. (1986) Martinis and Fox (1997) Lincecum and Martinis (2005) Xu et al. (2004) Karkhanis et al. (2007)
IleRS	CP1	Valine Leucine Threonine Cysteine	Homocysteine Homoserine	Jakubowski and Fersht (1981)
ValRS	CP1	Threonine Cysteine Serine	L-α Aminobutyrate Homocysteine	Jakubowski and Fersht (1981)
MetRS	Aminoacylation active site		Homocysteine Norleucine Ethionine Selenomethionine	Jakubowski and Fersht (1981), Budisa et al. (1995)

(continued)

Table 1 (continued)

AARS	Editing domain	Misactivated amino acids		References
		Standard	Nonstandard	
Class II				
AlaRS	C-terminal Domain	Glycine Serine Cysteine	L-α-Aminobutyrate	Tsui and Fersht (1981)
	AlaX (post)	Glycine Serine		Ahel et al. (2003)
PheRS	B3/B4 domain	Isoleucine Leucine Methionine Tyrosine	*p*-Aminophenylalanine	Roy et al. (2004)
ThrRS	N2 domain	Serine Valine		Dock-Bregeon et al. (2000)
	ThrRS-ed (post)	Serine		Korencic et al. (2004)
ProRS	INS (post)	Alanine		Beuning and Musier-Forsyth (2000)
	ProX (post)	Alanine		Ahel et al. (2003)
	YbaK (post)	Cysteine		An and Musier-Forsyth (2004)
	Aminoacylation active site (pre)	Alanine		Hati et al. (2006)
LysRS	Aminoacylation active site	Cysteine Threonine Alanine	Homocysteine Homoserine Ornithine	Jakubowski (1999)

For the an explanation of the acroynms used for the aminoacyl transfer RNA (tRNA) synthetases (*AARS*), see the text.

of the tRNA could be modeled to fit into the CP1 domain rather than into the Rossmann-fold-based aminoacylation site. On the basis of this structure, a translocation mechanism was proposed, wherein the tRNA acceptor end shuttles the noncognate amino acid from the aminoacylation active site to the editing pocket.

A novel "triple sieve" mechanism that requires an independently folded freestanding editing protein appears to be necessary for complete fidelity by class II prolyl-tRNA synthetase (ProRS) (Fig. 3b) (An and Musier-Forsyth 2004, 2005). Cognate proline as well as the smaller amino acids alanine and cysteine pass through the first sieve and are activated in the aminoacylation active site of ProRS. The insertion domain of bacterial ProRS carries out alanine-specific editing in *cis* and functions as the second sieve. However, this editing active site fails to target Cys-$tRNA^{Pro}$ for hydrolysis. Instead, in some organisms, a ternary complex composed of ProRS, tRNA, and the separate freestanding editing protein targets Cys-$tRNA^{Pro}$ for hydrolytic editing in *trans* (An and Musier-Forsyth 2005).

2.2 Editing Aminoacyl-tRNA Synthetases

Of the 20 AARSs, nine enzymes that represent both classes are known to possess editing activities that hydrolyze noncognate amino acids (Table 1). These include leucyl-tRNA synthetase (LeuRS) (Englisch et al. 1986), IleRS (Baldwin and Berg 1966), ValRS (Fersht and Kaethner 1976) and methionyl-tRNA synthetase (MetRS) (Fersht and Dingwall 1979a) from class I; alanyl-tRNA synthetase (AlaRS) (Tsui and Fersht 1981), phenylalanyl-tRNA synthetase (PheRS) (Yarus 1972), threonyl-tRNA synthetase (ThrRS) (Dock-Bregeon et al. 2000), ProRS (Beuning and Musier-Forsyth 2000) and lysyl-tRNA synthetase (LysRS) (Jakubowski 1999) from class II. Each of these enzymes discriminates against standard and in some cases nonstandard amino acids that are structurally similar to their cognate amino acid (Table 1).

The crystal structures of all the editing enzymes have been solved in the presence and/or absence of various substrates and substrate analogs (Table 2). The structures for IleRS (Nureki et al. 1998), ValRS (Fukai et al. 2000), and LeuRS (Cusack et al. 2000) with noncognate amino acids or editing substrate analogs bound clearly localize the editing active site to a threonine-rich region within the CP1 domain. The class I MetRS and the class II LysRS edit via a cyclization reaction that occurs in the synthetic site. Most class II AARSs that edit rely on hydrolytic modules and freestanding domains that are structurally dissimilar. Among class II enzymes, ThrRS (Dock-Bregeon et al. 2004; Hussain et al. 2006) and PheRS (Kotik-Kogan et al. 2005) have been cocrystallized with substrates bound in their editing domains. Very recently, the structure of a bacterial ProRS with a *cis*-editing domain was solved in the absence of bound substrate (Crépin et al. 2006). A substantial amount of structural information (Table 2) has been complemented by a wealth of mutational data to provide additional molecular insights into the editing mechanisms of both classes of AARSs.

Table 2 X-ray crystal structures of AARS editing enzymes

AARS	Complex	Org.	Substrate/analogue	Protein Data Bank code	References
Class I					
LeuRS	Adenylate	tt	Leu-AMS	1H3N	Cusack et al. (2000)
	Pre-transfer editing	tt	Norvalyl sulphamoyl adenosine	1OBH	Lincecum et al. (2003)
	Post-transfer editing	tt	Norvalyl aminodeoxyadenosine	1OBC	Lincecum et al. (2003)
	apo	ph	–	1WKB	Fukunaga and Yokoyama (2005c)
	Aminoacylation	ph	$tRNA^{Leu}$	1WZ2	Fukunaga and Yokoyama (2005a)
	Exit complex	tt	Norvaline, 2AD, $tRNA^{Leu}$	2BTE	Tukalo et al. (2005)
	Post-transfer editing	tt	$tRNA^{Leu}$	2BYT	Tukalo et al. (2005)
	CP1	ec	–	2AJG	Liu et al. (2006)
			Isoleucine	2AJI	Liu et al. (2006)
			Methionine	2AJH	Liu et al. (2006)
	With inhibitor	tt	$tRNA^{Leu}$, AN2690	2V0G	Rock et al. (2007)
			AMP, AN2690	2V0C	Rock et al. (2007)
IleRS	Amino acid bound	tt	L-Isoleucine, L-valine, Zn^{2+}	1ILE	Nureki et al. (1998)
	tRNA and inhibitor bound	sa	Mupirocin, ec-$tRNA^{Ile}$	1FFY	Silvian et al. (1999)
				1QU2	Silvian et al. (1999)
				1QU3	Silvian et al. (1999)
	Adenylate and inhibitor	tt	Ile-AMS	1JZQ	Nakama et al. (2001)
			Mupirocin	1JZS	Nakama et al. (2001)
	CP1 with amino acid	tt	*apo*	1UDZ	Fukunaga et al. (2004)
			L-Valine	1UE0	Fukunaga et al. (2004)
	CP1	tt	Val-AMS	1WK8	Fukunaga and Yokoyama (2006)
			–	1WNY	Fukunaga and Yokoyama (2006)
			Valyl amino deoxyadenosine	1WNZ	Fukunaga and Yokoyama (2006)
ValRS	Adenylation complex	tt	Val-AMS, $tRNA^{Val}$	1GAX	Fukai et al. (2000)
	Adenylation complex	ta	Val-AMS, $tRNA^{Val}$	1IVS	Fukai et al. (2003)
	apo	tt	–	1IYW	Fukai et al. (2003)
	CP1 domain	tt	Threonyl sulfamoyl adenosine	1WK9	Fukunaga and Yokoyama (2005b)
			pH 8.0	1WKA	Fukunaga and Yokoyama (2005b)

MetRS	*apo*	ec	Zn^{2+}	1QQT	Mechulam et al. (1999)
	apo	tt	–	1A8H	Sugiura et al. (2000)
	Amino acid bound	ec	Methionine	1F4L	Serre et al. (2001)
	C-terminal domain	pa	–	1MKH	Crépin et al. (2002)
	MetRS	ec	Trifluoromethionine	1PFW	Crépin et al. (2003)
			Difluoromethionine	1PFV	Crépin et al. (2003)
			Methionine phosphinate	1PFU	Crépin et al. (2003)
			Methionine phosphonate	1P7P	Crépin et al. (2003)
			Methionyl sulphamoyl adenosine	1PFY	Crépin et al. (2003)
			Met-AMP	1PG0	Crépin et al. (2003)
			Methionine, adenosine	1PG2	Crépin et al. (2003)
	apo	pa	Zn^{2+}	1RQG	Crépin et al. (2004)
	Adenylation	aa	$tRNA^{Met}$	2CSX	Nakanishi et al. (2005)
			$tRNA^{Met}$, Met-AMS	2CT8	Nakanishi et al. (2005)
	apo	tt	Y225F mutant	1WOY	Unpublished
Class II					
AlaRS	*apo*	aa	–	1RIQ	Swairjo et al. (2004)
	ATP	aa	ATP, Mg^{2+}	1YFR	Swairjo and Schimmel (2005)
	Catalytic fragment		Glycine	1YFT	Swairjo and Schimmel (2005)
	Amino acid bound		Alanine	1YFS	Swairjo and Schimmel (2005)
	Catalytic fragment		Serine	1YGB	Swairjo and Schimmel (2005)
	AlaX	ph	Serine	1WNU	Sokabe et al. (2005)
			Zn^{2+}	1WXO	Sokabe et al. (2005)
			Selenomethionine	1V7O	Sokabe et al. (2005)
	apo	ph	Zn^{2+}	1V4P	Ishijima et al. (2006)
	AlaX	ph	–	2E1B	Fukunaga and Yokoyama (2007)
PheRS	*apo*	tt	–	1PYS	Mosyak et al. (1995)
	Aminoacylation	tt	$tRNA^{Phe}$	1EIY	Goldgur et al. (1997)

(*continued*)

Table 2 (continued)

AARS	Complex	Org.	Substrate/analogue	Protein Data Bank code	References
	Adenylation complex	ta	PheOH-AMP	1B7Y	Reshetnikova et al. (1999)
	Adenylation complex	tt	Phenylalanine	1B70	Reshetnikova et al. (1999)
			Phe-AMP	1JJC	Fishman et al. (2001)
		tt	Tyrosine	2AMC	Kotik-Kogan et al. (2005)
			Tyrosyl sulphamoyl adenosine	2ALY	Kotik-Kogan et al. (2005)
			p-Chlorophenylalanine	2AKW	Kotik-Kogan et al. (2005)
	Adenylation complex	tt	$tRNA^{Phe}$, PheOH-AMP	2IY5	Moor et al. (2006)
	N-terminal PheRS-β fragment	ph	–	2CXI	Sasaki et al. (2006)
ThrRS	Adenylation complex	ec	AMP, $tRNA^{Thr}$	1QF6	Sankaranarayanan et al. (1999)
	ΔN truncation	ec	Thr-AMS	1EVL	Sankaranarayanan et al. (2000)
			Threonine	1EVK	Sankaranarayanan et al. (2000)
	ΔN truncation	ec	Ser-AMS	1FYF	Dock-Bregeon et al. (2000)
	Messenger RNA operator domain	ec	Threonyl sulfamoyl adenosine	1KOG	Torres-Larios et al. (2002)
	Adenylation	sa	Threonyl sulfamoyl adenosine	1NYQ	Torres-Larios et al. (2003)
			Threonine, ATP	1NYR	Torres-Larios et al. (2003)
	Editing domain	ec	–	1TJE	Dock-Bregeon et al. (2004)
			Serine	1TKE	Dock-Bregeon et al. (2004)
			Ser-AMS	1TKG	Dock-Bregeon et al. (2004)
			Serine-3′-aminoadenosine	1TKY	Dock-Bregeon et al. (2004)
	Editing domain	pa	–	1Y2Q	Dwivedi et al. (2005)
	Editing domain	pa	Serine-3′-aminoadenosine	2HL0	Hussain et al. (2006)
			Serine-3′-aminoadenosine	2HL1	Hussain et al. (2006)
			Ser-AMS	2HL2	Hussain et al. (2006)
			L-Serine	2HKZ	Hussain et al. (2006)
ProRS	Adenylation	tt	–	1HC7	Yaremchuk et al. (2001)
			$tRNA^{Pro}$, prolinol, ATP	1H4Q	Yaremchuk et al. (2001)
			$tRNA^{Pro}$, Pro-AMS	1H4S	Yaremchuk et al. (2001)

			Proline	1H4T	Yaremchuk et al. (2001)
	apo	mj	–	1NJ8	Kamtekar et al. (2003)
		mt	Cys-AMS	1NJ1	Kamtekar et al. (2003)
			–	1NJ2	Kamtekar et al. (2003)
			Pro-AMS	1NJ5	Kamtekar et al. (2003)
			Ala-AMS	1NJ6	Kamtekar et al. (2003)
	apo	rp	–	2I4L	Crépin et al. (2006)
			Pro-AMS	2I4M	Crépin et al. (2006)
			Cys-AMS	2I4N	Crépin et al. (2006)
			ATP	2I4O	Crépin et al. (2006)
	Adenylation	ef	ATP, prolinol	2J3L	Crépin et al. (2006)
			Pro-AMS	2J3M	Crépin et al. (2006)
LysRS	Amino acid bound (LysU)	ec	Lysine	1LYL	Onesti et al. (1995)
	Adenylation (LysU)	ec	Lysine	1E1O	Desogus et al. (2000)
			Lys-AMP	1E1T	Desogus et al. (2000)
			AMP-PCP, lysine	1E22	Desogus et al. (2000)
			Lysine, ATP, Mn^{2+}	1E24	Desogus et al. (2000)
	Amino acid bound (LysS)	ec	–	1BBW	Onesti et al. (2000)
			Lysine	1BBU	Onesti et al. (2000)
	apo (LysRS-I)	ph	Zn^{2+}	1IRX	Terada et al. (2002)

CP1, connective polypeptide 1; *INS*, novel insertion domain; *AA-AMS*, non hydrolyzable sulphamoyl adenylate of the corresponding amino acid (AA); *2AD*, 2-aminoadenosine; *PheOH-AMP*, unhydrolyzable phenyl adenylate, AMP-PCP, β,γ-methyleneadenosine 5 -triphosphate; *tt, Thermus thermophilus; ta, Thermus aquaticus; ec, Escherichia coli; pa, Pyrococcus abyssi; sa, Saccharomyces cerevisiae; mj, Methanocaldococcus jannaschii; aa, Aquifex aeolicus; mt, Methanothermobacter thermoautotrophicus; mk, Methanopyrus kandleri; ph, Pyrococcus horikoshii; rp, Rhodopseudomonas palustri; ef, Enterococcus faecalis*

2.3 Redundant Amino Acid Editing Pathways

Amino acid editing has also been referred to as proofreading or "kinetic proofreading" (Hopfield 1974; Hopfield et al. 1976; Yamane and Hopfield 1977). Although some authors have limited the definition of kinetic proofreading to a nonenzymatic pathway involving dissociation of the noncognate adenylate followed by solution hydrolysis (Fig. 4, pathway 2) (Fersht 1977b; Perona 2005), we have chosen to use the more general definition that includes enzyme-catalyzed hydrolysis (Hopfield et al. 1976; Yamane and Hopfield 1977). Thus, amino acid editing generally requires (1) an initial substrate binding step, (2) a second driving step involving ATP hydrolysis, and (3) discrimination between cognate and noncognate substrates using one or more rejection pathways (Fig. 4).

Numerous amino acid editing pathways have been proposed (Fig. 4). The linkages targeted for hydrolysis consist of a mixed phosphoanhydride (pre-transfer editing) and an aminoacyl ester bond (post-transfer editing) (Fig. 4). Pre-transfer editing, wherein

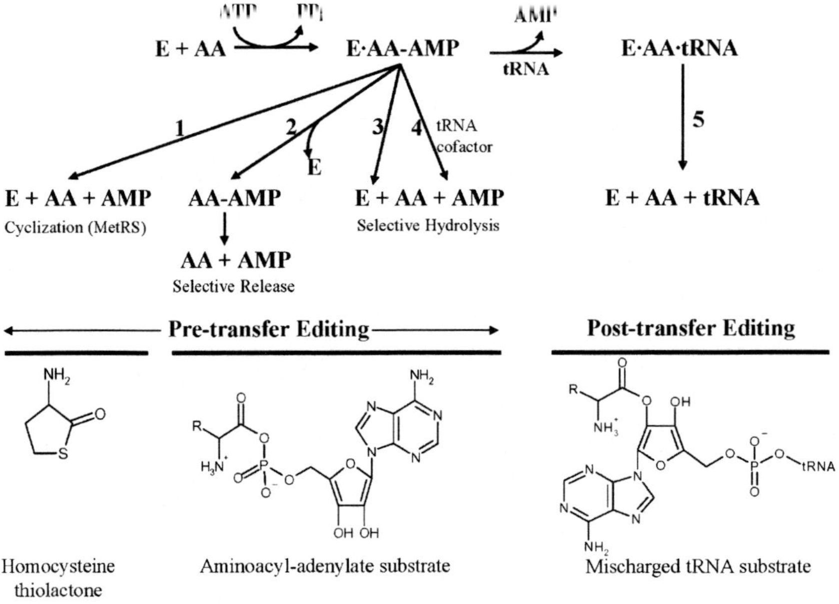

Fig. 4 Proposed editing pathways of the AARSs. Editing of noncognate amino acids can occur after transfer to the tRNA and hydrolyzes the mischarged tRNA (pathway 5) or it can target the adenylate intermediate (*AA-AMP*) in a tRNA-independent (pathways *1–3*) or dependent (pathway *4*) manner. The aminoacyl adenylate intermediate could be hydrolyzed while bound to the enzyme (pathways *3* and *4*) or could be selectively released to the aqueous environment where it is rapidly hydrolyzed (pathway *2*). In methionyl-tRNA synthetase (*MetRS*), misactivated homocysteine cyclizes during editing to form a thiolactone (pathway *1*). Cyclization can also occur for lysyl-tRNA synthetase misactivated amino acids

misactivated adenylate intermediates are hydrolyzed, may occur either in the absence (pathways 1–3 in Fig. 4) or in the presence (pathway 4 in Fig. 4) of a tRNA cofactor (Fersht and Dingwall 1979c). The first tRNA-independent pathway illustrates the case of cyclization of homocysteinyl adenylate by MetRS to form homocysteinyl-thiolactone, followed by expulsion from the active site (Fersht and Dingwall 1979a; Jakubowski and Fersht 1981; Jakubowski 1991). Alternatively, "selective release" of the noncognate adenylate intermediate from the amino acid activation active site would be followed by solution hydrolysis (pathway 2) in the aqueous environment (Hati et al. 2006; Splan et al. 2007). Finally, selective hydrolysis of the adenylate may occur in a tRNA-independent (pathway 3) or a tRNA-dependent (pathway 4) fashion, either in the aminoacylation active site (Gruic-Sovulj et al. 2005) or following selective release and translocation to a separate editing active site (Nomanbhoy et al. 1999; Nomanbhoy and Schimmel 2000; Bishop et al. 2002; Bishop et al. 2003). If a noncognate amino acid is attached to a tRNA, this mischarged tRNA is cleared via post-transfer editing (pathway 5) (Eldred and Schimmel 1972).

X-ray crystallography and biochemical investigations have suggested that the pre- and post-transfer editing sites could overlap (Lincecum et al. 2003; Dock-Bregeon et al. 2004; Fukunaga et al. 2004; Fukunaga and Yokoyama 2005b). In these cases, the common adenine and amino acid moieties of the pre- and post-transfer editing active sites are accommodated within two subsites of the editing active site (Fig. 5). However, as discussed below, it is not clear how a small adenylate molecule could be translocated from the synthetic active site to this remote amino acid editing site for hydrolysis.

Many AARSs have been reported to carry out both pre- and post-transfer editing, although in some cases, one pathway clearly seems to dominate. Interestingly, mutations that can simultaneously abolish both pre- and post-transfer editing activities have been identified (Lincecum et al. 2003), while other mutations can selectively inactivate one mechanism (Hendrickson et al. 2000) and in some cases activate the alternate mechanism (Williams and Martinis 2006). Though all the class I editing enzymes occupy the same subclass Ia, each appears to rely on a different dominant editing pathway. For example, MetRS edits homocysteine via the cyclization reaction described above (Fersht and Dingwall 1979a; Jakubowski and Fersht 1981; Jakubowski 1991). ValRS primarily employs post-transfer editing (Fersht and Kaethner 1976; Jakubowski and Fersht 1981), while IleRS depends mainly on pre-transfer editing (Fersht 1977b). For LeuRS, the *E. coli* enzyme exclusively uses post-transfer editing, whereas *Saccharomyces cerevisiae* cytoplasmic LeuRS primarily relies upon a pre-transfer editing mechanism (Englisch et al. 1986). Though these editing reactions are conserved through evolution, a dramatic shift between mechanisms may be achieved with relative ease (Hendrickson et al. 2002; Williams and Martinis 2006). The redundancy of these editing mechanisms suggests their importance to protein synthesis and fidelity.

Some class II enzymes also appear to use redundant editing mechanisms, although the relative contribution of each pathway toward overall fidelity is not yet clear in all cases. ProRS performs pre-transfer editing of alanyl adenylate in a tRNA-independent reaction that involves primarily enzymatic hydrolysis (Splan et al.

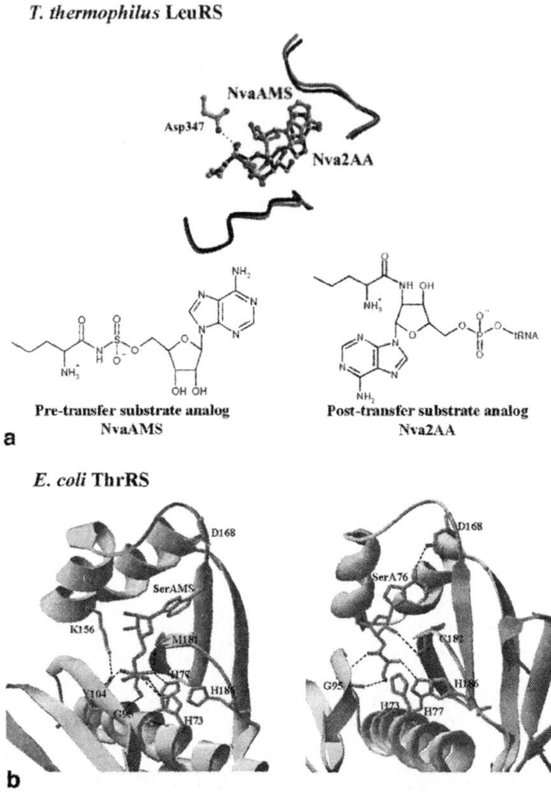

Fig. 5 Editing active sites of *Thermus thermophilus* leucyl-tRNA synthetase (*LeuRS*) and *Escherichia coli* threonyl-tRNA synthetase (*ThrRS*). **a** Overlapping pre-transfer (*light gray*) and post-transfer (*black*) editing active sites of *T. thermophilus* LeuRS with the superimposed pre-transfer analog (*NvaAMS*) and post-transfer (*Nva2AA*) editing analog (Lincecum et al. 2003). The universally conserved aspartate that is shared by the connective polypeptide 1 domains of IleRS, LeuRS and valyl-tRNA synthetase (ValRS) forms a salt bridge with the α-amino group of misactivated amino acid. **b** The pre- and post-transfer analogs bound to the editing active site of *E. coli* ThrRS (Dock-Bregeon et al. 2004). The pre-transfer (5′-*O*-[*N*-(L-seryl)-sulfamoyl]adenosine) and post-transfer analog (seryl-3′-aminoadenosine) form different sets of hydrogen bonds with residues of overlapping active sites

2007), with selective release of the adenylate followed by solution hydrolysis constituting a minor pathway (Hati et al. 2006). ProRS also clears Ala-tRNA^{Pro} via post-transfer editing (Beuning and Musier-Forsyth 2000). PheRS possesses both pre- and post-transfer editing capabilities to correct misactivation of noncognate tyrosine (Lin et al. 1983, 1984; Roy et al. 2004; Kotik-Kogan et al. 2005). Likewise, AlaRS performs pre- and post-transfer editing to hydrolyze misactivated or mischarged glycine and serine (Tsui and Fersht 1981; Beebe et al. 2003a, b). LysRS discriminates homocysteine, homoserine, and ornithine (Jakubowski 1999) using a pre-transfer cyclization mechanism similar to that described for MetRS (Jakubowski 1997). Post-transfer editing is the major pathway for ThrRS editing of Ser-$tRNA^{Thr}$, as this enzyme appears to lack a pre-transfer editing capability (Dock-Bregeon et al. 2000, 2004).

2.4 Translocation of Editing Substrates

Post-transfer editing involves hydrolysis of the aminoacyl-tRNA bond using an active site that is distinct from the site of aminoacylation; thus, conformational changes in both the enzyme and tRNA occur during the translocation process. A cocrystal structure of *T. thermophilus* LeuRS in the post-transfer editing orientation shows that the presence of tRNA causes the CP1 editing domain to rotate by approximately 35° relative to its *apo* state (Tukalo et al. 2005) (Fig. 6). This work also identified an "exit complex" where the $tRNA^{Leu}$ acceptor end is near the CP1 domain, but dissociated from LeuRS (Tukalo et al. 2005). In the aminoacylation complex of *Pyrococcus horikoshii* LeuRS with $tRNA^{Leu}$, the CP1 editing domain swings approximately 20° with respect to the canonical core of the enzyme when compared with the *apo* state. This orientation prevents a clash between the 5′ terminus of tRNA and the CP1 domain (Fukunaga and Yokoyama 2005c). Another LeuRS crystal structure depicts an intermediate state with the tRNA acceptor end halfway between the aminoacylation and editing active sites (Fukunaga and Yokoyama 2005c). More recently, a boron containing inhibitor called AN2690 (5-fluoro-1,3-dihydro-1-hydroxy-2,1-benzoxaborole) that binds in the LeuRS editing active site covalently trapped the uncharged 3′ end of the tRNA in the editing complex (Rock et al. 2007). This suggested that the uncharged 3′ end of the tRNA sweeps through the editing site enroute to binding to the aminoacylation active site.

Comparison of the *apo*-IleRS (Nureki et al. 1998) and the editing complex of IleRS with *E. coli* $tRNA^{Ile}$ (Silvian et al. 1999) also showed two distinct orientations of the editing domain that differed by an angle of 47° (Silvian et al. 1999). In addition, the $tRNA^{Ile}$ acceptor end adopted a stacked helical conformation that would allow insertion of A76 into the editing active site for proofreading and/or hydrolysis (Silvian et al. 1999). A hairpin conformation of the tRNA 3′ end would be required to propel the 3′-terminal adenosine into the aminoacylation active site. Molecular modeling and the cocrystal structures of *T. thermophilus* ValRS and $tRNA^{Val}$ suggested that its editing domain adopts different orientations during editing to shuttle the flexible CCA-3′ end of tRNA from the aminoacylation active site to the editing site (Fukai et al. 2000). A 3′ end shuttling mechanism is also consistent with structural data based on the class II ThrRS system (Dock-Bregeon et al. 2000, 2004). Interestingly, in contrast to the class Ia enzymes, the single-stranded CCA-3′ end of the tRNA changes from a helical to a bent conformation to reach the editing active site.

The mechanism of translocation of aminoacyl adenylates remains much less clear. Some pre-transfer editing reactions take place in the aminoacylation active

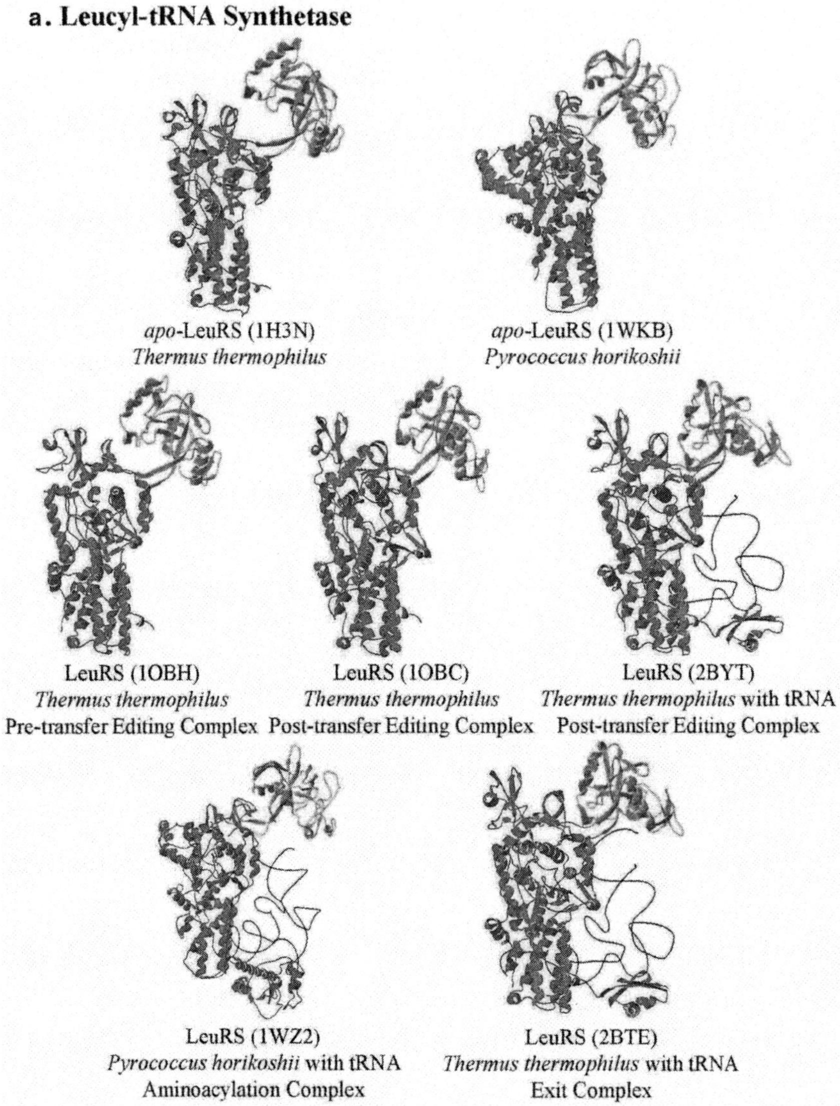

Fig. 6 Crystal structures of class I editing AARSs. **a** Crystal and cocrystal structures of LeuRS (Cusack et al. 2000; Lincecum et al. 2003; Fukunaga and Yokoyama 2005a, c; Tukalo et al. 2005). **b** Crystal and cocrystal structures of IleRS (Nureki et al. 1998; Silvian et al. 1999; Nakama et al. 2001). **c** Cocrystal structures of ValRS with $tRNA^{Val}$ in editing conformation (Fukai et al. 2000, 2003). The main body of the enzymes is shown in *dark gray*, with the inserted editing domain in *light gray*. In the cocrystal structures, cognate tRNA is shown in *black*. Protein Data Bank codes are indicated

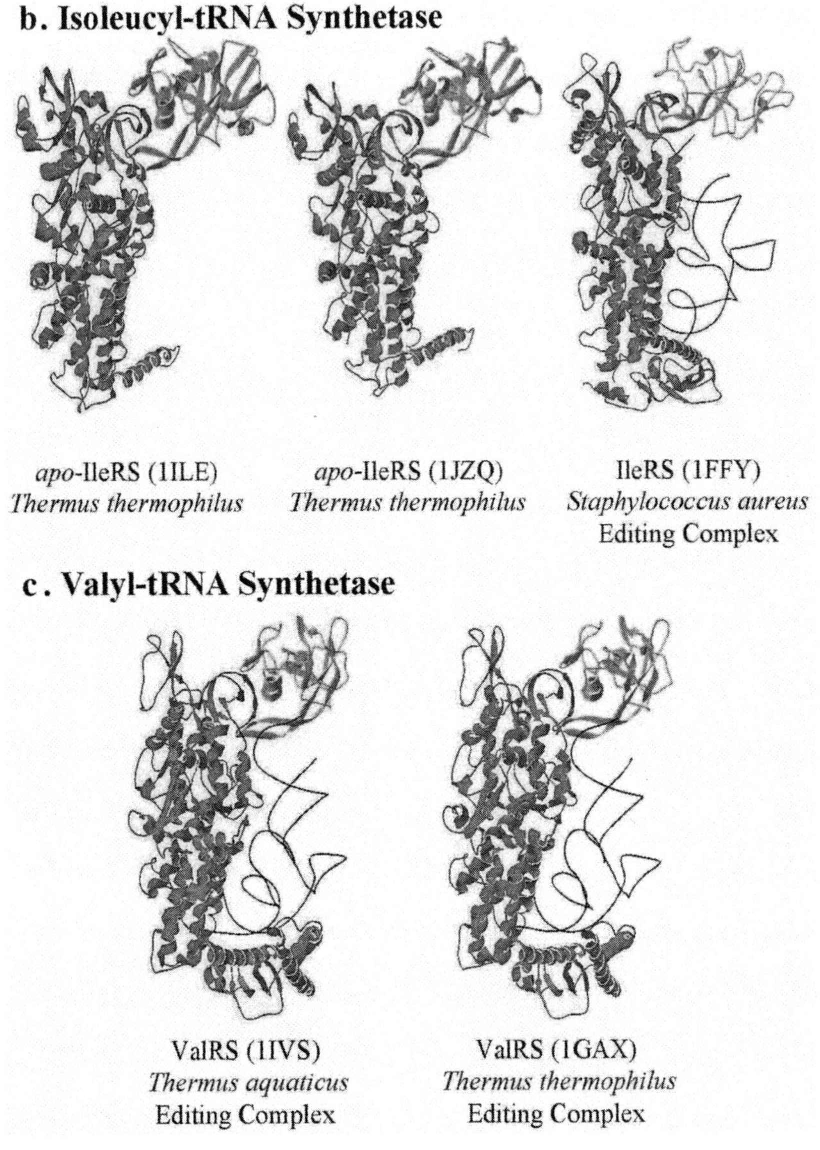

Fig. 6 (Continued)

site, obviating the need for a translocation mechanism (Jakubowski 1993a; Gruic-Sovulj et al. 2005; Perona 2005; Hati et al. 2006). However, fluorescence-based assays and mutational analysis of class I synthetases suggest that misactivated amino acids are translocated from the synthetic site to the editing domain in a tRNA-dependent manner (Nomanbhoy et al. 1999; Nomanbhoy and Schimmel

2000, 2003; Bishop et al. 2003). A fluorescence energy transfer assay based on the $tRNA^{Ile}$-dependent evacuation of fluorescent adenine analogs from the IleRS aminoacylation active site indicated that translocation of noncognate valine, and not hydrolysis, was the rate-limiting step during amino acid editing (Nomanbhoy et al. 1999). Likewise, ValRS translocation rates of misactivated amino acids for editing were similar for threonine, cysteine, and α-aminobutyrate (Nomanbhoy and Schimmel 2000). Interestingly, a single post-transfer editing event has been postulated to initiate conformational changes for translocation of aminoacyl adenylates during pre-transfer editing by IleRS and ValRS (Bishop et al. 2002; Nordin and Schimmel 2003).

Translocation determinants for pre-transfer editing in IleRS and LeuRS have also been identified. For IleRS, mutation of Lys183 and Trp421 within the CP1 domain did not significantly affect aminoacylation and editing activities, but reduced rates of translocation (Bishop et al. 2003). This lysine "hinge" is also conserved in LeuRS and influences translocation of aminoacyl adenylate when the dominant post-transfer editing activity was inactivated in the *E. coli* enzyme (Williams and Martinis 2006). In LeuRS, a flexible peptide at the surface of the CP1 domain is in close proximity to the lysine "hinge" and was proposed to form a precise molecular interface between the CP1 domain and the canonical core of the enzyme that is important to translocation. Mutations within this CP1-based peptide also impacted fidelity, suggesting that a translocation pathway for pre-transfer editing was activated (Williams and Martinis 2006).

3 CP1 Domain Based Editing – IleRS, ValRS, and LeuRS

Among the editing class I enzymes, LeuRS, IleRS, ValRS, and MetRS belong to subclass Ia, and are highly homologous to one another (Eriani et al. 1990; Burbaum and Schimmel 1991; O'Donoghue and Luthey-Schulten 2003). In LeuRS, IleRS, and ValRS, the CP1 domain is inserted into the main body of the enzyme and is connected via two β-strand linkers (Fig. 6). The CP1 domain houses the editing active site, which has diverged to accommodate each enzyme's specificity for amino acid editing (Schmidt and Schimmel 1994, 1995; Lin and Schimmel 1996; Nureki et al. 1998; Silvian et al. 1999; Mursinna and Martinis 2002; Zhai et al. 2007). The related MetRS has a small CP1 domain that is not functional in editing.

The location of the editing active site was first localized by cross-linking experiments, using *N*-bromoacetyl-Val-$tRNA^{Ile}$ (Schmidt and Schimmel 1995). Mutations of conserved residues (H401Q and Y403F) within the CP1 domain of IleRS, considerably decreased the editing activity of the enzyme (Schmidt and Schimmel 1995), but not aminoacylation. During deletion analysis, this domain also appeared to be dispensable for amino acid activation and charging (Starzyk et al. 1987). Conversely, mutations that abolished the aminoacylation ability of the enzyme, but did not affect editing, have also been identified (Schmidt and Schimmel 1994; Lin and Schimmel 1996). Finally, the CP1 domains for IleRS, ValRS, and LeuRS were isolated, independent of the enzyme's canonical aminoacylation core, and actively hydrolyzed their respective mischarged cognate tRNAs (Lin and Schimmel 1996; Chen et al. 2000). More recent data suggest that the connective β-linkers are necessary for viable post-transfer editing activity of the isolated CP1 domain from *E. coli* LeuRS (Betha et al. 2007). Collectively, these and other results provided strong experimental support for the original double sieve model (Fersht 1977a) that proposed exclusive separation of the aminoacylation and editing active site.

Interestingly, the CP1 domain in bacterial and mitochondrial LeuRSs appears to have a novel point of insertion into the catalytic domain of the enzyme. Primary sequence analyses show that the insertion of the CP1 for IleRS, ValRS, as well as archaeal and eukaryotic cytoplasmic LeuRS splits a zinc-binding site (ZN-1), but the bacterial and mitochondrial LeuRSs have the CP1 domain inserted after the ZN-1 domain (Cusack et al. 2000). As a result, the *T. thermophilus* IleRS and LeuRS structures show different orientations of the CP1 domains, with the LeuRS requiring approximately 180° difference in rotation to access the 3′ end of tRNA in its editing active site. Another polypeptide insert called the "leucine-specific domain" is unique to LeuRS. However, although this domain appears to be important for aminoacylation, it has little impact on editing (Vu and Martinis 2007).

A C-terminal domain with shared structural homology is also common in IleRS and ValRS as well as archaeal and eukaryal LeuRS. Bacterial LeuRSs have a C-terminal domain that is structurally unrelated. This C-terminal domain is necessary for both aminoacylation and amino acid editing in *T. thermophilus* IleRS and ValRS (Fukunaga and Yokoyama 2007) as well as *E. coli* LeuRS (Hsu et al. 2006). Conversely, the C-terminal truncated *P. horikoshii* LeuRS edits Ile-$tRNA^{Leu}$ efficiently (Fukunaga and Yokoyama 2005c) but also hydrolyzes Ile-$tRNA^{Ile}$ that is correctly formed by IleRS (Fukunaga and Yokoyama 2007). This suggests that the presence of the C-terminal domain *P. horikoshii* LeuRS is necessary to prevent misediting of Ile-$tRNA^{Ile}$. Surprisingly, deletion of this domain significantly enhanced both these activities in yeast mitochondrial LeuRS (Hsu et al. 2006). This likely reflects evolutionary compensation for its dual role as a splicing factor.

These enzymes also differ in whether they hydrolyze amino acids from the 2′ or 3′ hydroxyl of the A76 ribose of tRNA. In general, class I AARSs charge amino acids to the 2′ hydroxyl of tRNA. However, IleRS hydrolyzes the amino acid from the 3′ ribose hydroxyl, which suggests that the acylated amino acid migrates during translocation prior to editing (Nordin and Schimmel 2002). This mechanism is enzyme-specific as ValRS (Nordin and Schimmel 2002) and LeuRS (Lincecum et al. 2003) aminoacylate and edit at the 2′ hydroxyl.

Mutational and structural analysis of the class Ia editing enzymes have localized subsites of the editing active site that are responsible for interacting with the adenine, amino acid side chain, and central region of the editing substrate that is targeted for hydrolysis. In the central region, structures for LeuRS, IleRS, and ValRS show that a universally conserved aspartic acid residue forms a salt bridge with the amino group of the noncognate amino acid for both pre- and post-transfer editing substrates (Lincecum et al. 2003; Dock-Bregeon et al. 2004; Fukunaga et al. 2004; Fukunaga and Yokoyama 2005b) (Fig. 5). Mutation of this conserved aspartic acid abolishes amino acid editing in IleRS (Bishop et al. 2002), ValRS (Fukunaga and Yokoyama 2005b), and LeuRS (Lincecum et al. 2003).

In LeuRS, two neighboring conserved threonines, Thr247 and Thr248, within the editing active site also interact near the hydrolysis site (Lincecum et al. 2003). The hydroxyl group of Thr247 forms a hydrogen bond with both the carboxyl group of the bound amino acid and the 3′ hydroxyl group of the ribose ring. The amide and hydroxyl group of Thr248 form bonds with the 3′ hydroxyl group of the ribose ring. However, when each of these residues was mutated to alanine, the editing function was only slightly affected (Mursinna et al. 2004). A double mutation of both threonines to isosteric valines or alanines abolished editing activity, whereas substitution with serine maintained amino acid editing activity. It was proposed that these LeuRS threonines, which are also found in IleRS and ValRS, may be part of a hydrogen-bonding network that stabilizes the editing transition state (Zhai and Martinis 2005). Similar results with residues Thr228 and Thr230 in *T. thermophilus* IleRS (Fig. 7), which recognize the carboxyl group of valine, indicate their importance in post-transfer editing (Fukunaga et al. 2004).

This threonine-rich region that is conserved in IleRS, ValRS, and LeuRS also interacts with the bound amino acid and confers specificity to amino acid editing. Biochemical, X-ray crystallography and computational studies for LeuRS identified a conserved Thr252 in *E. coli* (Fig. 7) as a critical discriminator to block leucine at the editing active site (Mursinna et al. 2001, 2004; Lincecum et al. 2003; Zhai et al. 2007). The side chain of this residue lies at the bottom of the editing pocket and blocks the γ-methyl branch of leucine (Lincecum et al. 2003). Mutation to alanine or serine (Mursinna et al. 2001, 2004) uncouples specificity and hydrolyzes Leu-$tRNA^{Leu}$. Mutation of Thr252 to bulky residues fills in the amino acid binding pocket and blocks the site for editing (Mursinna and Martinis 2002; Tang and Tirrell 2002). Significantly, mutations at this site have already been capitalized upon to block leucine to incorporate novel amino acids during in vivo protein synthesis (Tang and Tirrell 2002; Turner et al. 2006).

Interestingly, mutational data indicate that the corresponding Thr233 in *T. thermophilus* IleRS does not take part in post-transfer editing (Hendrickson et al. 2002). However, mutation of a nearby Thr230 in the threonine-rich region uncouples specificity and causes hydrolysis of the correctly charged substrate (Fukunaga and Yokoyama 2006). Likewise, an H319A mutation in *E. coli* IleRS results in editing of correctly charged Ile-$tRNA^{Ile}$ (Hendrickson et al. 2002). X-ray structures show that the extra methyl group of isoleucine clashes with the His319 side chain, blocking the cognate amino acid from binding (Fukunaga et al. 2004). In *T. thermophilus* ValRS, Thr219, which is equivalent to the Thr252 specificity site in LeuRS, is part

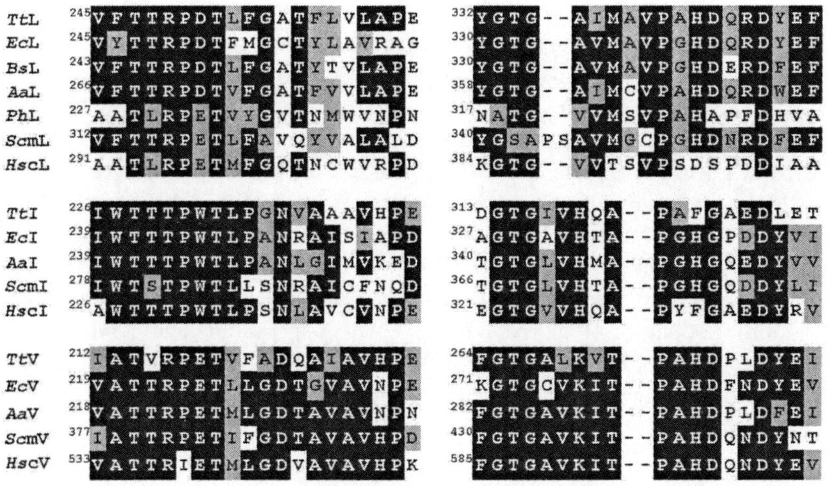

Fig. 7 Sequence alignment of the editing active site of LeuRS, IleRS, and ValRS. The most conserved residues within a particular synthetase are highlighted in *black* and homologous residues are shown in *gray*. *L*, LeuRS, *I*, IleRS; *V*, ValRS; *Tt*, *T. thermophilus; Ec, E. coli; Bs, Bacillus subtilus; Aa, Aquifex aeolicus; Ph, Pyrococcus horikoshii; Scm, Saccharomyces cerevisiae* (mitochondria); *Hsc, Homo sapiens* (cytoplasmic)

of the hydrogen-bonding network surrounding the editing substrate, but appears to play an indirect role (Fukunaga and Yokoyama 2005b). However, mutation of a highly conserved Lys277 within the editing site of *E. coli* ValRS results in deacylation of the cognate Val-tRNA^{Val} (Hountondji et al. 2002). This is likely because the amino acid binding pocket has been filled in to accommodate smaller amino acids as targets for hydrolysis.

Adenines of the aminoacyl adenylate and terminal tRNA residue of the editing substrates bind to a highly conserved GTG motif (Lincecum et al. 2003; Fukunaga and Yokoyama 2005b). In ValRS, this flexible loop undergoes conformational changes to accommodate differences in binding pre- and post-transfer editing substrates (Fukunaga and Yokoyama 2005b). A semiconserved tyrosine in *E. coli* LeuRS (Tyr330) has been proposed to position the terminal A76 of tRNA at the editing active site during this translocation process (Liu et al. 2006). In *T. thermophilus* ValRS, the homologous Phe264 undergoes a conformational change in the presence of tRNA^{Val} and its mutation to alanine affects editing activity. In addition, a nearby conserved Asp276 side chain forms a hydrogen bond with the 3′ hydroxyl group of A76 of tRNA^{Val} (Fukunaga and Yokoyama 2005b).

The CP1 domains from ValRS (Fukunaga and Yokoyama 2005b), LeuRS (Lincecum et al. 2003; Tukalo et al. 2005; Liu et al. 2006), and IleRS (Fukunaga and Yokoyama 2006) have evolved different strategies for recognition of their respective editing substrates. In *T. thermophilus* ValRS, a cleft formed by the side chains of Arg216, Thr219, Lys270, Thr272, and Asp279 positions threonine for binding (Fukunaga and Yokoyama 2005b) in the editing active site. The side chain hydroxyl of the threonine substrate is recognized by the side chains of Lys270, Thr272, and Asp279 to confer specificity. Similarly, in *E. coli* LeuRS, the side chains of noncognate methionine and isoleucine are accommodated via formation of a hydrogen-bond network with highly conserved Thr247, Met336, and Asp345 (Liu et al. 2006; Zhai et al. 2007). In *T. thermophilus* IleRS, the Trp227 and Thr230 side chains have been proposed to participate in substrate recognition during both pre- and post-transfer editing (Fukunaga and Yokoyama 2006). For both IleRS editing substrates (Val-AMP and Val-$tRNA^{Ile}$), valine-specific recognition is conferred by His319 and Thr233. Interestingly, the valine side chains are positioned in distinct orientations in the two different complexes.

3.1 tRNA Determinants for Editing

In ValRS and LeuRS, tRNA identity determinants for aminoacylation also impact editing activity. Mutational analysis of $tRNA^{Val}$ or transplant of its identity elements into the framework of heterologous tRNAs showed that there is a close correlation between aminoacylation with valine and stimulation of editing activity (Tardif and Horowitz 2002). Investigations of $tRNA^{Leu}$ deletion constructs determined that the anticodon stem loop and variable loop were dispensable for aminoacylation and overall editing by *E. coli* LeuRS and suggested that the D-loop was critical in tRNA identity (Larkin et al. 2002). Substitution and deletion mutations at U17, G18, G19, U54, U55, C56, and G59 within full-length *E. coli* $tRNA^{Leu}$ transcripts supported the conclusion that D/TψC-loop tertiary interaction affected both aminoacylation and editing (Du and Wang 2003).

In contrast, in the IleRS system, the $tRNA^{Ile}$ nucleotide determinants that are required for aminoacylation are distinct from those that confer overall editing, which is comprised of pre- and post-transfer editing activities (Hale et al. 1997). Chimeric $tRNA^{Ile}$/$tRNA^{Val}$ molecules determined that three D-loop residues G16, D20, and D21 were essential to overall editing activity (Hale et al. 1997), although surprisingly, they do not affect the chemical step of deacylation (Farrow et al. 1999). The lack of stimulation of editing by a minihelix derived from the acceptor stem of $tRNA^{Ile}$ is also consistent with the importance of specific D-loop nucleotides in the overall editing reaction (Nordin and Schimmel 1999). It was proposed that these D-loop nucleotides may induce a conformational change during translocation of the mischarged tRNA product from the aminoacylation to the editing active site (Farrow et al. 1999). These data led to a model that misacylated tRNA is a crucial intermediate in overall editing, and that pre-transfer editing is initiated only after one post-transfer editing event (Bishop et al. 2002).

Pre-transfer editing is stimulated by the addition of tRNA in a number of class I and II AARS systems. In the case of class I IleRS and ValRS, it has been suggested that cognate tRNA triggers translocation of the misactivated amino acid from the aminoacylation active site to the editing active site (Lin and Schimmel 1996; Nomanbhoy et al. 1999). Interestingly, a 61-nucleotide DNA aptamer was selected that could stimulate ATP hydrolysis in the presence of IleRS and noncognate valine, but not isoleucine (Hale and Schimmel 1996); thus, the DNA aptamer appears to mimic the tRNA and this suggests that an acceptor hydroxyl group is not completely necessary for stimulation of pre-transfer editing.

Early work showed that alterations of the $tRNA^{Ile}$ 3′ CCA end that destroyed isoleucine acceptor activity also eliminated editing of Val-AMP by IleRS (Baldwin and Berg 1966). More extensive analysis of 3′ end variants determined that the only tRNAs that could stimulate ATPase activity in IleRS were those that were active in post-transfer editing (Nordin and Schimmel 2003). Likewise, mutation of the 3′-terminal A76 in $tRNA^{Val}$ to C or U produces tRNA molecules that can be aminoacylated with valine (Tamura et al. 1994; Tardif et al. 2001), but are defective in post-transfer editing, as well as in stimulation of ATP hydrolysis (Tardif et al. 2001). Thus, in both ValRS and IleRS, mutations that differentially affect aminoacylation and editing can be isolated and post-transfer editing appears to be a requirement for overall editing. This result is also consistent with the cocrystal of *T. thermophilus* ValRS with $tRNA^{Val}$ that identifies specific interactions between functional groups on A76 and the editing active site (Fukai et al. 2000).

The initial site of aminoacylation for most class I enzymes is the 2′ hydroxyl of A76 (Cramer et al. 1975). A catalytic role in editing was originally proposed for the 3′ hydroxyl of misacylated Val-$tRNA^{Ile}$, which was hypothesized to assist in activation of a catalytic water molecule (von der Haar and Cramer 1976; Freist and Cramer 1983). More recent studies using a variety of 3′ end modified substrates showed that IleRS deacylates a mischarged 2′-dA76 substrate, but cannot hydrolyze Val-3′-dA76 $tRNA^{Ile}$ (Nordin and Schimmel 1999). Thus, under normal conditions, transacylation from the 2′ to the 3′ hydroxyl appears to be required for deacylation of Val-$tRNA^{Ile}$ by IleRS. In contrast, ValRS deacylates mischarged 3′-dA76 $tRNA^{Val}$, albeit with a tenfold reduced rate compared with Thr-$tRNA^{Val}$.

More limited studies of tRNA determinants for editing have been performed for class II synthetases. Efficient editing by class II AlaRS is dependent on a covalently continuous two-domain tRNA structure rather than specific sequence elements. Whereas a minihelix derived from the acceptor stem of *E. coli* $tRNA^{Ala}$ failed to stimulate editing, chimeric tRNAs with sequence alterations in the D, anticodon, and TψC stem loops were efficiently deacylated by *E. coli* AlaRS (Beebe et al. 2003a). In the *E. coli* PheRS system, 3′ end modified tRNAs were tested for stimulation of tyrosine-dependent ATPase activity (Roy et al. 2004). Only 3′-dA $tRNA^{Phe}$ retained phenylalanylation activity, but was unable to stimulate editing. Additional 3′ end substitutions revealed an essential function for the 3′ hydroxyl group in hydrolysis by PheRS (Ling et al. 2007). Although the exact role is still not clear, this group is hypothesized to function as a hydrogen-bond donor, helping with water molecule positioning or with substrate binding.

4 Class II Synthetase Editing Mechanisms

4.1 Triple Sieves and Freestanding Editing Domains

As evidence accumulated for proofreading by class I synthetases, more detailed investigations into similar editing functions in class II AARSs were carried out. Amino acid editing within this class was first observed with PheRS, which can deacylate Ile-tRNA^{Phe} (Yarus 1973). Similar activities were reported with AlaRS, which targets glycyl adenylate (Tsui and Fersht 1981), ThrRS, which hydrolyzes serine via post-transfer editing (Dock-Bregeon et al. 2000), and ProRS, which clears alanine through both pre- and post-transfer editing pathways (Beuning and Musier-Forsyth 2000). In accordance with Fersht's original double sieve hypothesis for the class I IleRS and ValRS, a similar paradigm with well-separated active sites exists for class II AARSs. However, in contrast to class I enzymes, which rely primarily on the CP1 domain for editing, structurally dissimilar domains have been implicated for editing by class II enzymes.

Mutagenesis analyses localized the ThrRS editing active site to a cleft in an N-terminal domain (N2) (Dock-Bregeon et al. 2000). In bacteria and eukaryotes, N2 is highly conserved, but a distinct N-terminal editing domain that performs the analogous function is present in the majority of archaeal ThrRSs (Beebe et al. 2004; Korencic et al. 2004). The editing domain of AlaRS is weakly related to the ThrRS N2 domain, but is located internally (Dock-Bregeon et al. 2000; Beebe et al. 2003b). A novel insertion domain (INS) confers editing to prokaryotic ProRS (Wong et al. 2002), and a fourth distinct editing domain is present in PheRS (Roy and Ibba 2006). Freestanding editing proteins that are homologs of editing domains found in ProRS (Wong et al. 2003), AlaRS (Ahel et al. 2003), and ThrRS (Korencic et al. 2004) have been discovered. These may have evolved to further enhance the accuracy of protein synthesis (Pezo et al. 2004), or to carry out other, still unknown, cellular functions (Geslain and Ribas de Pouplana 2004).

4.2 Prolyl-tRNA Synthetase

Most bacterial ProRS enzymes possess a post-transfer editing activity to distinguish proline from alanine (Beuning and Musier-Forsyth 2000; Ahel et al. 2002). A unique domain inserted between motifs 2 and 3 of bacterial ProRS (INS) is responsible for hydrolyzing Ala-tRNA^{Pro} (Wong et al. 2002, 2003). Lower eukaryotic ProRSs, possess a weakly homologous INS-like editing domain appended to the N-terminus. Whereas *Plasmodium falciparum* ProRS was shown to catalyze editing of *E. coli* Ala-tRNA^{Pro}, *S. cerevisiae* ProRS lacked this activity (Ahel et al. 2003). The *S. cerevisiae* enzyme also fails to deacylate yeast Ala-tRNA^{Pro} (Sternjohn et al. 2007).

Alanine-scanning mutagenesis of highly conserved residues in the INS domain of *E. coli* ProRS identified key residues required for post-transfer editing (Wong et al. 2002). Mutation to alanine of a highly conserved lysine residue, Lys279, selectively decreased post-transfer editing activity by 95% without affecting aminoacylation with proline. As expected, K279A ProRS mischarged alanine onto tRNA^{Pro}. An H369A mutation altered editing substrate specificity. With this variant, deacylation of Ala-tRNA^{Pro} was decreased by 80% with concomitant increase of Pro-tRNA^{Pro} hydrolysis (Wong et al. 2002).

The INS domain was found to be dispensable for pre-transfer editing by *E. coli* ProRS. A ΔINS-ProRS construct with an optimized 16-residue glycine/serine linker displayed approximately 1,200-fold reduced proline activation efficiency and failed to activate alanine, but aminoacylated tRNA^{Pro} with an overall k_{cat}/K_M that was comparable with that of wild-type ProRS (Hati et al. 2006). Although the ΔINS construct did not exhibit pre-transfer editing activity against alanine, this was not unexpected owing to the lack of alanyl adenylate formation. In contrast, the ΔINS variant did exhibit significant pre-transfer editing of substrates that were reasonably well activated, such as *cis*-4-hydroxyproline and even proline (Hati et al. 2006). Studies also showed that a separate editing domain is not required for pre-transfer editing by ProRS. Enzymatic hydrolysis of the noncognate adenylate in the aminoacylation active site is the major pre-transfer editing pathway, with selective release of the noncognate adenylate into solution contributing a minor component in the case of some ProRS species (Hati et al. 2006).

ProRSs from all kingdoms of life also misactivate and mischarge noncognate cysteine with catalytic efficiencies that are comparable to that for cognate proline in vitro (Beuning and Musier-Forsyth 2001; Ahel et al. 2002); however, these enzymes lack a cysteine-specific editing activity (Beuning and Musier-Forsyth 2001; Ahel et al. 2002; An and Musier-Forsyth 2004). The discovery of a single-domain protein (*Haemophilus influenzae* YbaK) with significant sequence homology to the INS editing domain (Wolf et al. 1999; Zhang et al. 2000) solved the mystery of how cells clear Cys-tRNA^{Pro}. The *H. influenzae* YbaK was initially shown to display weak hydrolysis of Ala-tRNA^{Pro} in *trans* (Wolf et al. 1999; Zhang et al. 2000; Wong et al. 2003), but was later demonstrated to efficiently deacylate Cys-tRNA^{Pro} (An and Musier-Forsyth 2004; Ruan and Söll 2005). Mischarged Cys-tRNA^{Pro} was also shown to form in *E. coli* cells and to be hydrolyzed by the YbaK protein in vivo (Ruan and Söll 2005). Taken together, these data support a "triple sieve" mechanism of proofreading in the ProRS system (Fig. 3b). Interestingly, an INS editing domain paralog, ProX from the bacterium *Clostridium sticklandii*, was reported to efficiently hydrolyze Ala-tRNA^{Pro}, but not Cys-tRNA^{Pro} (Ahel et al. 2003). The ProRS from this species lacks an INS domain. Thus, there exist at least two subclasses of freestanding INS-like proteins with different specificities.

The crystal structure of *H. influenzae* YbaK (Zhang et al. 2000) shows that the protein contains a highly curved seven-stranded b-sheet surrounded by six short a-helices. An oxyanion hole, similar to that of a serine protease, and a strictly conserved lysine residue (Lys46) are located in a putative ligand binding pocket. This residue is essential for post-transfer editing by *H. influenzae* YbaK (An and Musier-Forsyth 2004), consistent with the critical role of the equivalent Lys279 residue in *E. coli* ProRS (Wong et al. 2002). The first structure of a ProRS with an intact INS domain from *Enterococcus faecalis* has recently been solved (Crépin et al. 2006) and shows close structural similarity to *H. influenzae* YbaK (Fig. 8a). Structural genomics studies have identified additional proteins with structural homology to the YbaK protein. These include single domain proteins from *Caulobacter crescentus* (1VJF), *Agrobacterium tumefaciens* (1VKI), and *Aeropyrum pernix* (1WDV) (Murayama et al. 2005).

How is tRNA recognition achieved by freestanding editing modules that lack tRNA-specific elements such as anticodon binding domains, and how do these small proteins compete for binding with EF-Tu? Answers to these questions are also beginning to emerge. Studies with *H. influenzae* YbaK showed that binding to uncharged tRNA is relatively weak ($K_d > 5\mu M$) and nonspecific in the absence of ProRS (S. An and K. Musier-Forysth, unpublished data), and that YbaK preferentially deacylates any Cys-tRNA sequence, including Cys-$tRNA^{Cys}$, independent of specific acceptor stem sequence elements (An and Musier-Forsyth 2005; Ruan and Söll 2005). However, in the presence of ProRS, specific deacylation of Cys-$tRNA^{Pro}$ is observed. YbaK does not deacylate Cys-$tRNA^{Cys}$ when CysRS is included, and in fact stimulates cognate charging activity (An and Musier-Forsyth 2005). In the presence of EF-Tu, deacylation activity is abolished, suggesting that YbaK functions at a step prior to tRNA release from the synthetase. Binding and cross-linking studies confirmed that ProRS·YbaK·tRNA associate closely and specifically. Thus, specific *trans*-editing of Cys-$tRNA^{Pro}$ by *H. influenzae* YbaK appears to be ensured through the formation of a novel ternary complex with ProRS (An and Musier-Forsyth 2005).

4.3 Threonyl-tRNA Synthetase

An N-terminal domain (N2) is responsible for editing in bacterial and eukaryotic ThrRSs (Dock-Bregeon et al. 2000) and shares weak sequence homology with an internal editing domain of AlaRS (Sankaranarayanan et al. 1999; Dock-Bregeon et al. 2000; Beebe et al. 2003b). Crystal structures of *E. coli* ThrRS obtained in the absence and presence of cognate tRNA showed that a zinc ion in the aminoacylation active site interacts with the hydroxyl group of threonine and serine to effectively discriminate against valine during amino acid selection (Sankaranarayanan et al. 1999, 2000). Thus, hydrolytic editing activity is required to clear just serine, which has been observed via a post-transfer editing mechanism (Dock-Bregeon et al. 2000, 2004).

High-resolution crystal structures of the *E. coli* ThrRS editing domain have been solved in the absence and presence of serine and pre- and post-transfer editing

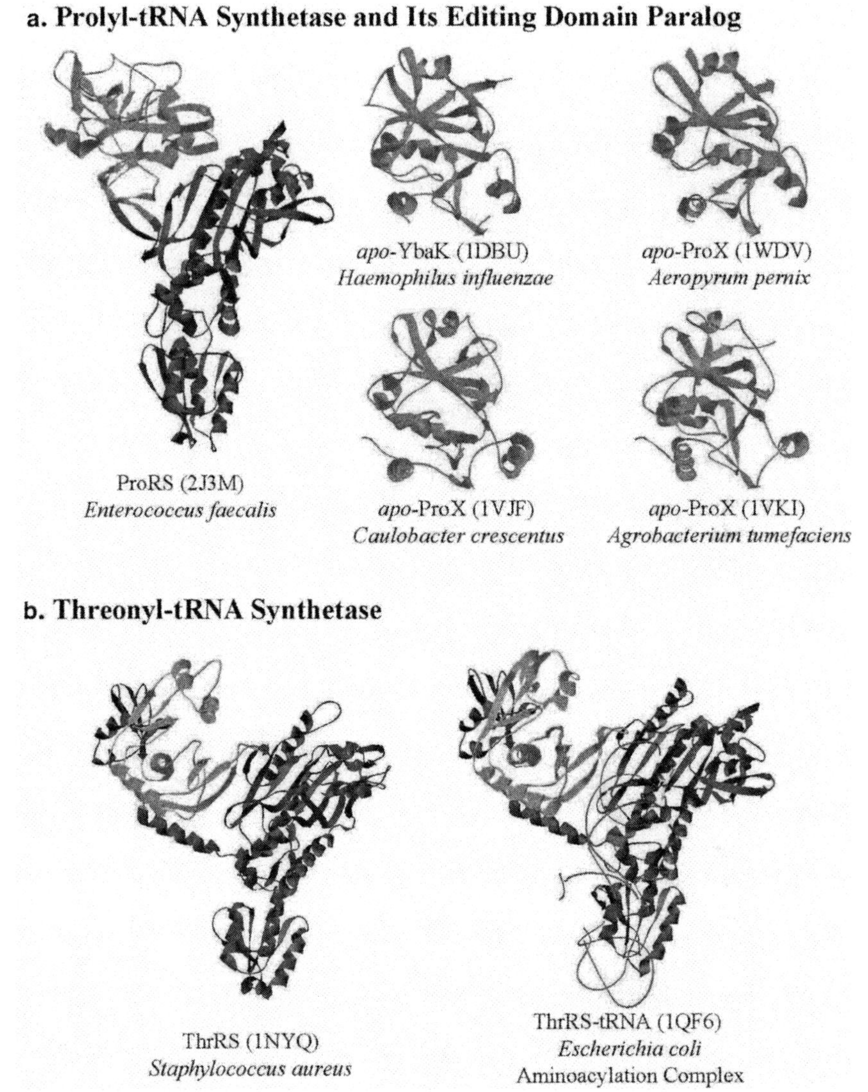

Fig. 8 Crystal structures of class II editing synthetases. **a** Crystal structure of ProRS (Crépin et al. 2006) and its editing domain paralogs (Zhang et al. 2000; Murayama et al. 2005) (1VJF, 1VKI; unpublished data). **b** Crystal structures of ThrRS (Sankaranarayanan et al. 1999; Torres-Larios et al. 2003). **c** Crystal structure of alanyl-tRNA synthetase (*AlaRS*) (Swairjo et al. 2004) and its editing domain paralog AlaX (Sokabe et al. 2005). **d** Crystal structures of phenylalanyl-tRNA synthetase (*PheRS*) (Mosyak et al. 1995; Goldgur et al. 1997; Sasaki et al. 2006). The main body of the enzymes is shown in *dark gray*, with the inserted editing domain in *light gray*. In the cocrystal structures (e.g., ThrRS–tRNA), cognate tRNA is shown in backbone presentation. Protein Data Bank codes are indicated

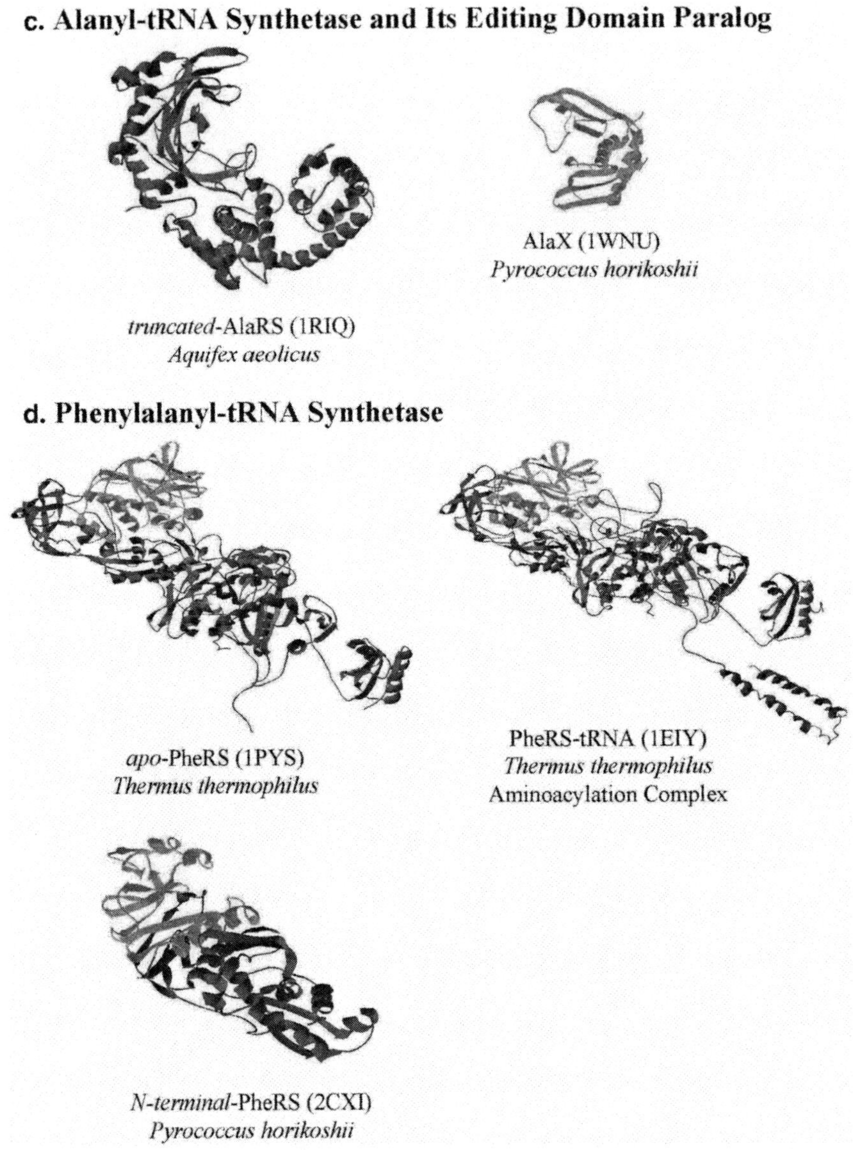

Fig. 8 (Continued)

substrate analogs (5′-*O*-[*N*-(L-seryl)-sulfamoyl]adenosine and seryl-3′-aminoadenosine) (Fig. 5b). Comparison of this structural information led to a proposed hydrolytic mechanism that involved two catalytic water molecules (W1 and W2). His73, one of the two histidines of the conserved HxxxH motif, stabilizes W1 and is hypothesized to nucleophilically attack the carboxyl carbon of the seryl group. The second water molecule, W2, was proposed to donate a proton to O3′ of A76, to cleave the C-O3′ (i.e., aminoacyl ester) bond for post-transfer editing (Dock-Bregeon et al. 2004). Three residues, His77, Tyr104, and Asp180, block cognate threonine from binding and mediate hydrolysis via the catalytic water molecules. Interestingly, the 5′-*O*-[*N*-(L-seryl)-sulfamoyl]adenosine phosphate group replaces a key catalytic water molecule upon binding, providing a structural basis for the lack of pre-transfer editing of seryl adenylate.

In contrast to the N2 domain of bacterial and eukaryotic ThrRSs, most archaeal ThrRS enzymes possess a distinct N-terminal editing domain that was also demonstrated to clear Ser-$tRNA^{Thr}$ (Beebe et al. 2004; Korencic et al. 2004). Freestanding modules that resemble the editing domain of archaeal ThrRSs were also identified in several species (Beebe et al. 2004; Korencic et al. 2004). The *Sulfolobus solfataricus* editing domain, which is fused to a C-terminal anticodon binding domain (ThrRS-ed), hydrolyzes Ser-$tRNA^{Thr}$ in *trans* (Korencic et al. 2004). This crenarchaeal species also has an additional *thrS*-related gene that encodes a bacterial-type catalytic domain fused to an anticodon binding domain (ThrRS-cat). Not surprisingly, the latter protein exhibited aminoacylation activity, but lacked editing activity.

The X-ray crystal structure of the N-terminal editing domain of archaeal *Pyrococcus abyssi* ThrRS (Pab-NTD) has significant structural homology with *E. coli* D-Tyr-$tRNA^{Tyr}$ deacylase (DTD), even though they only share 14% sequence identity (Dwivedi et al. 2005). Notably, although Pab-NTD binds D-serine and D-threonine, it only interacts with L-serine amongst the corresponding L-amino acids that were tested (Dwivedi et al. 2005). Recently, insights into the enantioselectivity of Pab-NTD were elucidated via a combination of site-directed mutagenesis and X-ray crystallography (Hussain et al. 2006). Mutation to methionine of a single invariant lysine residue in Pab-NTD decreased L-serine binding, but retained the D-amino acid binding affinity. Conversely, mutation of an invariant Met129 residue to lysine in DTD facilitated L-amino acid binding. Thus, a single residue plays a critical role in the inversion of enantioselectivity in both enzymes.

4.4 Alanyl-tRNA Synthetase

AlaRS clears mischarged serine and glycine through a post-transfer editing pathway (Tsui and Fersht 1981; Beebe et al. 2003b). As indicated earlier, a C-terminal editing domain of AlaRS is weakly homologous to the N2-terminal editing domain of bacterial and eukaryotic ThrRS (Beebe et al. 2003b, 2004). Although crystal structures of an active fragment of *Aquifex aeolicus* AlaRS have been solved in the presence of bound amino acids, the editing active site was not present in this truncated domain (Swairjo and Schimmel 2005). However, the cocrystal structure with serine bound in the aminoacylation active site showed significant conformational changes compared with the structures of complexes with alanine or glycine. In particular, in the alanine- and glycine-bound AlaRS structures, the side chain amino group of Asn194 interacts with the carboxylate group of the bound amino acids. In contrast, when serine binds, a conformational change occurs to accommodate the larger side chain. In addition, the amide of the Asn194 side chain forms a hydrogen bond with the hydroxyl group of the bound serine. These results suggested a mechanism whereby serine can bypass steric exclusion of the AlaRS active site (Swairjo and Schimmel 2005).

A freestanding module called AlaX was also identified, and shares weak homology to the editing domains of AlaRS. AlaX from *Methanosarcina barkeri* and *S. solfataricus* deacylates mischarged Ser-tRNA^{Ala} and Gly-tRNA^{Ala}, but not Ser-tRNA^{Thr} or correctly charged Ala-tRNA^{Ala} and Ser-tRNA^{Ser}, demonstrating that AlaX specifically edits in *trans* (Ahel et al. 2003).

A crystal structure of the *P. horikoshii* AlaX protein was solved in the absence and in the presence of an editing substrate, suggesting a mechanism for serine and alanine discrimination (Sokabe et al. 2005). The side chain hydroxyl group of serine is recognized by a hydrogen-bonding network that is composed of Thr30, Asp20, and a water molecule. The hydroxyl group of Thr30 is located at the entrance of the hydrophilic pocket and near the γ-methylene group of the bound serine. The hydrogen-bond interactions between Thr30 and the bound serine likely facilitate binding of the noncognate amino acid for editing, whereas the γ-methyl group of alanine would be expected to clash with Thr30 in binding. Structural analyses, along with mutagenesis and sequence alignments, suggest that the serine discrimination model that is dependent on Thr30 is comparable to the AlaRS editing mechanism (Sokabe et al. 2005).

4.5 Phenylalanyl-tRNA Synthetase

E. coli PheRS possesses both pre- and post-transfer editing. The latter has been localized to the B3/B4 domain of the β-subunit of the enzyme (Roy et al. 2004). Crystal structures of *T. thermophilus* PheRS showed that this unique hydrolytic domain is about 35–40 Å away from the aminoacylation active site that resides in the β-subunit (Kotik-Kogan et al. 2005). In addition, since recent evidence demonstrates that neither EF-Tu nor the ribosome discriminates between Tyr-tRNA^{Phe} and Phe-tRNA^{Phe}, PheRS appears to provide the sole proofreading mechanism that is necessary to maintain translational accuracy (Ling et al. 2007). However, the PheRS editing domain lacks structural features reminiscent of other known synthetase editing domains.

Although PheRS mischarges several noncognate amino acids, only tyrosine is a substrate for post-transfer editing (Roy et al. 2005). X-ray crystal structures of *T. thermophilus* PheRS with tyrosine, Tyr-AMS (tyrosyl adenylate analog), and *p*-chlorophenylalanine revealed that the substrate binds at the interface between the B3 and the B4 domains (Kotik-Kogan et al. 2005). Molecular modeling studies suggest an analogous binding mode in the case of *E. coli* PheRS (Roy et al. 2004). Thus, specific discrimination of tyrosine versus phenylalanine occurs by anchoring the *p*-hydroxyl group of tyrosine at the B3/B4 domain interface within the β-subunit via a hydrogen bond with a conserved glutamic acid (Roy et al. 2004; Kotik-Kogan et al. 2005; Ling et al. 2007). A nucleophilic attack of water is substrate assisted by the free 3′ hydroxyl group of the terminal ribose (Ling et al. 2007).

A recent crystal structure of an N-terminal fragment of the archaeal/eukaryal-type *P. horikoshii* PheRS β-subunit revealed a different orientation for the editing domain relative to that observed in the bacterial system (Sasaki et al. 2006). Structure-based alignment studies also suggested that residues postulated to be critical for *T. thermophilus* PheRS editing (Kotik-Kogan et al. 2005) are not conserved in the archaeal/eukaryal-type synthetase (Sasaki et al. 2006). Alanine-scanning mutagenesis of the archaeal enzyme identified an asparagine, which is important for catalysis, as well as key residues that facilitated *p*-hydroxyl group recognition of tyrosine (Sasaki et al. 2006).

Interestingly, while yeast cytosolic PheRS possesses editing activity for Tyr-$tRNA^{Phe}$ (Igloi et al. 1978; Lin et al. 1984; Roy et al. 2005), the yeast mitochondrial enzyme appears to have lost this function (Roy et al. 2005). Unlike for other class II editing synthetases, however, efforts to identify a putative *trans*-editing module in the yeast PheRS system were not successful (Roy et al. 2005).

5 Single Site Editing and Cyclization

5.1 *Methionyl-tRNA Synthetase*

MetRS is highly homologous to class Ia LeuRS, IleRS, and ValRS, but possesses a much smaller CP1 domain that does not have an editing function. Rather, MetRS edits within the synthetic active site in the canonical core via an altered tRNA-independent pre-transfer editing mechanism (Fersht and Dingwall 1979a). MetRS edits smaller, unnatural amino acids on the basis of size selection (Fersht and Dingwall 1979a; Jakubowski 1991), and primarily targets homocysteine, which differs from methionine by a single methyl group (Fig. 2). Interestingly, homocysteine is the immediate precursor of methionine biosynthesis in the cell.

Editing of homocysteine involves its intramolecular cyclization to form homocysteine thiolactone subsequent to the activation step (Fig. 4). The side chain of homocysteine performs a nucleophilic attack on the activated homocysteinyl adenylate during the editing reaction (Jakubowski and Fersht 1981). The active site of MetRS cyclizes cognate methionine to form *S*-methylhomocysteine thiolactone in vitro (Jakubowski 1993a), but this reaction is minimized in vivo (Jakubowski 1993b).

Mutagenesis data suggest that the synthetic active site for aminoacylation and editing pathways is molecularly partitioned (Kim et al. 1993). Likewise, structural studies have demonstrated a conformational change induced by methionine binding that is absent when noncognate homocysteine binds (Serre et al. 2001). In *E. coli* MetRS, residues Trp305 and Tyr15 anchor the terminal methyl group and sulfur atom in the side chain of the bound methionine. Mutations W305A and Y15A impair discrimination and confer editing against the cognate methionine as well as homocysteine (Ghosh et al. 1991b; Kim et al. 1993). Conserved residues Asp52 and Arg233 interact with the main chain amino and carboxyl groups of methionine (Ghosh et al. 1991a) and also appear to play a role in the cyclization mechanism (Kim et al. 1993).

5.2 *Lysyl-tRNA Synthetase*

Unlike other AARSs, a LysRS representative has been found in each class (Ibba et al. 1997; Ambrogelly et al. 2002). With rare exception, LysRSs from both classes are not found together (Ibba et al. 1997; Polycarpo et al. 2003). LysRS-I and LysRS-II both use the same identity elements for $tRNA^{Lys}$ recognition, but differences in their active sites result in divergent mechanisms for amino acid activation (Ibba et al. 1999). LysRS-II relies on electrostatic interactions for binding lysine in an acidic active site (Ataide and Ibba 2004). Aminoacylation studies suggest that similar to the class I glutaminyl-tRNA synthetase (GlnRS) and glutamyl-tRNA synthetase (GluRS), LysRS-I has a high level of initial substrate discrimination at the active site and does not require additional proofreading (Levengood et al. 2004).

LysRS-II however, has a more promiscuous active site (Levengood et al. 2004) and edits via a cyclization mechanism similar to MetRS (Jakubowski 1999). For example, homocysteine, homoserine, and ornithine are cyclized to form a thiolactone, lactone, and lactam respectively (Jakubowski 1997, 1999). Within the aminoacylation active site of *E. coli* LysRS-II, mutagenesis suggests that Phe426 interacts with the lysine side chain to promote specificity (Ataide and Ibba 2004). In addition, aspartic acid residues at positions 240, 278, and 428 provide a negatively charged active site which contributes to substrate discrimination (Ataide and Ibba 2004).

6 Loss of Editing Function in AARSs

Some tRNA synthetases, primarily mitochondrial enzymes, have lost their editing activities. As mentioned earlier, this includes *S. cerevisiae* mitochondrial PheRS (Roy et al. 2005) and human mitochondrial ProRS (Musier-Forsyth et al. 1997), which are completely missing a post-transfer editing domain. All eukaryotic-like cytoplasmic ProRS enzymes also lack post-transfer editing activity, as these enzymes do not possess the novel insertion domain that is responsible for this editing activity in most bacterial ProRSs (Beuning and Musier-Forsyth 2001). At least some of these post-transfer editing-deficient enzymes appear to have been compensated with a higher initial selectivity at the amino acid activation step (Beuning and Musier-Forsyth 2001).

The *S. cerevisiae* ProRS possesses an N-terminal extension with weak homology to the bacterial post-transfer editing active site, but does not perform post-transfer editing in vitro (Ahel et al. 2003). Recently, it was shown that replacement of the N-terminal domain of *S. cerevisiae* ProRS with the *E. coli* INS domain confers post-transfer editing activity to this chimeric enzyme, with specificity for yeast Ala-tRNA^{Pro} (Sternjohn et al. 2007). In contrast, the isolated INS domain displays only weak editing activity and lacks tRNA sequence specificity. These results demonstrate how in evolution a weak editing activity can be converted to a more robust state through fusion to the body of a synthetase. In this manner, a single editing module can be distributed to different synthetases, and simultaneously acquire specificity and enhanced activity.

In another example, human mitochondrial LeuRS, which is homologous to bacterial LeuRSs (Bullard et al. 2000), has acquired substitutions within its editing active site rendering it inactive (Lue and Kelley 2005). However, kinetic analysis of aminoacylation suggests that greater stringency in the synthetic active site maintains the thresholds of fidelity that are sufficient for protein synthesis (Lue and Kelley 2005). Interestingly, yeast mitochondrial LeuRS has a viable amino acid editing active site, but when it is inactivated there appears to be little consequence to the cell (Karkhanis et al. 2006). This suggests that the mitochondria may simply tolerate a lower level of fidelity. It is also possible that other mechanisms of fidelity, such as freestanding editing proteins in these organelles, could edit in *trans*.

7 Role of AARS Editing In Vivo

The prevalence of the amino acid editing reaction for many of the AARSs suggests its importance to the cell. Recently, a modest defect in the AlaRS editing reaction was linked to mammalian disease (Lee et al. 2006). A missense mutation in the editing domain of mouse AlaRS produced low levels of mischarged Ser-tRNA^{Ala}. Ribosomal misincorporation of serine in place of alanine yielded misfolded proteins that were concentrated in Purkinje neural cells and resulted in progressive neurodegenerative disease. Incorporation of editing-defective ValRS into mammalian cells also resulted in ribosomal mistranslation, which correlated to decreases in cell viability (Nangle et al. 2006). An editing-defective IleRS statistically enhances the mutation rate in aging bacteria (Bacher and Schimmel 2007).

Many unimolecular organisms require the amino acid editing reaction for viability, particularly in the presence of excess noncognate amino acid. This includes fidelity mechanisms for LeuRS (Xu et al. 2004; Karkhanis et al. 2006, 2007), PheRS (Roy et al. 2004), and ValRS (Nangle et al. 2002) in *E. coli*. In addition, although hydrolysis of Ser-tRNA^{Thr} in *trans* by the freestanding archaeal *S. solfataricus* editing domain (ThrRS-ed) was dispensable in vivo when challenged by serine levels normally found in cells (Korencic et al. 2004), in the presence of elevated serine concentrations, an accumulation of Ser-tRNA^{Thr} resulted in a growth defect of a *S. solfataricus* ThrRS-ed deletion strain. Interestingly, intracellular studies using editing-defective LeuRSs indicated that nonstandard amino acids, rather than standard amino acids, may pose a greater threat to the fidelity of LeuRS (Karkhanis et al. 2007).

Under some cellular conditions, editing-defective AARSs are tolerated and can even be used to expand the genetic code. For example, a ValRS editing defect was used to incorporate aminobutyrate into *E. coli* proteins at high levels (Döring et al. 2001). In addition, editing-defective LeuRSs in the yeast mitochondria appeared to have little or no consequence to cellular viability (Karkhanis et al. 2006). Interestingly, norvaline can be globally incorporated into proteins at leucine positions during high expression conditions of recombinant proteins in *E. coli*, suggesting that it escapes the LeuRS amino acid editing mechanism (Apostol et al. 1997).

Tirrell and coworkers have capitalized upon *E. coli* amino acid auxotrophs and wild-type AARSs for global incorporation of unnatural amino acids (see the chapter by Beatty and Tirrell, this volume). In these systems, the *E. coli* AARS stably aminoacylates its cognate wild-type tRNA with a nonstandard amino acid, which is then inserted into proteins at the canonical codons. This technique has been successfully used to incorporate several phenylalanine analogs by wild-type and mutant PheRSs (Kast and Hennecke 1991; Sharma et al. 2000; Datta et al. 2002; Kirshenbaum et al. 2002; Beatty et al. 2005). Unsaturated methionine and isoleucine analogs were also incorporated via MetRS and IleRS (van Hest and Tirrell 1998; Kiick et al. 2000, 2001; Cirino et al. 2003; Wang et al. 2003; Link et al. 2004; Mock et al. 2006). In one case, the unsaturated isoleucine analog 2-amino-3-methyl-4-pentenoic acid was not edited by IleRS owing to its large size (Mock et al. 2006). Likewise, unsaturated, ketone-containing, and fluorinated leucine analogs were introduced into proteins via LeuRS mischarging activities (Tang and Tirrell 2002) (Table 3).

8 Orthogonal AARS and tRNAs

The development of orthogonal AARS–tRNA pairs has facilitated stable aminoacylation of nonstandard amino acids that can be used efficiently in vivo during protein synthesis. Orthogonal systems have the following requirements: (1) the orthogonal AARS must not aminoacylate the endogenous tRNAs; (2) the orthogonal tRNA must not be aminoacylated by the endogenous AARS; (3) the orthogonal AARS must preferentially and stably charge the nonstandard amino acid versus

Table 3 Modification of AARS's amino acid fidelity for applications in protein engineering

AARS	Mutation/ modification	Result	References
LeuRS	T252A	Incorporates fluorescent amino acid 2-amino-3-[5-(dimethylamino) naphthalene-1-sulfonamide] propanoic acid into yeast	Summerer et al. (2006)
LeuRS	T252Y	Incorporates norleucine, norvaline, allylglycine, homoallylglycine, homopropargylglycine, and 2-butynylglycine into *E. coli* proteins	Tang and Tirrell (2002)
ValRS	T222P	Incorporates aminobutyrate into *E. coli* proteins	Döring et al. (2001)
PheRS	A294G	Incorporates *p*-bromo phenylalanine, *p*-iodophenylalanine, *p*-cyanophenylalanine, *p*-ethynylphenylalanine, and *p*-azidophenylalanine and 2-pyridylalanine, 3-pyridylalanine, and 4-pyridylalanine into target mouse DHFR protein	Kast and Hennecke (1991) Sharma et al. (2000) Kirshenbaum et al. (2002)
PheRS	A294G, T251G	Incorporates *p*-acetylphenylalanine into target mouse DHFR protein	Datta et al. (2002)

DHFR dehydrofolate reductase

standard endogenous amino acids. Several systems have already been developed to incorporate a variety of unnatural amino acids (see the chapter by Köhrer and RajBhandary, this volume).

The RajBhandary laboratory has designed orthogonal AARS–tRNA pairs that have been applicable in bacteria, yeast, and mammalian cells. *E. coli* GlnRS and mutated human $tRNA^{fMet}_{CUA}$ were successfully used in *S. cerevisiae* for site-specific amber suppression (Kowal et al. 2001). In *E. coli*, yeast TyrRS was mutated so that it no longer interacted with *E. coli* $tRNA^{Pro}$, but specifically aminoacylated an *E. coli* $tRNA^{fMet}_{2}$ mutant to suppress an amber mutation (Kowal et al. 2001). Subsequent data showed that an amber suppressor tRNA (*supF*), derived from *E. coli* $tRNA^{Tyr}_{1}$, can be specifically aminoacylated by *E. coli* TyrRS and then imported into COS1 mammalian cells (Köhrer et al. 2001).

The laboratory of Peter Schultz has site-specifically incorporated a large number of nonstandard amino acids in vivo via orthogonal AARS–tRNA pairs in bacteria and eukarya. One of their first systems paired a mutated *M. jannaschii* $tRNA^{Tyr}_{CUA}$ with a modified *M. jannaschii* TyrRS that contained five substituted residues within its aminoacylation active site. This orthogonal pair facilitated incorporation of *o*-methyltyrosine into chloroamphenicol acetyltransferase and dihydrofolate reductase (Wang et al. 2001). In addition, the pair has been evolved to incorporate more than 20 nonstandard amino acids into proteins in *E. coli* (Xie and Schultz 2005). These include glycosylated and aliphatic side chains, as well as tyrosine and phenylalanine analogs. Additional orthogonal AARS–tRNA pairs have been generated for *E. coli* expression based on yeast GlnRS (Liu and Schultz 1999), *P. horikoshiii* GluRS (Santoro et al. 2003), and *P. horikoshii* LysRS (Anderson et al. 2004). Likewise, *E. coli* TyrRS has been expressed in yeast (Chin et al. 2003; Deiters et al. 2003) and *Bacillus subtilis* tryptophanyl-tRNA synthetase in mammalian cells (Zhang et al. 2004) to develop promising orthogonal systems. More recently, a new pair consisting of yeast PheRS (T415G) and mutant amber suppressor ($tRNA^{Phe}_{CUA\_UG}$) were introduced into an *E. coli* expression strain for the site specific incorporation of tryptophan analogs (Kwon and Tirrell 2007). Nonstandard amino acid incorporation in yeast has also been engineered using *E. coli* LeuRS (Wu et al. 2004; Turner et al. 2006), which was optimized by an editing mutation (T252A) (Mursinna et al. 2001) that hydrolyzed correctly charged Leu-$tRNA^{Leu}$ (Summerer et al. 2006).

9 Adapting AARS Editing Sites for Protein Engineering Applications

Canonical and orthogonal AARSs provide a powerful tool for the incorporation of novel amino acids into proteins. Significantly, those that are promiscuous and require amino acid editing activities can be exploited to expand the genetic code. Inactivation of the editing active site would facilitate stable production of mischarged tRNAs for ribosomal protein synthesis. For example, in LeuRS, IleRS, and ValRS, a single mutation of a universally conserved aspartic acid that was described earlier effectively abolishes amino acid editing to yield mischarged tRNAs (Bishop et al. 2002). In ProRS, substitution of a single highly conserved lysine to alanine also eliminated amino acid editing to produce Ala-$tRNA^{Pro}$ (Wong et al. 2002, 2003).

Another strategy relied upon rational design to fill in the amino acid binding pocket with bulky amino acids at the Thr252 position in LeuRS. This resulted in blocked editing activity (Mursinna and Martinis 2002) and facilitated in vivo incorporation of nonstandard amino acids (Tang and Tirrell 2002). There are many additional mechanisms that can be used to inactivate amino acid editing, including the insertion of peptides (Chen et al. 2000), prolines (Hendrickson et al. 2000), or deletion mutations (Beebe et al. 2003b) to grossly disrupt the editing active site topology, while maintaining aminoacylation activity. Mutational disruption of a network of hydrogen-bonding interactions among highly conserved residues in the editing active site of ThrRS also yielded mischarged Ser-$tRNA^{Thr}$ (Dock-Bregeon et al. 2000, 2004).

Mutations in both the aminoacylation and editing active sites have been engineered to redefine the specificity in vitro for a number of standard AARSs, in addition to those orthogonal AARSs that have been evolutionarily selected (described briefly earlier). For example, an A294G mutation in the PheRS α-subunit allows aminoacylation of *para*-substituted phenylalanine analogs, including tyrosine and *p*-fluoro-*l*-phenylalanine (Ibba et al. 1994). This mutation in combination with editing mutants in the β-subunits (E334A and A356W), allowed production of stably aminoacylated Tyr-$tRNA^{Phe}$ (Roy et al. 2004). In *E. coli* AlaRS, mutation of Cys666 to alanine in the editing active site facilitated stable mischarging of glycine and serine onto $tRNA^{Ala}$. A second Q584H mutation enhanced the mischarging of C666A to produce Ser-$tRNA^{Ala}$ and Gly-$tRNA^{Ala}$ (Beebe et al. 2003b).

Altering the editing active site specificity could be capitalized upon to hydrolyze correctly charged standard amino acids and enhance discrimination of a novel AARS's activities. As described before, an *E. coli* LeuRS T252A mutation (Mursinna et al. 2001) suppressed cognate Leu-$tRNA^{Leu}$ formation in an orthogonal AARS (Summerer et al. 2006). In the editing site of *E. coli* IleRS, a His333A mutation hydrolyzed both Ile-$tRNA^{Ile}$ and Val-$tRNA^{Ile}$ at a rate similar to that of wild-type IleRS deacylation of Val-$tRNA^{Ile}$ (Hendrickson et al. 2002). In a similar manner, the *E. coli* K277A mutant ValRS deacylates cognate charged Val-$tRNA^{Val}$ (Hountondji et al. 2002). Likewise, for class II AARSs, H369A and H369C mutations in the insertion domain of *E. coli* ProRS resulted in an enzyme that readily deacylates Pro-$tRNA^{Pro}$ (Wong et al. 2002). In addition, editing of correctly charged Ala-$tRNA^{Ala}$ was conferred by introducing mutations into *P. horikoshii* AlaRS (Q633M) as well as a parallel mutation in the AlaX *trans*-editing domain (T30V) (Sokabe et al. 2005).

A growing foundation of molecular information based on structural and biochemical analyses has provided significant insight into amino acid specificities of the AARSs and their fidelity mechanisms. These remarkable enzymes in the first step of protein synthesis that commit an amino acid for site-specific incorporation into the polypeptide chain are poised for adaptation to enable custom synthesis of novel proteins. Rational design and in vivo selection of mutations in the aminoacylation and editing active sites promise to yield a greatly expanded genetic code that has enormous medical and technology applications.

Acknowledgments This work was supported by grants from the National Institutes of Health (GM063789-05 to S.A.M. and GM049928 to K.M.-F.).

References

Ahel I, Stathopoulos C, Ambrogelly A, Sauerwald A, Toogood H, Hartsch T, Söll D (2002) Cysteine activation is an inherent in vitro property of prolyl-tRNA synthetases. J Biol Chem 277:34743–34748

Ahel I, Korencic D, Ibba M, Söll D (2003) *Trans*-editing of mischarged tRNAs. Proc Natl Acad Sci USA 100:15422–15427

Ambrogelly A, Korencic D, Ibba M (2002) Functional annotation of class I lysyl-tRNA synthetase phylogeny indicates a limited role for gene transfer. J Bacteriol 184:4594–4600

An S, Musier-Forsyth K (2004) *Trans*-editing of Cys-$tRNA^{Pro}$ by *Haemophilus influenzae* YbaK protein. J Biol Chem 279:42359–42362

An S, Musier-Forsyth K (2005) Cys-tRNA(Pro) editing by *Haemophilus influenzae* YbaK via a novel synthetase·YbaK·tRNA ternary complex. J Biol Chem 280:34465–34472

Anderson JC, Wu N, Santoro SW, Lakshman V, King DS, Schultz PG (2004) An expanded genetic code with a functional quadruplet codon. Proc Natl Acad Sci USA 101:7566–7571

Apostol I, Levine J, Lippincott J, Leach J, Hess E, Glascock CB, Weickert MJ, Blackmore R (1997) Incorporation of norvaline at leucine positions in recombinant human hemoglobin expressed in *Escherichia coli*. J Biol Chem 272:28980–28988

Ataide SF, Ibba M (2004) Discrimination of cognate and noncognate substrates at the active site of class II lysyl-tRNA synthetase. Biochemistry 43:11836–11841

Bacher JM, Schimmel P (2007) An editing-defective aminoacyl-tRNA synthetase is mutagenic in aging bacteria via the SOS response. Proc Natl Acad Sci USA 104:1907–1912

Baldwin AN, Berg P (1966) Transfer ribonucleic acid-induced hydrolysis of valyladenylate bound to isoleucyl ribonucleic acid synthetase. J Biol Chem 241:839–845

Beatty KE, Xie F, Wang Q, Tirrell DA (2005) Selective dye-labeling of newly synthesized proteins in bacterial cells. J Am Chem Soc 127:14150–14151

Beebe K, Merriman E, Schimmel P (2003a) Structure-specific tRNA determinants for editing a mischarged amino acid. J Biol Chem 278:45056–45061

Beebe K, Ribas De Pouplana L, Schimmel P (2003b) Elucidation of tRNA-dependent editing by a class II tRNA synthetase and significance for cell viability. EMBO J 22:668–675

Beebe K, Merriman E, Ribas De Pouplana L, Schimmel P (2004) A domain for editing by an archaebacterial tRNA synthetase. Proc Natl Acad Sci USA 101:5958–5963

Betha AK, Williams AM, Martinis SA (2007) Isolated CP1 domain of *Escherichia coli* leucyl-tRNA synthetase is dependent on flanking hinge motifs for amino acid editing activity. Biochemistry 46:6258–6267

Beuning PJ, Musier-Forsyth K (2000) Hydrolytic editing by a class II aminoacyl-tRNA synthetase. Proc Natl Acad Sci USA 97:8916–8920

Beuning PJ, Musier-Forsyth K (2001) Species-specific differences in amino acid editing by class II prolyl-tRNA synthetase. J Biol Chem 276:30779–30785

Bishop AC, Nomanbhoy TK, Schimmel P (2002) Blocking site-to-site translocation of a misactivated amino acid by mutation of a class I tRNA synthetase. Proc Natl Acad Sci USA 99:585–590

Bishop AC, Beebe K, Schimmel PR (2003) Interstice mutations that block site-to-site translocation of a misactivated amino acid bound to a class I tRNA synthetase. Proc Natl Acad Sci USA 100:490–494

Bullard JM, Cai YC, Spremulli LL (2000) Expression and characterization of the human mitochondrial leucyl-tRNA synthetase. Biochim Biophys Acta 1490:245–258

Budisa N, Steipe B, Demange P, Eckerskorn C, Kellermann J, Huber R (1995) High-level biosynthetic substitution of methionine in proteins by its analogs 2-aminohexanoic acid, selenomethionine, telluromethionine and ethionine in *Escherichia coli*. Eur J Biochem 230:788–796

Burbaum JJ, Schimmel P (1991) Structural relationships and the classification of aminoacyl-tRNA synthetases. J Biol Chem 266:16965–16968

Carter CW Jr (1993) Cognition, mechanism, and evolutionary relationships in aminoacyl-tRNA synthetases. Annu Rev Biochem 62:715–748

Chen JF, Guo NN, Li T, Wang ED, Wang YL (2000) CP1 domain in *Escherichia coli* leucyl-tRNA synthetase is crucial for its editing function. Biochemistry 39:6726–6731

Chin JW, Cropp TA, Anderson JC, Mukherji M, Zhang Z, Schultz PG (2003) An expanded eukaryotic genetic code. Science 301:964–967

Cirino PC, Tang Y, Takahashi K, Tirrell DA, Arnold FH (2003) Global incorporation of norleucine in place of methionine in cytochrome P450 BM-3 heme domain increases peroxygenase activity. Biotechnol Bioeng 83:729–734

Cramer F, Faulhammer H, von der Haar F, Sprinzl M, Sternbach H (1975) Aminoacyl-tRNA synthetases from baker's yeast: reacting site of enzymatic aminoacylation is not uniform for all tRNAs. FEBS Lett 56:212–214

Crépin T, Schmitt E, Blanquet S, Mechulam Y (2002) Structure and function of the C-terminal domain of methionyl-tRNA synthetase. Biochemistry 41:13003–13011

Crépin T, Schmitt E, Mechulam Y, Sampson PB, Vaughan MD, Honek JF, Blanquet S (2003) Use of analogues of methionine and methionyl adenylate to sample conformational changes during catalysis in *Escherichia coli* methionyl-tRNA synthetase. J Mol Biol 332:59–72

Crépin T, Schmitt E, Blanquet S, Mechulam Y (2004) Three-dimensional structure of methionyl-tRNA synthetase from *Pyrococcus abyssi*. Biochemistry 43:2635–2644

Crépin T, Schmitt E, Mechulam Y, Sampson PB, Vaughan MD, Honek JF, Blanquet S (2003) Use of analogues of methionine and methionyl adenylate to sample conformational changes during catalysis in Escherichia coli methionyl-tRNA synthetase. J Mol Biol 332:59–72

Crépin T, Yaremchuk A, Tukalo M, Cusack S (2006) Structures of two bacterial prolyl-tRNA synthetases with and without a *cis*-editing domain. Structure 14:1511–1525

Cusack S (1995) Eleven down and nine to go. Nat Struct Biol 2:824–831

Cusack S, Berthet-Colominas C, Härtlein M, Nassar N, Leberman R (1990) A second class of synthetase structure revealed by X-ray analysis of *Escherichia coli* seryl-tRNA synthetase at 2.5 Å. Nature 347:249–255

Cusack S, Yaremchuk A, Tukalo M (2000) The 2 Å crystal structure of leucyl-tRNA synthetase and its complex with a leucyl-adenylate analogue. EMBO J 19:2351–2361

Datta D, Wang P, Carrico IS, Mayo SL, Tirrell DA (2002) A designed phenylalanyl-tRNA synthetase variant allows efficient in vivo incorporation of aryl ketone functionality into proteins. J Am Chem Soc 124:5652–5653

Deiters A, Cropp TA, Mukherji M, Chin JW, Anderson JC, Schultz PG (2003) Adding amino acids with novel reactivity to the genetic code of *Saccharomyces cerevisiae*. J Am Chem Soc 125:11782–11783

Desogus G, Todone F, Brick P, Onesti S (2000) Active site of lysyl-tRNA synthetase: structural studies of the adenylation reaction. Biochemistry 39:8418–8425

Dock-Bregeon A, Sankaranarayanan R, Romby P, Caillet J, Springer M, Rees B, Francklyn CS, Ehresmann C, Moras D (2000) Transfer RNA-mediated editing in threonyl-tRNA synthetase. The class II solution to the double discrimination problem. Cell 103:877–884

Dock-Bregeon AC, Rees B, Torres-Larios A, Bey G, Caillet J, Moras D (2004) Achieving error-free translation; the mechanism of proofreading of threonyl-tRNA synthetase at atomic resolution. Mol Cell 16:375–386

Döring V, Mootz HD, Nangle LA, Hendrickson TL, de Crécy-Lagard V, Schimmel P, Marliere P (2001) Enlarging the amino acid set of *Escherichia coli* by infiltration of the valine coding pathway. Science 292:501–504

Du X, Wang ED (2003) Tertiary structure base pairs between D- and TpsiC-loops of *Escherichia coli* tRNA(Leu) play important roles in both aminoacylation and editing. Nucleic Acids Res 31:2865–2872

Dwivedi S, Kruparani SP, Sankaranarayanan R (2005) A D-amino acid editing module coupled to the translational apparatus in archaea. Nat Struct Mol Biol 12:556–557

Eldred EW, Schimmel PR (1972) Rapid deacylation by isoleucyl transfer ribonucleic acid synthetase of isoleucine-specific transfer ribonucleic acid aminoacylated with valine. J Biol Chem 247:2961–2964

Englisch S, Englisch U, von der Haar F, Cramer F (1986) The proofreading of hydroxy analogues of leucine and isoleucine by leucyl-tRNA synthetases from *E. coli* and yeast. Nucleic Acids Res 14:7529–7539

Eriani G, Delarue M, Poch O, Gangloff J, Moras D (1990) Partition of tRNA synthetases into two classes based on mutually exclusive sets of sequence motifs. Nature 347:203–206

Eriani G, Cavarelli J, Martin F, Ador L, Rees B, Thierry JC, Gangloff J, Moras D (1995) The class II aminoacyl-tRNA synthetases and their active site: evolutionary conservation of an ATP binding site. J Mol Evol 40:499–508

Farrow MA, Nordin BE, Schimmel P (1999) Nucleotide determinants for tRNA-dependent amino acid discrimination by a class I tRNA synthetase. Biochemistry 38:16898–16903

Fersht AR (1977a) Enzyme structure and mechanism. W.H. Freeman and Company Limited, pp 283

Fersht AR (1977b) Editing mechanisms in protein synthesis. Rejection of valine by the isoleucyl-tRNA synthetase. Biochemistry 16:1025–1030

Fersht AR, Dingwall C (1979a) An editing mechanism for the methionyl-tRNA synthetase in the selection of amino acids in protein synthesis. Biochemistry 18:1250–1256

Fersht AR, Dingwall C (1979b) Cysteinyl-tRNA synthetase from *Escherichia coli* does not need an editing mechanism to reject serine and alanine. High binding energy of small groups in specific molecular interactions. Biochemistry 18:1245–1249

Fersht AR, Dingwall C (1979c) Establishing the misacylation/deacylation of the tRNA pathway for the editing mechanism of prokaryotic and eukaryotic valyl-tRNA synthetases. Biochemistry 18:1238–1245

Fersht AR, Kaethner MM (1976) Enzyme hyperspecificity. Rejection of threonine by the valyl-tRNA synthetase by misacylation and hydrolytic editing. Biochemistry 15:3342–3346

Fersht AR, Shindler JS, Tsui WC (1980) Probing the limits of protein-amino acid side chain recognition with the aminoacyl-tRNA synthetases. Discrimination against phenylalanine by tyrosyl-tRNA synthetases. Biochemistry 19:5520–5524

First EA (2005) Catalysis of the tRNA aminoacylation reaction. In: Aminoacyl-tRNA synthetases, Landes Bioscience/Eurekah.com, Georgetown, TX, pp 328–352

Fishman R, Ankilova V, Moor N, Safro M (2001) Structure at 2.6 Å resolution of phenylalanyl-tRNA synthetase complexed with phenylalanyl-adenylate in the presence of manganese. Acta Crystallogr D Biol Crystallogr 57(Pt 11):1534–1544

Freist W (1989) Mechanisms of aminoacyl-tRNA synthetases: a critical consideration of recent results. Biochemistry 28:6787–6795

Freist W, Cramer F (1983) Isoleucyl-tRNA synthetase from Baker's yeast. Catalytic mechanism, 2′,3′-specificity and fidelity in aminoacylation of $tRNA^{Ile}$ with isoleucine and valine investigated with initial-rate kinetics using analogs of tRNA, ATP and amino acids. Eur J Biochem 131:65–80

Fukai S, Nureki O, Sekine S, Shimada A, Tao J, Vassylyev DG, Yokoyama S (2000) Structural basis for double-sieve discrimination of L-valine from L-isoleucine and L-threonine by the complex of tRNA(Val) and valyl-tRNA synthetase. Cell 103:793–803

Fukai S, Nureki O, Sekine S, Shimada A, Vassylyev DG, Yokoyama S (2003) Mechanism of molecular interactions for tRNA(Val) recognition by valyl-tRNA synthetase. RNA 9:100–111

Fukunaga R, Yokoyama S (2005b) Structural basis for non-cognate amino acid discrimination by the valyl-tRNA synthetase editing domain. J Biol Chem 280:29937–29945

Fukunaga R, Yokoyama S (2005c) Crystal structure of leucyl-tRNA synthetase from the archaeon *Pyrococcus horikoshii* reveals a novel editing domain orientation. J Mol Biol 346:57–71

Fukunaga R, Yokoyama S (2006) Structural basis for substrate recognition by the editing domain of isoleucyl-tRNA synthetase. J Mol Biol 359:901–912

Fukunaga R, Yokoyama S (2007) The C-terminal domain of the archaeal leucyl-tRNA synthetase prevents misediting of isoleucyl-tRNA(Ile). Biochemistry 46:4985–4996

Fukunaga R, Fukai S, Ishitani R, Nureki O, Yokoyama S (2004) Crystal structures of the CP1 domain from *Thermus thermophilus* isoleucyl-tRNA synthetase and its complex with L-valine. J Biol Chem 279:8396–8402

Fukunaga R, Yokoyama S (2005a) Aminoacylation complex structures of leucyl-tRNA synthetase and tRNA(Leu) reveal two modes of discriminator-base recognition. Nat Struct Mol Biol 12:915–922

Geslain R, Ribas de Pouplana L (2004) Regulation of RNA function by aminoacylation and editing? Trends Genet 20:604–610

Ghosh G, Kim HY, Demaret JP, Brunie S, Schulman LH (1991a) Arginine-395 is required for efficient in vivo and in vitro aminoacylation of tRNAs by *Escherichia coli* methionyl-tRNA synthetase. Biochemistry 30:11767–11774

Ghosh G, Pelka H, Schulman LH, Brunie S (1991b) Activation of methionine by *Escherichia coli* methionyl-tRNA synthetase. Biochemistry 30:9569–9575

Giegé R, Sissler M, Florentz C (1998) Universal rules and idiosyncratic features in tRNA identity. Nucleic Acids Res 26:5017–5035

Goldgur Y, Mosyak L, Reshetnikova L, Ankilova V, Lavrik O, Khodyreva S, Safro M (1997) The crystal structure of phenylalanyl-tRNA synthetase from *Thermus thermophilus* complexed with cognate $tRNA^{Phe}$. Structure 5:59–68

Gruic-Sovulj I, Uter N, Bullock T, Perona JJ (2005) tRNA-dependent aminoacyl-adenylate hydrolysis by a nonediting class I aminoacyl-tRNA synthetase. J Biol Chem 280:23978–23986
Hale SP, Schimmel P (1996) Protein synthesis editing by a DNA aptamer. Proc Natl Acad Sci USA 93:2755–2758
Hale SP, Auld DS, Schmidt E, Schimmel P (1997) Discrete determinants in transfer RNA for editing and aminoacylation. Science 276:1250–1252
Hati S, Ziervogel B, Sternjohn J, Wong FC, Nagan MC, Rosen AE, Siliciano PG, Chihade JW, Musier-Forsyth K (2006) Pre-transfer editing by class II prolyl-tRNA synthetase: role of aminoacylation active site in "selective release" of noncognate amino acids. J Biol Chem 281: 27862–27872
Hendrickson TL, Schimmel P (2003) Transfer RNA-dependent amino acid discrimination by aminoacyl-tRNA synthetases. In: translation mechanisms. Kluwer Academic/Plenum Publishers, pp 35–64
Hendrickson TL, Nomanbhoy TK, Schimmel P (2000) Errors from selective disruption of the editing center in a tRNA synthetase. Biochemistry 39:8180–8186
Hendrickson TL, Nomanbhoy TK, de Crécy-Lagard V, Fukai S, Nureki O, Yokoyama S, Schimmel P (2002) Mutational separation of two pathways for editing by a class I tRNA synthetase. Mol Cell 9:353–362
Hendrickson TL, de Crécy-Lagard V, Schimmel P (2004) Incorporation of nonnatural amino acids into proteins. Annu Rev Biochem 73:147–176
Hopfield JJ (1974) Kinetic proofreading: a new mechanism for reducing errors in biosynthetic processes requiring high specificity. Proc Natl Acad Sci USA 71:4135–4139
Hopfield JJ, Yamane T, Yue V, Coutts SM (1976) Direct experimental evidence for kinetic proofreading in aminoacylation of $tRNA^{Ile}$. Proc Natl Acad Sci USA 73:1164–1168
Hountondji C, Dessen P, Blanquet S (1986) Sequence similarities among the family of aminoacyl-tRNA synthetases. Biochimie 68:1071–1078
Hountondji C, Lazennec C, Beauvallet C, Dessen P, Pernollet JC, Plateau P, Blanquet S (2002) Crucial role of conserved lysine 277 in the fidelity of tRNA aminoacylation by *Escherichia coli* valyl-tRNA synthetase. Biochemistry 41:14856–14865
Hsu JL, Rho SB, Vanella KM, Martinis SA (2006) Functional divergence of a unique C-terminal domain of leucyl-tRNA synthetase to accommodate its splicing and aminoacylation roles. J Biol Chem 281:23075–23082
Hussain T, Kruparani SP, Pal B, Dock-Bregeon AC, Dwivedi S, Shekar MR, Sureshbabu K, Sankaranarayanan R (2006) Post-transfer editing mechanism of a D-aminoacyl-tRNA deacylase-like domain in threonyl-tRNA synthetase from archaea. EMBO J 25:4152–4162
Ibba M, Söll D (2000) Aminoacyl-tRNA synthesis. Annu Rev Biochem 69:617–650
Ibba M, Kast P, Hennecke H (1994) Substrate specificity is determined by amino acid binding pocket size in *Escherichia coli* phenylalanyl-tRNA synthetase. Biochemistry 33:7107–7112
Ibba M, Morgan S, Curnow AW, Pridmore DR, Vothknecht UC, Gardner W, Lin W, Woese CR, Söll D (1997) A euryarchaeal lysyl-tRNA synthetase: resemblance to class I synthetases. Science 278:1119–1122
Ibba M, Losey HC, Kawarabayasi Y, Kikuchi H, Bunjun S, Söll D (1999) Substrate recognition by class I lysyl-tRNA synthetases: a molecular basis for gene displacement. Proc Natl Acad Sci USA 96:418–423
Ibba M, Becker HD, Stathopoulos C, Tumbula DL, Söll D (2000) Author correction. Trends Biochem Sci 25:380
Igloi GL, von der Haar F, Cramer F (1978) Aminoacyl-tRNA synthetases from yeast: generality of chemical proofreading in the prevention of misaminoacylation of tRNA. Biochemistry 17:3459–3468
Ishijima J, Uchida Y, Kuroishi C, Tuzuki C, Takahashi N, Okazaki N, Yutani K, Miyano M (2006) Crystal structure of alanyl-tRNA synthetase editing-domain homolog (PH0574) from a hyperthermophile, *Pyrococcus horikoshii* OT3 at 1.45 Å resolution. Proteins 62: 1133–1137

Jakubowski H (1991) Proofreading in vivo: editing of homocysteine by methionyl-tRNA synthetase in the yeast *Saccharomyces cerevisiae*. EMBO J 10:593–598

Jakubowski H (1993a) Proofreading and the evolution of a methyl donor function. Cyclization of methionine to S-methyl homocysteine thiolactone by *Escherichia coli* methionyl-tRNA synthetase. J Biol Chem 268:6549–6553

Jakubowski H (1993b) Energy cost of proofreading in vivo: The charging of methionine tRNAs in *Escherichia coli*. FASEB J 7:168–172

Jakubowski H (1997) Aminoacyl thioester chemistry of class II aminoacyl-tRNA synthetases. Biochemistry 36:11077–11085

Jakubowski H (1999) Misacylation of $tRNA^{Lys}$ with noncognate amino acids by lysyl-tRNA synthetase. Biochemistry 38:8088–8093

Jakubowski H, Fersht AR (1981) Alternative pathways for editing non-cognate amino acids by aminoacyl-tRNA synthetases. Nucleic Acids Res 9:3105–3117

Kamtekar S, Kennedy WD, Wang J, Stathopoulos C, Söll D, Steitz TA (2003) The structural basis of cysteine aminoacylation of $tRNA^{Pro}$ by prolyl-tRNA synthetases. Proc Natl Acad Sci USA 100:1673–1678

Karkhanis VA, Boniecki MT, Poruri K, Martinis SA (2006) A viable amino acid editing activity in the leucyl-tRNA synthetase CP1-splicing domain is not required in the yeast mitochondria. J Biol Chem 281:33217–33225

Karkhanis VA, Mascarenhas AP, Martinis SA (2007) Amino acid toxicities of *Escherichia coli* that are prevented by leucyl-tRNA synthetase amino acid editing. J Bacteriol 189: 8765–8768

Kast P, Hennecke H (1991) Amino acid substrate specificity of *Escherichia coli* phenylalanyl-tRNA synthetase altered by distinct mutations. J Mol Biol 222:99–124

Kiick KL, van Hest JC, Tirrell DA (2000) Expanding the scope of protein biosynthesis by altering the methionyl-tRNA synthetase activity of a bacterial expression host. Angew Chem Int Ed Engl 39:2148–2152

Kiick KL, Weberskirch R, Tirrell DA (2001) Identification of an expanded set of translationally active methionine analogues in *Escherichia coli*. FEBS Lett 502:25–30

Kim HY, Ghosh G, Schulman LH, Brunie S, Jakubowski H (1993) The relationship between synthetic and editing functions of the active site of an aminoacyl-tRNA synthetase. Proc Natl Acad Sci USA 90:11553–11557

Kirshenbaum K, Carrico IS, Tirrell DA (2002) Biosynthesis of proteins incorporating a versatile set of phenylalanine analogues. Chembiochem 3:235–237

Köhrer C, Xie L, Kellerer S, Varshney U, RajBhandary UL (2001) Import of amber and ochre suppressor tRNAs into mammalian cells: a general approach to site-specific insertion of amino acid analogues into proteins. Proc Natl Acad Sci USA 98:14310–14315

Korencic D, Ahel I, Schelert J, Sacher M, Ruan B, Stathopoulos C, Blum P, Ibba M, Söll D (2004) A freestanding proofreading domain is required for protein synthesis quality control in Archaea. Proc Natl Acad Sci USA 101:10260–10265

Kotik-Kogan O, Moor N, Tworowski D, Safro M (2005) Structural basis for discrimination of L-phenylalanine from L-tyrosine by phenylalanyl-tRNA synthetase. Structure 13:1799–1807

Kowal AK, Köhrer C, RajBhandary UL (2001) Twenty-first aminoacyl-tRNA synthetase-suppressor tRNA pairs for possible use in site-specific incorporation of amino acid analogues into proteins in eukaryotes and in eubacteria. Proc Natl Acad Sci USA 98:2268–2273

Kwon I, Tirrell DA (2007) Site-specific incorporation of tryptophan analogues into recombinant proteins in bacterial cells. J Am Chem Soc 129:10431–10437

Landès C, Perona JJ, Brunie S, Rould MA, Zelwer C, Steitz TA, Risler JL (1995) A structure-based multiple sequence alignment of all class I aminoacyl-tRNA synthetases. Biochimie 77:194–203

Larkin DC, Williams AM, Martinis SA, Fox GE (2002) Identification of essential domains for *Escherichia coli* tRNA(leu) aminoacylation and amino acid editing using minimalist RNA molecules. Nucleic Acids Res 30:2103–2113

Lee JW, Beebe K, Nangle LA, Jang J, Longo-Guess CM, Cook SA, Davisson MT, Sundberg JP, Schimmel P, Ackerman SL (2006) Editing-defective tRNA synthetase causes protein misfolding and neurodegeneration. Nature 443:50–55

Levengood J, Ataide SF, Roy H, Ibba M (2004) Divergence in noncognate amino acid recognition between class I and class II lysyl-tRNA synthetases. J Biol Chem 279:17707–17714

Lin L, Schimmel P (1996) Mutational analysis suggests the same design for editing activities of two tRNA synthetases. Biochemistry 35:5596–5601

Lin SX, Baltzinger M, Remy P (1983) Fast kinetic study of yeast phenylalanyl-tRNA synthetase: an efficient discrimination between tyrosine and phenylalanine at the level of the aminoacyladenylate-enzyme complex. Biochemistry 22:681–689

Lin SX, Baltzinger M, Remy P (1984) Fast kinetic study of yeast phenylalanyl-tRNA synthetase: role of $tRNA^{Phe}$ in the discrimination between tyrosine and phenylalanine. Biochemistry 23:4109–4116

Lin L, Hale SP, Schimmel P (1996) Aminoacylation error correction. Nature 384:33–34

Lincecum TL, Jr., Tukalo M, Yaremchuk A, Mursinna RS, Williams AM, Sproat BS, Van Den Eynde W, Link A, Van Calenbergh S, Grotli M, Martinis SA, Cusack S (2003) Structural and mechanistic basis of pre- and posttransfer editing by leucyl-tRNA synthetase. Mol Cell 11:951–963

Ling J, Roy H, Ibba M (2007) Mechanism of tRNA-dependent editing in translational quality control. Proc Natl Acad Sci USA 104:72–77

Link AJ, Vink MK, Tirrell DA (2004) Presentation and detection of azide functionality in bacterial cell surface proteins. J Am Chem Soc 126:10598–10602

Liu DR, Schultz PG (1999) Progress toward the evolution of an organism with an expanded genetic code. Proc Natl Acad Sci USA 96:4780–4785

Liu Y, Liao J, Zhu B, Wang ED, Ding J (2006) Crystal structures of the editing domain of *Escherichia coli* leucyl-tRNA synthetase and its complexes with Met and Ile reveal a lock-and-key mechanism for amino acid discrimination. Biochem J 394:399–407

Loftfield RB (1963) The frequency of errors in protein biosynthesis. Biochem J 89:82–92

Loftfield RB, Vanderjagt D (1972) The frequency of errors in protein biosynthesis. Biochem J 128:1353–1356

Ludmerer SW, Schimmel P (1987) Construction and analysis of deletions in the amino-terminal extension of glutamine tRNA synthetase of *Saccharomyces cerevisiae*. J Biol Chem 262:10807–10813

Lue SW, Kelley SO (2005) An aminoacyl-tRNA synthetase with a defunct editing site. Biochemistry 44:3010–3016

Martinis SA, Schimmel P (1999) Aminoacyl tRNA synthetases. *Escherichia coli* and Salmonella: cellular and molecular biology. Neidhardt FC. Washington DC. ASM Press, pp 887–901

Martinis SA, Plateau P, Cavarelli J, Florentz C (1999a) Aminoacyl-tRNA synthetases: a family of expanding functions. EMBO J 18:4591–4596

Martinis SA, Plateau P, Cavarelli J, Florentz C (1999b) Aminoacyl-tRNA synthetases: A new image for a classical family. Biochimie 81:683–700

Mechulam Y, Schmitt E, Maveyraud L, Zelwer C, Nureki O, Yokoyama S, Konno M, Blanquet S (1999) Crystal structure of *Escherichia coli* methionyl-tRNA synthetase highlights species-specific features. J Mol Biol 294:1287–1297

Mock ML, Michon T, van Hest JC, Tirrell DA (2006) Stereoselective incorporation of an unsaturated isoleucine analogue into a protein expressed in *E. coli*. ChemBioChem 7:83–87

Moor N, Kotik-Kogan O, Tworowski D, Sukhanova M, Safro M (2006) The crystal structure of the ternary complex of phenylalanyl-tRNA synthetase with $tRNA^{Phe}$ and a phenylalanyl-adenylate analogue reveals a conformational switch of the CCA end. Biochemistry 45: 10572–10583

Moras D (1992) Structural and functional relationships between aminoacyl-tRNA synthetases. Trends Biochem Sci 17:159–164

Mosyak L, Reshetnikova L, Goldgur Y, Delarue M, Safro MG (1995) Structure of phenylalanyl-tRNA synthetase from *Thermus thermophilus*. Nat Struct Biol 2:537–547

Murayama K, Kato-Murayama M, Katsura K, Uchikubo-Kamo T, Yamaguchi-Hirafuji M, Kawazoe M, Akasaka R, Hanawa-Suetsugu K, Hori-Takemoto C, Terada T, Shirouzu M, Yokoyama S (2005) Structure of a putative *trans*-editing enzyme for prolyl-tRNA synthetase from *Aeropyrum perinix* K1 at 1.7 Å resolution. Acta Crystallogr F Struc Biol Cryst Comm F61:26–29

Mursinna RS, Martinis SA (2002) Rational design to block amino acid editing of a tRNA synthetase. J Am Chem Soc 124:7286–7287

Mursinna RS, Lincecum TL, Jr., Martinis SA (2001) A conserved threonine within *Escherichia coli* leucyl-tRNA synthetase prevents hydrolytic editing of leucyl-tRNA^{Leu}. Biochemistry 40: 5376–5381

Mursinna RS, Lee KW, Briggs JM, Martinis SA (2004) Molecular dissection of a critical specificity determinant within the amino acid editing domain of leucyl-tRNA synthetase. Biochemistry 43:155–165

Musier-Forsyth K, Stehlin C, Burke B, Liu H (1997) Understanding species-specific differences in substrate recognition by *Escherichia coli* and human prolyl-tRNA synthetases. Nucleic Acids Symp Ser 36:5–7

Nakama T, Nureki O, Yokoyama S (2001) Structural basis for the recognition of isoleucyl-adenylate and an antibiotic, mupirocin, by isoleucyl-tRNA synthetase. J Biol Chem 276:47387–47393

Nangle LA, De Crecy Lagard V, Döring V, Schimmel P (2002) Genetic code ambiguity. Cell viability related to the severity of editing defects in mutant tRNA synthetases. J Biol Chem 277:45729–45733

Nangle LA, Motta CM, Schimmel P (2006) Global effects of mistranslation from an editing defect in mammalian cells. Chem Biol 13:1091–1100

Newberry KJ, Hou YM, Perona JJ (2002) Structural origins of amino acid selection without editing by cysteinyl-tRNA synthetase. EMBO J 21:2778–2787

Nomanbhoy TK, Schimmel PR (2000) Misactivated amino acids translocate at similar rates across surface of a tRNA synthetase. Proc Natl Acad Sci USA 97:5119–5122

Nomanbhoy TK, Hendrickson TL, Schimmel P (1999) Transfer RNA-dependent translocation of misactivated amino acids to prevent errors in protein synthesis. Mol Cell 4:519–528

Nordin BE, Schimmel P (1999) RNA determinants for translational editing. Mischarging a minihelix substrate by a tRNA synthetase. J Biol Chem 274:6835–6838

Nordin BE, Schimmel P (2002) Plasticity of recognition of the 3′-end of mischarged tRNA by class I aminoacyl-tRNA synthetases. J Biol Chem 277:20510–20517

Nordin BE, Schimmel P (2003) Transiently misacylated tRNA is a primer for editing of misactivated adenylates by class I aminoacyl-tRNA synthetases. Biochemistry 42:12989–12997

Nureki O, Vassylyev DG, Tateno M, Shimada A, Nakama T, Fukai S, Konno M, Hendrickson TL, Schimmel P, Yokoyama S (1998) Enzyme structure with two catalytic sites for double-sieve selection of substrate. Science 280:578–582

Onesti S, Desogus G, Brevet A, Chen J, Plateau P, Blanquet S, Brick P (2000) Structural studies of lysyl-tRNA synthetase: conformational changes induced by substrate binding. Biochemistry 39:12853–12861

Onesti S, Miller AD, Brick P (1995) The crystal structure of the lysyl-tRNA synthetase (LysU) from *Escherichia coli*. Structure 3:163–176

O'Donoghue P, Luthey-Schulten Z (2003) On the evolution of structure in aminoacyl-tRNA synthetases. Microbiol Mol Biol Rev 67:550–573

Pauling L (1958) The probability of errors in protein synthesis. In: Festschrift Arthur Stoll Siebzigsten Geburtstag, pp 597–602

Perona JJ (2005) Two-step pathway to aminoacylated tRNA. Structure 13:1397–1398

Pezo V, Metzgar D, Hendrickson TL, Waas WF, Hazebrouck S, Döring V, Marliere P, Schimmel P, De Crécy-Lagard V (2004) Artificially ambiguous genetic code confers growth yield advantage. Proc Natl Acad Sci USA 101:8593–8597

Polycarpo C, Ambrogelly A, Ruan B, Tumbula-Hansen D, Ataide SF, Ishitani R, Yokoyama S, Nureki O, Ibba M, Söll D (2003) Activation of the pyrrolysine suppressor tRNA requires formation of a ternary complex with class I and class II lysyl-tRNA synthetases. Mol Cell 12:287–294

Reshetnikova L, Moor N, Lavrik O, Vassylyev DG (1999) Crystal structures of phenylalanyl-tRNA synthetase complexed with phenylalanine and a phenylalanyl-adenylate analogue. J Mol Biol 287:555–568

Ribas de Pouplana L, Schimmel P (2001) Two classes of tRNA synthetases suggested by sterically compatible dockings on tRNA acceptor stem. Cell 104:191–193

Rock FL, Mao W, Yaremchuk A, Tukalo M, Crépin T, Zhou H, Zhang YK, Hernandez V, Akama T, Baker SJ, Plattner JJ, Shapiro L, Martinis SA, Benkovic SJ, Cusack S, Alley MR (2007) An antifungal agent inhibits an aminoacyl-tRNA synthetase by trapping tRNA in the editing site. Science 316:1759–1761

Roy H, Ibba M (2006) Phenylalanyl-tRNA synthetase contains a dispensable RNA-binding domain that contributes to the editing of noncognate aminoacyl-tRNA. Biochemistry 45: 9156–9162

Roy H, Ling J, Irnov M, Ibba M (2004) Post-transfer editing in vitro and in vivo by the beta subunit of phenylalanyl-tRNA synthetase. EMBO J 23:4639–4648

Roy H, Ling J, Alfonzo J, Ibba M (2005) Loss of editing activity during the evolution of mitochondrial phenylalanyl-tRNA synthetase. J Biol Chem 280:38186–38192

Ruan B, Söll D (2005) The bacterial YbaK protein is a Cys-$tRNA^{Pro}$ and Cys-$tRNA^{Cys}$ deacylase. J Biol Chem 280:25887–25891

Sankaranarayanan R, Dock-Bregeon AC, Romby P, Caillet J, Springer M, Rees B, Ehresmann C, Ehresmann B, Moras D (1999) The structure of threonyl-tRNA synthetase-tRNA(Thr) complex enlightens its repressor activity and reveals an essential zinc ion in the active site. Cell 97:371–381

Sankaranarayanan R, Dock-Bregeon AC, Rees B, Bovee M, Caillet J, Romby P, Francklyn CS, Moras D (2000) Zinc ion mediated amino acid discrimination by threonyl-tRNA synthetase. Nat Struct Biol 7:461–465

Santoro SW, Anderson JC, Lakshman V, Schultz PG (2003) An archaebacteria-derived glutamyl-tRNA synthetase and tRNA pair for unnatural amino acid mutagenesis of proteins in *Escherichia coli*. Nucleic Acids Res 31:6700–6709

Sasaki HM, Sekine S, Sengoku T, Fukunaga R, Hattori M, Utsunomiya Y, Kuroishi C, Kuramitsu S, Shirouzu M, Yokoyama S (2006) Structural and mutational studies of the amino acid-editing domain from archaeal/eukaryal phenylalanyl-tRNA synthetase. Proc Natl Acad Sci USA 103:14744–14749

Schimmel P, Giegé R, Moras D, Yokoyama S (1993) An operational RNA code for amino acids and possible relationship to genetic code. Proc Natl Acad Sci USA 90:8763–8768

Schmidt E, Schimmel P (1994) Mutational isolation of a sieve for editing in a transfer RNA synthetase. Science 264:265–267

Schmidt E, Schimmel P (1995) Residues in a class I tRNA synthetase which determine selectivity of amino acid recognition in the context of tRNA. Biochemistry 34:11204–11210

Schreier AA, Schimmel PR (1972) Transfer ribonucleic acid synthetase catalyzed deacylation of aminoacyl transfer ribonucleic acid in the absence of adenosine monophosphate and pyrophosphate. Biochemistry 11:1582–1589

Serre L, Verdon G, Choinowski T, Hervouet N, Risler JL, Zelwer C (2001) How methionyl-tRNA synthetase creates its amino acid recognition pocket upon L-methionine binding. J Mol Biol 306:863–876

Sharma N, Furter R, Kast P, Tirrell DA (2000) Efficient introduction of aryl bromide functionality into proteins in vivo. FEBS Lett 467:37–40

Silvian LF, Wang J, Steitz TA (1999) Insights into editing from an ile-tRNA synthetase structure with $tRNA^{ile}$ and mupirocin. Science 285:1074–1077

Sokabe M, Okada A, Yao M, Nakashima T, Tanaka I (2005) Molecular basis of alanine discrimination in editing site. Proc Natl Acad Sci USA 102:11669–11674

Splan KE, Ignatov ME, Musier-Forsyth K (2008) Transfer RNA modulates the editing mechanism used by class II prolyl-tRNA synthetase. J Biol Chem 283:7128–7134

Starzyk RM, Webster TA, Schimmel P (1987) Evidence for dispensable sequences inserted into a nucleotide fold. Science 237:1614–1618

Sternjohn J, Hati S, Siliciano PG, Musier-Forsyth K (2007) Restoring species-specific posttransfer editing activity to a synthetase with a defunct editing domain. Proc Natl Acad Sci USA 104:2127–2132

Sugiura I, Nureki O, Ugaji-Yoshikawa Y, Kuwabara S, Shimada A, Tateno M, Lorber B, Giegé R, Moras D, Yokoyama S, Konno M (2000) The 2.0 Å crystal structure of *Thermus thermophilus* methionyl-tRNA synthetase reveals two RNA-binding modules. Structure Fold Des 8:197–208

Summerer D, Chen S, Wu N, Deiters A, Chin JW, Schultz PG (2006) A genetically encoded fluorescent amino acid. Proc Natl Acad Sci USA 103:9785–9789

Swairjo MA, Otero FJ, Yang XL, Lovato MA, Skene RJ, McRee DE, Ribas de Pouplana L, Schimmel P (2004) Alanyl-tRNA synthetase crystal structure and design for acceptor-stem recognition. Mol Cell 13:829–841

Swairjo MA, Schimmel PR (2005) Breaking sieve for steric exclusion of a noncognate amino acid from active site of a tRNA synthetase. Proc Natl Acad Sci USA 102:988–993

Tamura K, Nameki N, Hasegawa T, Shimizu M, Himeno H (1994) Role of the CCA terminal sequence of tRNA(Val) in aminoacylation with valyl-tRNA synthetase. J Biol Chem 269:22173–22177

Tang Y, Tirrell DA (2002) Attenuation of the editing activity of the *Escherichia coli* leucyl-tRNA synthetase allows incorporation of novel amino acids into proteins in vivo. Biochemistry 41:10635–10645

Tardif KD, Horowitz J (2002) Transfer RNA determinants for translational editing by *Escherichia coli* valyl-tRNA synthetase. Nucleic Acids Res 30:2538–2545

Tardif KD, Liu M, Vitseva O, Hou YM, Horowitz J (2001) Misacylation and editing by *Escherichia coli* valyl-tRNA synthetase: Evidence for two tRNA binding sites. Biochemistry 40: 8118–8125

Terada T, Nureki O, Ishitani R, Ambrogelly A, Ibba M, Söll D, Yokoyama S (2002) Functional convergence of two lysyl-tRNA synthetases with unrelated topologies. Nat Struct Biol 9: 257–262

Torres-Larios A, Dock-Bregeon AC, Romby P, Rees B, Sankaranarayanan R, Caillet J, Springer M, Ehresmann C, Ehresmann B, Moras D (2002) Structural basis of translational control by *Escherichia coli* threonyl tRNA synthetase. Nat Struct Biol 9:343–347

Torres-Larios A, Sankaranarayanan R, Rees B, Dock-Bregeon AC, Moras D (2003) Conformational movements and cooperativity upon amino acid, ATP and tRNA binding in threonyl-tRNA synthetase. J Mol Biol 331:201–211

Tsui WC, Fersht AR (1981) Probing the principles of amino acid selection using the alanyl-tRNA synthetase from *Escherichia coli*. Nucleic Acids Res 9:4627–4637

Tukalo M, Yaremchuk A, Fukunaga R, Yokoyama S, Cusack S (2005) The crystal structure of leucyl-tRNA synthetase complexed with tRNA(Leu) in the post-transfer-editing conformation. Nat Struct Mol Biol 12:923–930

Turner JM, Graziano J, Spraggon G, Schultz PG (2006) Structural plasticity of an aminoacyl-tRNA synthetase active site. Proc Natl Acad Sci USA 103:6483–6488

van Hest JC, Tirrell DA (1998) Efficient introduction of alkene functionality into proteins in vivo. FEBS Lett 428:68–70

von der Haar F, Cramer F (1976) Hydrolytic action of aminoacyl-tRNA synthetases from baker's yeast: "chemical proofreading" preventing acylation of tRNA(Ile) with misactivated valine. Biochemistry 15:4131–4138

Vu MT, Martinis SA (2007) A unique insert of leucyl-tRNA synthetase is required for aminoacylation and not amino acid editing. Biochemistry 46:5170–5176

Wang L, Brock A, Herberich B, Schultz PG (2001) Expanding the genetic code of *Escherichia coli*. Science 292:498–500

Wang P, Tang Y, Tirrell DA (2003) Incorporation of trifluoroisoleucine into proteins in vivo. J Am Chem Soc 125:6900–6906

Webster T, Tsai H, Kula M, Mackie GA, Schimmel P (1984) Specific sequence homology and three-dimensional structure of an aminoacyl transfer RNA synthetase. Science 226:1315–1317

Williams AM, Martinis SA (2006) Mutational unmasking of a tRNA-dependent pathway for preventing genetic code ambiguity. Proc Natl Acad Sci USA 103:3586–3591

Wolf YI, Aravind L, Grishin NV, Koonin EV (1999) Evolution of aminoacyl-tRNA synthetases – analysis of unique domain architectures and phylogenetic trees reveals a complex history of horizontal gene transfer events. Genome Res 9:689–710

Wong FC, Beuning PJ, Nagan M, Shiba K, Musier-Forsyth K (2002) Functional role of the prokaryotic proline-tRNA synthetase insertion domain in amino acid editing. Biochemistry 41:7108–7115

Wong FC, Beuning PJ, Silvers C, Musier-Forsyth K (2003) An isolated class II aminoacyl-tRNA synthetase insertion domain is functional in amino acid editing. J Biol Chem 278:52857–52864

Wu N, Deiters A, Cropp TA, King D, Schultz PG (2004) A genetically encoded photocaged amino acid. J Am Chem Soc 126:14306–14307

Xie J, Schultz PG (2005) Adding amino acids to the genetic repertoire. Curr Opin Chem Biol 9:548–554

Xu MG, Li J, Du X, Wang ED (2004) Groups on the side chain of T252 in *Escherichia coli* leucyl-tRNA synthetase are important for discrimination of amino acids and cell viability. Biochem Biophys Res Commun 318:11–16

Yamane T, Hopfield JJ (1977) Experimental evidence for kinetic proofreading in the aminoacylation of tRNA by synthetase. Proc Natl Acad Sci USA 74:2246–2250

Yaremchuk A, Tukalo M, Grotli M, Cusack S (2001) A succession of substrate induced conformational changes ensures the amino acid specificity of *Thermus thermophilus* prolyl-tRNA synthetase: comparison with histidyl-tRNA synthetase. J Mol Biol 309:989–1002

Yarus M (1972) Phenylalanyl-tRNA synthetase and isoleucyl-tRNA^{Phe}: A possible verification mechanism for aminoacyl-tRNA. Proc Natl Acad Sci USA 69:1915–1919

Yarus M (1973) Verification of misacylated tRNA^{phe} is apparently carried out only by phenylalanyl-tRNA synthetase. Nat New Biol 245:5–6

Zhai Y, Martinis SA (2005) Two conserved threonines collaborate in the *Escherichia coli* leucyl-tRNA synthetase amino acid editing mechanism. Biochemistry 44:15437–15443

Zhai Y, Nawaz MH, Lee KW, Kirkbride E, Briggs JM, Martinis SA (2007) Modulation of substrate specificity within the amino acid editing site of leucyl-tRNA synthetase. Biochemistry 46:3331–3337

Zhang H, Huang K, Li Z, Banerjei L, Fisher KE, Grishin NV, Eisenstein E, Herzberg O (2000) Crystal structure of YbaK protein from *Haemophilus influenzae* (HI1434) at 1.8 Å resolution: Functional implications. Proteins 40:86–97

Zhang Z, Alfonta L, Tian F, Bursulaya B, Uryu S, King DS, Schultz PG (2004) Selective incorporation of 5-hydroxytryptophan into proteins in mammalian cells. Proc Natl Acad Sci USA 101:8882–8887

Specialized Components of the Translational Machinery for Unnatural Amino Acid Mutagenesis: tRNAs, Aminoacyl-tRNA Synthetases, and Ribosomes

C. Köhrer(✉) and U.L. RajBhandary

Contents

Abstract Site-specific incorporation of unnatural amino acids (amino acid analogues) into proteins adds a new dimension to studies of protein structure and function. Here, we describe in detail the development of methods for site-specific incorporation of unnatural amino acids with novel chemical, physical and biological properties using specialized suppressor tRNAs alongside engineered aminoacyl-tRNA synthetases and ribosomes.

1 Introduction

Incorporation of unnatural amino acids (amino acid analogues) with novel chemical, physical, and biological properties into proteins has added a new dimension to protein engineering and has provided the molecular tools for (1) studies of protein folding,

C. Köhrer
Department of Biology, Massachusetts Institute of Technology, 77 Massachusetts Avenue, Cambridge, MA 02139, USA, e-mail: koehrer@mit.edu

C. Köhrer and U.L. RajBhandary (eds.) *Protein Engineering*.
Nucleic Acids and Molecular Biology 22,

structure, and function and (2) design of proteins with new or enhanced characteristics. Ribosome-based incorporation of unnatural amino acids, often termed "unnatural amino acid mutagenesis," has expanded the repertoire of building blocks available for protein synthesis beyond the 20 naturally occurring amino acids. The most commonly used unnatural amino acids to date include those that are fluorescent or photoactivatable, those that carry heavy atoms (e.g., iodine) or reactive side chains (e.g., keto amino acids), and those that mimic post-translational modifications such as phosphorylation or glycosylation. Amino acid analogues that can be used as affinity or spectroscopic probes and those that introduce backbone modifications have opened up many new applications of protein engineering. Unnatural amino acid mutagenesis offers an important alternative to conventional chemical synthesis and modification of peptides/proteins, which are often limited by the size of the peptide/proteins that can be synthesized, sample heterogeneity, and/or low yield.

Different strategies for global (residue-specific) and site-specific incorporation of unnatural amino acids into proteins have been developed, using both prokaryotic and eukaryotic systems. A detailed description of global amino acid replacement and its application can be found in the chapters by Beatty and Tirrell and Mascarenhas et al. in this volume. This chapter will focus on methods for site-specific incorporation of unnatural amino acids using specialized suppressor transfer RNAs (tRNAs) alongside engineered aminoacyl-tRNA synthetases and ribosomes. We provide an introduction to the basic principles and highlight recent studies that have brought about the optimization of individual components of the translational apparatus leading to improved systems that are now available for use.

2 Basic Principles of Unnatural Amino Acid Mutagenesis

2.1 Expansion of the Genetic Code

The most common strategy for site-specific insertion of unnatural amino acids, both in vitro and in vivo, relies on the readthrough of an amber (UAG) stop codon in a messenger RNA (mRNA) by an amber suppressor tRNA that is aminoacylated with the desired unnatural amino acid. Thus, the standard genetic code is "expanded" by reassigning the amber stop codon to a specific unnatural amino acid (Noren et al. 1989) (Fig. 1a). To avoid incorporation of natural amino acids at the designated site in the target protein, the key requirement for this approach is that the amber suppressor tRNA be "orthogonal" and not be a substrate for any of the endogenous aminoacyl-tRNA synthetases present in the respective expression system. In the beginning, efforts were limited almost exclusively to the use of various amber suppressor tRNAs. More recently, the genetic code has been expanded further by identifying orthogonal ochre and opal suppressor tRNAs, which translate ochre (UAA) and opal (UGA) codons, respectively, and frameshift suppressor tRNAs, which translate four- or five-base codons. Yet another approach to expansion of the

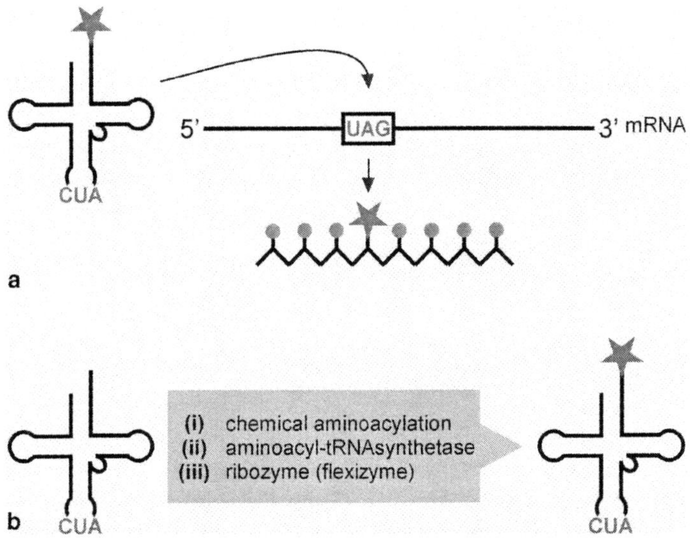

Fig. 1 Transfer RNA (tRNA)-mediated protein engineering. **a** General approach to site-specific incorporation of unnatural amino acids. **b** Different approaches can be used to attach the desired unnatural amino acid to the suppressor tRNA. *i* chemical aminoacylation of suppressor tRNAs, i.e., ligation of a truncated tRNA lacking the last two 3′-terminal nucleotides to a chemically synthesized dinucleotide carrying the unnatural amino acid, as pioneered by Hecht and coworkers (Heckler et al. 1984). Alternatively, tRNAs can be aminoacylated chemically using amino acid thioesters linked to a peptide nucleic acid (peptide nucleic acid thioesters; Ninomiya et al. 2004). Recently, Sisido and coworkers described a simplified protocol for acylation of dinucleotides in cationic micelles (Ninomiya 2003). *ii* use of a wild-type (Hartman et al. 2006, 2007) or engineered aminoacyl-tRNA synthetase that aminoacylates the suppressor tRNA with the desired unnatural amino acid (Wang et al. 2001; Sakamoto et al. 2002; Chin et al. 2003). *iii* ribozyme (e.g., Flexizyme)-catalyzed aminoacylation (Saito et al. 2001; Saito and Suga 2001; Murakami et al. 2002, 2003, 2006)

genetic code involves the introduction of unnatural base pairs (e.g., *isoG:isoC*), an idea first developed by Benner and colleagues (Bain et al. 1992).

Much work on unnatural amino acid mutagenesis relies on cell-free protein synthesis systems using orthogonal suppressor tRNAs that are aminoacylated (1) chemically (Heckler et al. 1984; Ninomiya 2003, 2004; detailed in the chapter by Hecht, this volume), (2) enzymatically using wild-type or mutant aminoacyl-tRNA synthetases, or (3) with the help of engineered ribozymes (Saito et al. 2001; Saito and Suga 2001; Murakami et al. 2002, 2003, 2006) (Fig. 1b). Recently, Szostak and coworkers have compiled a list of over 90 amino acid analogues – including β-amino acids, *N*-methyl amino acids, and α,α-disubstituted amino acids – that are substrates for wild-type aminoacyl-tRNA synthetases in an in vitro reaction (Hartman et al. 2006; Hartman et al. 2007). Pre-aminoacylated suppressor tRNAs have also been imported into *Xenopus* oocytes and mammalian cells by means of microinjection, electroporation, or transfection (Nowak et al. 1995; Köhrer et al. 2001, 2003; Ilegems et al. 2002, 2004; Monahan et al. 2003). The chapter by Dougherty in this volume describes the highly successful application of this approach to studies of ion channels, GABA, and nicotinic receptors expressed in *Xenopus* oocytes.

The use of pre-aminoacylated suppressor tRNAs provides the most general approach to incorporation of unnatural amino acids, since virtually any unnatural amino acid can be attached to the tRNA using the methods mentioned above. However, this approach is often limited by the low yields of target protein being synthesized, as the availability of the unnatural amino acid for incorporation is defined by the amount of the pre-aminoacylated tRNA that is supplied. In contrast, the use of orthogonal suppressor tRNA/aminoacyl-tRNA synthetase pairs allows the continuous regeneration of the aminoacyl-tRNA during the course of protein synthesis and thereby makes possible the efficient synthesis in vivo of proteins containing unnatural amino acids. These pairs, which function alongside the native cellular tRNA/aminoacyl-tRNA synthetase pairs, consist of an orthogonal suppressor tRNA and an orthogonal aminoacyl-tRNA synthetase that aminoacylates only the suppressor tRNA but no other tRNA in the cell. Once such a pair has been identified for use in *Escherichia coli*, yeast, or mammalian cells, the orthogonal aminoacyl-tRNA synthetase has to be modified in such a way that it activates a specific unnatural amino acid but not the normal amino acid and attaches it to the orthogonal suppressor tRNA.

2.2 *Orthogonal Suppressor tRNA/Aminoacyl-tRNA Synthetase Pairs*

The identification of most orthogonal suppressor tRNA/aminoacyl-tRNA synthetase pairs exploits the kingdom-specific recognition of certain tRNAs by their cognate aminoacyl-tRNA synthetases and requires the import of both the suppressor tRNA and the aminoacyl-tRNA synthetase from a heterologous organism. For example, bacterial $tRNA^{Tyr}$ or an amber suppressor tRNA derived from it is not recognized by eukaryotic tyrosyl-tRNA synthetase (TyrRS); hence, an amber suppressor tRNA derived from *E. coli* or *Bacillus stearothermophilus* $tRNA^{Tyr}$ is not active in suppression in yeast (Edwards and Schimmel 1990; Chin et al. 2003) or in mammalian cells (Sakamoto et al. 2002) unless the bacterial TyrRS is coexpressed alongside the suppressor tRNA. It was also confirmed that the bacterial TyrRS is orthogonal in eukaryotic cells, i.e., does not recognize the eukaryotic $tRNA^{Tyr}$ or any other eukaryotic tRNA (Clark and Eyzaguirre 1962). Similarly, a suppressor tRNA/aminoacyl-tRNA synthetase pair derived from the archaeal *Methanococcus jannaschii* $tRNA^{Tyr}$/TyrRS pair was shown to be orthogonal in *E. coli* (Wang et al. 2001); in this case, the archaeal suppressor tRNA had to be mutated to eliminate some cross-reactivity with bacterial synthetases (Fig. 2a).

An example of the first orthogonal suppressor tRNA/aminoacyl-tRNA synthetase pair developed for use in mammalian cells (Fig. 2b) comes from the use of an amber suppressor tRNA derived from *E. coli* $tRNA^{Gln}$ (*hsup2am*) and *E. coli* glutaminyl-tRNA synthetase (GlnRS) (Drabkin et al. 1996). The suppressor tRNA was expressed in mammalian cells and its activity as a suppressor tRNA in readthrough of amber

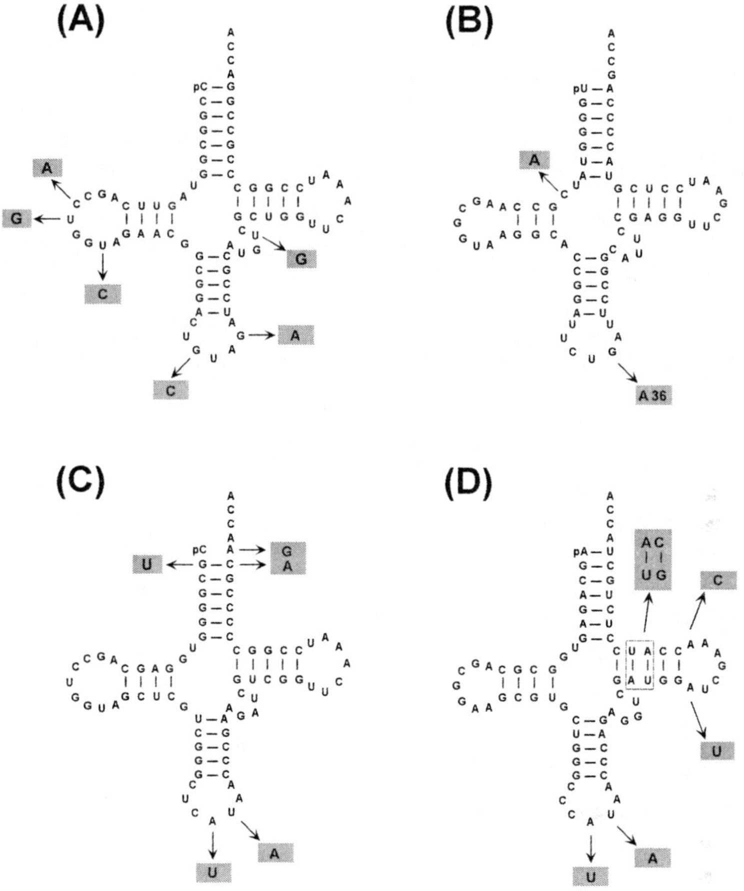

Fig. 2 Orthogonal suppressor tRNA/aminoacyl-tRNA synthetase pairs. **a** Suppressor tRNA derived from *Methanococcus jannaschii* tRNA^{Tyr} for use in *Escherichia coli* along with *M. jannaschii* TyrRS (Wang et al. 2001). **b** Suppressor tRNA derived from *E. coli* tRNA^{Gln} (*hsup2am*) for use in mammalian cells and yeast along with *E. coli* GlnRS (Drabkin et al. 1996; Köhrer et al. 2004). **c** Suppressor tRNA derived from *E. coli* tRNA$_2^{fMet}$ (*fMam*) for use in *E. coli* along with *Saccharomyces cerevisiae* TyrRS (Lee and RajBhandary 1991; Kowal et al. 2001). **d** Suppressor tRNA derived from human initiator tRNA$_i^{Met}$ (*hM2am*) for use in mammalian cells and yeast along with *E. coli* GlnRS (Drabkin et al. 1998; Kowal et al. 2001; Köhrer et al. 2004)

codons in a reporter gene was shown to be strictly dependent on coexpression of *E. coli* GlnRS. Based on this orthogonal suppressor tRNA/aminoacyl-tRNA synthetase pair, a complete set of highly active orthogonal amber, ochre, and opal suppressor tRNA/*E. coli* GlnRS pairs has been generated (Köhrer et al. 2004).

In an alternative approach, orthogonal suppressor tRNA/aminoacyl-tRNA synthetase pairs were developed for *E. coli*, mammalian cells, and yeast, containing a suppressor tRNA derived from the endogenous initiator tRNA (Lee and RajBhandary 1991; Drabkin et al. 1998; Kowal et al. 2001). A mutant of *E. coli* initiator tRNA$_2^{fMet}$

(*fMam*; U35A36/U2:A71/G72) is orthogonal in *E. coli* (Fig. 2c). Mutations in the anticodon abolish recognition by *E. coli* methionyl-tRNA synthetase, but allow recognition by GlnRS. Subsequent mutations in the acceptor stem abolish recognition by GlnRS and initiation factor IF2 (RajBhandary 1994). The mutant tRNA is, however, recognized by *Saccharomyces cerevisiae* TyrRS, and acts as an amber suppressor in yeast (Lee and RajBhandary 1991) and in *E. coli* cells expressing *S. cerevisiae* TyrRS (Chow and RajBhandary 1993; Kowal et al. 2001). Another amber suppressor tRNA, orthogonal in mammalian cells and yeast, was derived from the human initiator $tRNA_i^{Met}$ (Drabkin et al. 1998; Kowal et al. 2001) (Fig. 2d). The tRNA mutant (*hM2am*) containing mutations in the anticodon (U35A36) and TψC-arm (U50G51:C63A64/U54/C60) is a substrate for *E. coli* GlnRS, and allows suppression of amber alleles in the yeast tester strain *S. cerevisiae* HEY301-129 (*met8-1am, trp1-1am, his4-580am*) expressing *E. coli* GlnRS, as indicated by growth on minimal media lacking methionine, tryptophan, or histidine (Kowal et al. 2001).

It is important to point out that the transcription of tRNA genes in bacteria and mammalian cells differs greatly. In mammalian cells, intragenic transcriptional control elements, A and B boxes, direct transcription of tRNA genes by RNA polymerase III (Sprague 1994). This requirement limits the efficient synthesis of certain bacterial tRNAs in mammalian cells. For example, an amber suppressor derived from *E. coli* $tRNA^{Tyr}$ is poorly transcribed due to the lack of a proper A box (TRGCNNAGY; positions 8–16; Sprague 1994), while an amber suppressor derived from *B. stearothermophilus* $tRNA^{Tyr}$ contains a perfect A box and is, therefore, suitable for use in mammalian cells (Sakamoto et al. 2002). This problem can be avoided by placing the tRNA gene under control of the H1 promoter (Wang et al. 2007b). The H1 promoter drives the expression of the human H1RNA and is recognized by polymerase III; however, it works as an external promoter and does not require intragenic control elements (Myslinski et al. 2001).

Since the conception of the first orthogonal suppressor tRNA/aminoacyl-tRNA synthetase pairs in the 1990s, various such pairs have been identified. These include archaeal $tRNA^{Leu}$/leucyl-tRNA synthetase (LeuRS) (Anderson and Schultz 2003), $tRNA^{Glu}$/glutamyl-tRNA synthetase (GluRS) (Santoro et al. 2003), and $tRNA^{Lys}$/lysyl-tRNA synthetase (LysRS) (Anderson et al. 2004) for use in *E. coli*, and *E. coli* $tRNA^{Leu}$/LeuRS (Wu et al. 2004) and *B. subtilis* $tRNA^{Trp}$/tryptophanyl-tRNA synthetase (TrpRS) (Zhang et al. 2004) for use in yeast and mammalian cells (Table 1).

2.3 *Evolution of Orthogonal Aminoacyl-tRNA Synthetases That Accept Unnatural Amino Acids*

Having identified orthogonal suppressor tRNA/aminoacyl-tRNA synthetase pairs, the next step is to mutagenize the aminoacyl-tRNA synthetase in such a way that it selectively activates a specific unnatural amino acid and subsequently attaches it to the suppressor tRNA instead of the normal amino acid (reviewed in Hendrickson et al. 2004; Köhrer and RajBhandary 2005; Wang et al. 2006; Xie and Schultz

Table 1 Examples of orthogonal suppressor tRNA/aminoacyl-tRNA synthetase pairs described in this review

Suppressor tRNA/aminoacyl-tRNA synthetase pair	References
Developed for a prokaryotic host	
Escherichia coli $tRNA_2^{fMet}$/*Saccharomyces cerevisiae* TyrRS	Kowal et al. (2001)
Methanococcus jannaschii $tRNA^{Tyr}$/*M. jannaschii* TyrRS	Wang et al. (2001)
Halobacterium sp. $tRNA^{Leu}$/*Methanobacterium thermoautotrophicum* LeuRS	Anderson and Schultz (2003)
Archaeal[a] $tRNA^{Glu}$/*Pyrococcus horikoshii* GluRS	Santoro et al. (2003)
Archaeal[b] $tRNA^{Lys}$/*P. horikoshii* type I LysRS	Anderson et al. (2004)
Developed for a eukaryotic host	
E. coli $tRNA^{Gln}$/*E. coli* GlnRS	Drabkin et al. (1996), Köhrer et al. (2004)
Human $tRNA_i^{Met}$/*E. coli* GlnRS	Drabkin et al. (1998), Kowal et al. (2001)
E. coli or *Bacillus stearothermophilus* $tRNA^{Tyr}$/*E. coli* TyrRS	Sakamoto et al. (2002), Chin et al. (2003)
B. subtilis $tRNA^{Trp}$/*B. subtilis* TrpRS	Zhang et al. (2004)
E. coli $tRNA^{Leu}$/*E. coli* LeuRS	Wu et al. (2004)

tRNA transfer RNA, *TyrRS* tyrosyl-tRNA synthetase, *LeuRS* leucyl-tRNA synthetase, *GluRS* glutamyl-tRNA synthetase, *LysRS* lysyl-tRNA synthetase, *GlnRS* glutaminyl-tRNA synthetase.

[a] tRNA is derived from the consensus sequence obtained from a multiple-sequence alignment of archaeal $tRNA^{Glu}$ sequences.

[b] Frameshift suppressor tRNA derived from the archaeal $tRNA^{Lys}$.

2006). In this section, we provide three examples of such specialized aminoacyl-tRNA synthetases for site-specific incorporation of unnatural amino acids into proteins in *E. coli*, mammalian cells, and yeast. In general, two different approaches are taken to alter the amino acid specificity of an aminoacyl-tRNA synthetase: the first one uses libraries of synthetase variants that are subjected to appropriate selection and counterselection schemes (directed evolution; examples 1 and 2); the second one is based on rational design by introducing changes in the amino acid binding pocket in a site-specific manner (example 3).

2.3.1 Site-Specific Incorporation of Unnatural Amino Acids into Proteins in *E. coli*

Example 1: Directed evolution of an engineered aminoacyl-tRNA synthetase for incorporation of *O*-methyl-L-tyrosine (OMeTyr).

The first example of site-specific incorporation of an unnatural amino acid in *E. coli* came from the laboratory of Schultz and coworkers (Wang et al. 2001). The orthogonal suppressor tRNA/aminoacyl-tRNA synthetase pair used was derived from the *M. jannaschii* $tRNA^{Tyr}$/TyrRS system. *M. jannaschii* TyrRS fulfills the basic requirement of orthogonality in *E. coli* and also lacks a typical editing mechanism, thereby reducing the risk of rejecting the unnatural amino acid by pre- or post-transfer editing (Jakubowski and Fersht 1981; see also the chapter by Mascarenhas et al., this volume). While the amber suppressor tRNA derived from *M. jannaschii* $tRNA^{Tyr}$ showed high efficiency in translation in *E. coli*, it was a partial substrate for one or more *E. coli* aminoacyl-tRNA synthetases. To improve the orthogonality of this amber suppressor tRNA, 11 nucleotides that do not interact directly with *M. jannaschii* TyrRS were randomly mutagenized, and the resulting tRNA library was passed through positive and negative selections to isolate a suppressor tRNA that is not recognized by *E. coli* aminoacyl-tRNA synthetases but is efficiently aminoacylated by *M. jannaschii* TyrRS (Fig. 2a).

To alter the amino acid specificity of *M. jannaschii* TyrRS to aminoacylate the newly generated suppressor tRNA with the unnatural amino acid *O*-methyl-L-tyrosine (OMeTyr), a directed evolution approach was developed. On the basis of the crystal structure of the bacterial TyrRS homologue from *B. stearothermophilus* (Brick et al. 1989), a random library of mutants targeting five residues (Tyr32, Glu107, Asp158, Ile159, and Leu162) in the amino acid binding pocket of *M. jannaschii* TyrRS was generated. This "active-site" library of *M. jannaschii* TyrRS mutants was subjected to iterative rounds of stringent positive and negative selections, resulting in the successful isolation of a mutant TyrRS (Tyr32→Gln32, Glu107→Thr107, Asp158→Ala158, and Leu162→Pro162) that specifically attaches OMeTyr to the amber suppressor tRNA, thus allowing incorporation of the unnatural amino acid into a protein (Fig. 3). In similar experiments, more than 20 unnatural amino acids have been incorporated using the *M. jannaschii* $tRNA^{Tyr}$/TyrRS system (Xie and Schultz 2006); most recently, the evolution of *M. jannaschii* TyrRS mutants specific for 3-amino-L-tyrosine (an analogue suitable for UV–vis and electron paramagnetic resonance studies; Seyedsayamdost et al. 2007), *p*-carboxymethyl-L-phenylalanine (a phosphotyrosine mimetic; Xie et al. 2007), and 4-trifluoromethyl-L-phenylalanine (a ^{19}F-NMR label; Jackson et al. 2007) was reported.

2.3.2 Site-Specific Incorporation of Unnatural Amino Acids into Proteins in Eukaryotes

Example 2: Directed evolution of engineered aminoacyl-tRNA synthetases for incorporation of tyrosine analogues in yeast and mammalian cells.

The synthesis of numerous eukaryotic proteins of scientific, pharmacological, or industrial interest (e.g., antibodies, transcription factors, membrane proteins) necessitates the availability of eukaryotic expression systems that allow proper protein folding and post-translational modifications (e.g., phosphorylation, glycosylation), critical for the structure and function of such proteins. In 2003, an orthogonal suppressor tRNA/aminoacyl-tRNA synthetase pair was used for the site-specific incorporation of several unnatural amino acids into proteins in *S. cerevisiae* (Chin et al. 2003). A suppressor tRNA derived from *E. coli* $tRNA^{Tyr}$ was used alongside *E. coli* TyrRS. *E. coli* TyrRS mutants specific for various tyrosine analogues including the keto amino acids *p*-acetyl-L-phenylalanine (Acp) and *p*-benzoyl-L-phenylalanine

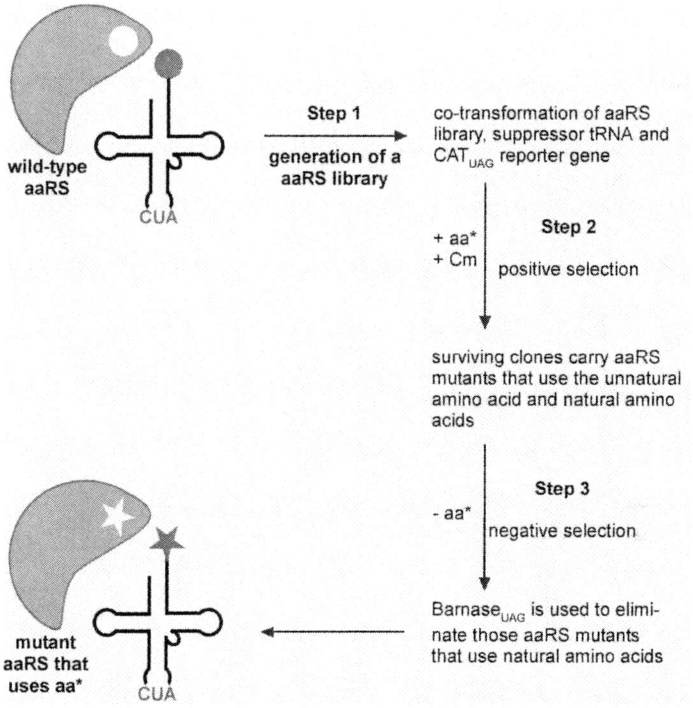

Fig. 3 Expanding the genetic code in *E. coli* by directed evolution. *Step 1*: Generation of an aminoacyl-tRNA synthetase (*aaRS*) library, in which critical residues in the amino acid binding pocket are randomized. Transformation of *E. coli* with the aminoacyl-tRNA synthetase library and the cognate suppressor tRNA. *Step 2*: Positive selection is based on suppression of an amber stop codon at a permissive site in chloramphenicol acetyltransferase (*CAT*) in the presence of the unnatural amino acid (*aa**). The resulting clones carry functional aminoacyl-tRNA synthetase mutants that use the unnatural amino acid or any of the natural amino acids. *Step 3*: Negative selection is based on suppression of one or more amber codons in the highly toxic barnase in the absence of the unnatural amino acid. Clones with aminoacyl-tRNA synthetase mutants that use natural amino acids are eliminated; only those clones with aminoacyl-tRNA synthetase mutants incorporating the desired unnatural amino acid, which is not present in the medium, survive, *Cm* chloramphenicol (Wang et al. 2001)

(Bzp) were generated through directed evolution by applying positive and negative selection processes, analogous to those developed in *E. coli* (see Sect. 2.3.1).

The combination of an amber suppressor tRNA derived from *B. stearothermophilus* $tRNA^{Tyr}$ with one of the TyrRS variants originally developed for use in yeast (Chin et al. 2003) was applied successfully for incorporation of the photoreactive keto amino acid Bzp into the adaptor protein Grb2 in mammalian cells (Hino et al. 2005). Bzp was incorporated at ten sequential positions in the ligand binding pocket of the SH2 domain of Grb2 to study in vivo cross-linking of Grb2 with the EGF receptor and other proteins. Studies such as these demonstrate the application of unnatural amino acid mutagenesis to analyses of protein–protein interactions in vivo in mammalian cells in the context of their native environment.

Additional reports showed that mammalian cells are highly suitable for selective and efficient incorporation of unnatural amino acids (Liu et al. 2007; Wang et al. 2007b). Recently, the laboratories of Sakmar and RajBhandary have applied unnatural amino acid mutagenesis to G-protein-coupled receptors (GPCRs) (Ye et al. 2008). Two keto amino acids, Acp and Bzp, were incorporated into the chemokine receptor CCR5 (major coreceptor for the human immunodeficiency virus) and rhodopsin (visual photoreceptor) in mammalian cells. Functional GPCRs that contained Acp or Bzp at various sites were obtained in high yield (Fig. 4). Rhodopsin containing Acp at three different sites was also purified from mammalian cells (0.5–2 μg/10^7 cells) and reacted with fluorescein hydrazide in vitro to produce fluorescently

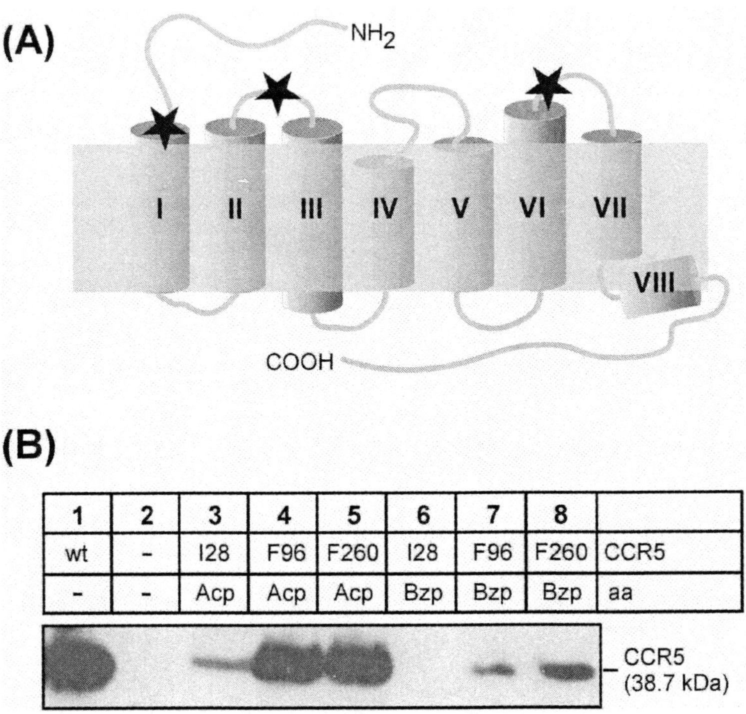

Fig. 4 Expansion of the genetic code in eukaryotes. Site-specific incorporation of keto amino acids into G-protein-coupled receptors. **a** The secondary structure of CCR5. Positions in CCR5 subjected to site-specific incorporation of unnatural amino acids are indicated. **b** Expression of functional CCR5 mutants containing *p*-acetyl-L-phenylalanine (*Acp*) or *p*-benzoyl-L-phenylalanine (*Bzp*) at positions 28, 96, or 260. HEK293T cells were transfected with plasmids carrying the genes for wild-type CCR5 or CCR5 mutant with an amber mutation at position I28, F96, or F260. Plasmids encoding a suppressor tRNA (derived from *Bacillus stearothermophilus* $tRNA^{Tyr}$), and *E. coli* TyrRS variants specific for Acp or Bzp were cotransfected, and the corresponding unnatural amino acids (Acp or Bzp) were provided in the cell media as indicated. Cell lysates were analyzed by immunoblot analysis using an antibody against the C-terminal portion of CCR5 (Ye et al. 2008)

labeled rhodopsin. Current dynamic studies of GPCRs rely on fluorescence techniques that are often hampered by inherent limitations; the two most common methods to introduce site-specific fluorescent labels are maleimide chemistry targeting cysteine residues and the use of green fluorescent protein fusions. The site-specific incorporation of reactive keto groups such as Acp or Bzp into GPCRs allows their reaction with different reagents to introduce a variety of fluorescent, spectroscopic, and other probes, thereby adding a new dimension to studies of GPCRs, including GPCR–G protein interactions.

Example 3: Rational design of engineered aminoacyl-tRNA synthetases for incorporation of unnatural amino acids in mammalian cells.

Yokoyama and colleagues (Kiga et al. 2002) developed an *E. coli* TyrRS mutant specific for 3-iodo-L-tyrosine by rational design. On the basis of the crystal structure of *B. stearothermophilus* TyrRS (Brick et al. 1989), three residues in the amino acid binding pocket were chosen for site-directed mutagenesis. A double-mutant of *E. coli* TyrRS (Tyr37→Val37, Gln195→Cys195) was found to recognize 3-iodo-L-tyrosine specifically and was used for efficient incorporation of the analogue into proteins in a wheat germ cell-free system (Kiga et al. 2002) and in mammalian cells (Sakamoto et al. 2002). A high-resolution crystal structure of the *E. coli* TyrRS Val37Cys195 mutant complexed with 3-iodo-L-tyrosine revealed the structural details of 3-iodo-L-tyrosine recognition (Kobayashi et al. 2005). Van der Waals interactions between the iodine atom and Val37 and Cys195 stabilize the position of 3-iodo-L-tyrosine in the amino acid binding pocket (Fig. 5).

A similar approach was taken to engineer an aminoacyl-tRNA synthetase specific for 5-hydroxy-L-tryptophan (Zhang et al. 2004). A single point mutation in *B. subtilis* TrpRS (Val144→Pro144) allowed incorporation of 5-hydroxy-L-tryptophan in mammalian cells when coexpressed with an opal suppressor tRNA derived from *B. subtilis* $tRNA^{Trp}$.

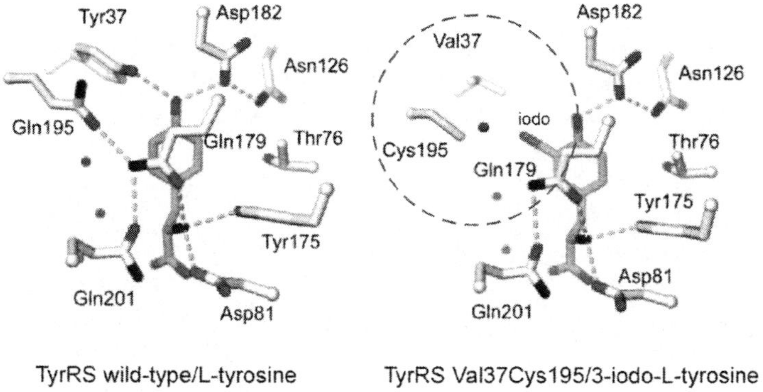

Fig. 5 Site-specific incorporation of 3-iodo-L-tyrosine in mammalian cells. Structural basis for recognition of 3-iodo-L-tyrosine by the mutant of *E. coli* TyrRS Val37Cys195 (Kobayashi et al. 2005)

3 Engineering the Perfect Host for Unnatural Amino Acid Mutagenesis: Optimization and Fine-Tuning of Unnatural Amino Acid Mutagenesis by Manipulating Individual Components of the Translational Machinery

It is well known that nonsense suppression depends upon several factors, including the context of the stop codon within the mRNA (Bossi and Roth 1980; Buckingham 1994), the nature of the suppressor tRNA, and the nature of the unnatural amino acid. Increased suppression efficiency is essential, particularly for the synthesis of proteins carrying two or more amino acid analogues, requiring two or more suppression events in the same mRNA. This section will showcase various approaches directed towards improving the efficiency of unnatural amino acid mutagenesis, by creating new suppressor tRNAs or modulating the activity and/or specificity of translation factors and the ribosome.

3.1 Suppressor tRNAs with Enhanced Activities

Early work on unnatural amino acid mutagenesis involved mainly the use of amber suppressor tRNAs. Köhrer et al. (2001, 2003) showed that the import into mammalian cells of ochre suppressor tRNAs, derived from bacterial tRNAs, led to specific suppression of an ochre codon in a reporter mRNA. It was further shown that import of a mixture of pre-aminoacylated amber and ochre suppressor tRNAs led to concomitant suppression of an amber and an ochre codon in an mRNA, suggesting that this approach could be used for the synthesis of proteins carrying two different unnatural amino acids in mammalian cells (Köhrer et al. 2003).

With the objective of expanding the nature and number of unnatural amino acids that can be introduced into proteins, the generation of a complete set of orthogonal amber, ochre, and opal suppressor tRNA/aminoacyl-tRNA synthetase pairs for use in mammalian cells was described (Köhrer et al. 2004). Amber, ochre, and opal suppressor tRNAs (*hsup2am, hsup2oc*, and *hsup2op*), derived from *E. coli* $tRNA^{Gln}$, were generated for use in mammalian cells. The activity of each suppressor tRNA was shown to be completely dependent upon the coexpression of *E. coli* GlnRS.

The activity of a suppressor tRNA in protein synthesis is affected by sequences in and around the anticodon loop and stem and by base modifications, especially those in the anticodon loop (Colby et al. 1976; Yanofsky and Soll 1977; Yarus 1982; Yarus et al. 1986; Agris 2004; Agris et al. 2007). For example, modification of A37 located next to the anticodon strengthens the interaction between codon and anticodon (Ericson and Björk 1991; Björk 1995). The enzyme responsible for modifying the A37 residue, the dimethylallyl diphosphate:tRNA dimethylallyltransferase, has been identified in *E. coli*, yeast, and mammalian cells; its minimal substrate requirements consists of a stretch of three As, A36-A37-A38, in the anticodon loop (Motorin et al. 1997). To improve the activity of the *hsup2am, hsup2oc*, and *hsup2op* suppressor tRNAs (Köhrer et al. 2004), mutations were introduced into the anticodon loop (U32C, U38A), creating an A36-A37-A38 recognition element and mimicking an anticodon loop configuration found in many strong suppressor tRNAs. The resulting tRNAs had significantly increased activities in suppression (36-fold for amber, 156-fold for ochre, and 200-fold for opal suppressor tRNA; Fig. 6), and have been used in combination to concomitantly suppress two or three termination codons in an mRNA. Important applications of such enhanced suppressor tRNAs, which are used alongside the appropriate aminoacyl-tRNA synthetases, include site-specific incorporation of two or, possibly, even three different unnatural amino acids and regulated suppression of amber, ochre, and opal codons in mammalian cells.

THG73, an amber suppressor tRNA^{Gln} from *Tetrahymena thermophila* (with a G73 mutation), has been used successfully for the incorporation of more than 100 different unnatural amino acids by injection of pre-aminoacylated tRNA into *Xenopus* oocytes (Beene et al. 2003; for details see the chapter by Dougherty, this volume). However, under certain conditions (e.g., increasing amounts of tRNA, extended incubation

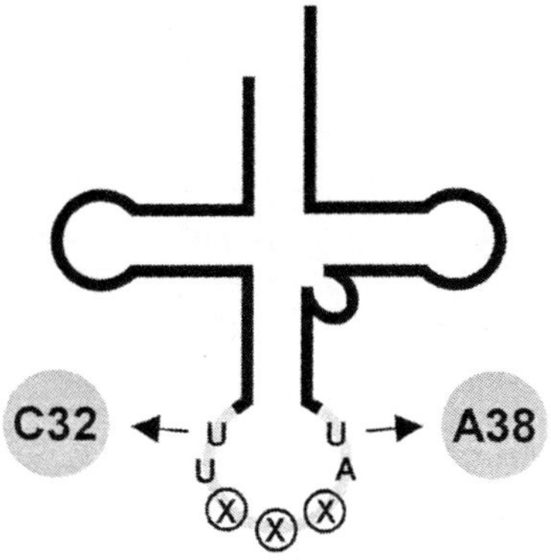

anticodon XXX	amber CUA	ochre UUA	opal UCA
U32U38	0.95%	0.03%	0.05%
	↓ **36 x**	↓ **156 x**	↓ **200 x**
C32A38	34.5%	4.54%	10.4%

Fig. 6 Complete set of orthogonal amber, ochre, and opal suppressor tRNAs for use in eukaryotic cells. Suppression efficiencies of individual suppressor tRNAs in mammalian cells, cotransfected with a plasmid for *E. coli* GlnRS and a luciferase reporter construct, are indicated (Köhrer et al. 2004)

times), THG73 is mis-aminoacylated by endogenous aminoacyl-tRNA synthetases with glutamine. A library of *T. thermophila* glutamine amber suppressor tRNAs with mutations in the acceptor stem was generated and analyzed in detail, along with an opal suppressor tRNAs, derived from the same tRNA (Rodriguez et al. 2007a, b). Acceptor stem mutants with improved orthogonality were isolated, while preserving high suppression efficiencies.

3.2 Systems with Reduced Release Factor Activity

Site-specific incorporation of unnatural amino acids into proteins by nonsense suppression has opened up exciting new possibilities for protein engineering. Nevertheless, while established as the method of choice, the strategy involving the use of nonsense codons to define the site of interest in the target protein has some inherent weaknesses. When a ribosome translating an mRNA reaches a stop codon, release factors bind to the "empty" ribosomal A-site and catalyze translation termination resulting in the release of the nascent polypeptide chain (Nakamura et al. 1996; Kisselev et al. 2003). In *E. coli* and other bacteria, this reaction is carried out by release factors 1 and 2 (RF-1, RF-2) recognizing amber/ochre and ochre/opal stop codons, respectively; in eukaryotes, eRF-1 recognizes all three stop codons. Consequently, a suppressor tRNA recognizing the stop codon has to compete against the release factor, leading to a reduction of readthrough efficiency by the suppressor tRNA.

Under optimal conditions, the readthrough efficiency obtained with certain orthogonal suppressor tRNAs can be as high as 25–75% in *E. coli* (Wang et al. 2006) and 30–40% in mammalian cells (Köhrer et al. 2004), but is often lower and drops drastically when two or more stop codons are used within the same mRNA for insertion of two or more unnatural amino acids. Several strategies have been developed to increase suppression efficiencies. These include (1) reducing the levels of active release factor, and most recently (2) the development of specialized ribosomes (ribo-X) for improved suppression efficiencies in *E. coli*.

3.2.1 Reduced Levels of Release Factor Lead to Increased Amber Suppression

In a study by Abelson, Miller, and colleagues (Kleina et al. 1990), 17 *E. coli* amber suppressor tRNA genes were constructed, yielding suppressor tRNAs of different efficiency. The same study also showed that the use of an *E. coli* host with a temperature-sensitive mutation in RF-1 markedly improved suppression efficiency of weaker suppressor tRNAs in vivo. Partial heat inactivation of the temperature-sensitive RF-1 resulted in an *E. coli* cell-free transcription/translation system with reduced levels of RF-1, increasing suppression efficiency more than tenfold (Short et al. 1999). Similarly, the omission of RF-1 from the fully reconstituted PURE system led to superior readthrough efficiency of nonsense codons (Shimizu et al. 2001; see also the chapter by Hirao et al., this volume).

In eukaryotes, eRF-1 recognizes all three stop codons and competes with amber, ochre, and opal suppressor tRNAs at the ribosomal A-site (Drugeon et al. 1997). Yarus, Leinwand, and coworkers (Carnes et al. 2003) showed that reduction of eRF-1 levels in mammalian cells allowed increased readthrough of a leaky stop codon in the absence of suppressor tRNAs. Small interfering RNAs were targeted against different portions of the eRF-1 coding region to increase readthrough efficiency approximately 2.5-fold. These results were confirmed in a study by Janzen and Geballe (2004), who analyzed the effects of depleting eukaryotic release factors, eRF-1 and eRF-3, on translation termination in various human cell lines. The idea of downregulating eRF-1 by RNA interference was subsequently applied to increase the efficiency of incorporation of unnatural amino acids into proteins in mammalian cells approximately fivefold (Ilegems et al. 2004).

It is evident, however, that complete depletion of release factor for further improvement of suppression efficiency is not feasible, as ribosomes would read through natural termination codons, causing cytotoxicity and ultimately cell death.

3.2.2 Specialized Ribosomes: Ribo-X

Recent work by Chin and colleagues (Wang et al. 2007a) revolutionized unnatural amino acid mutagenesis in *E. coli* by endowing cells with an additional set of highly specialized ribosomes that function side by side with natural ribosomes (Fig. 7). These specialized ribosomes, called ribo-X, are remarkable for two reasons: (1) they translate only the UAG-containing mRNA encoding the target protein; and (2) they have reduced affinity for −1.

First described by de Boer (Hui and de Boer 1987), Dahlberg (Jacob et al. 1987), and their colleagues in the late 1980s, specialized ribosomes have been and are being used extensively in studies of the translational machinery, such as the interaction between ribosome and mRNA, the binding of tRNA to the ribosome, and the function of individual components of the ribosome. The principle of ribosomes that are dedicated to translate a single mRNA species is based on the well-established roles of sequences on the mRNA (Shine–Dalgarno sequences) and sequences near the 3′ end of the 16S rRNA (anti-Shine–Dalgarno sequences). A ribosome can, therefore, be targeted to a particular mRNA by changing the Shine–Dalgarno sequence located seven to 13 nucleotides upstream of the AUG initiation codon in the mRNA and by concomitantly introducing the appropriate complementary changes in the anti-Shine–Dalgarno sequence in the 16S rRNA. An additional mutation (C1192U) in the 16S rRNA confers resistance to spectinomycin (Makosky and Dahlberg 1987). As a result, the spectinomycin-sensitive wild-type ribosomes can be "switched off" by addition of the antibiotic, while the specialized ribosomes remain unaffected.

Chin and colleagues have extended the principle of specialized ribosomes (Rackham and Chin 2005) and have isolated ribosomes with increased efficiency in

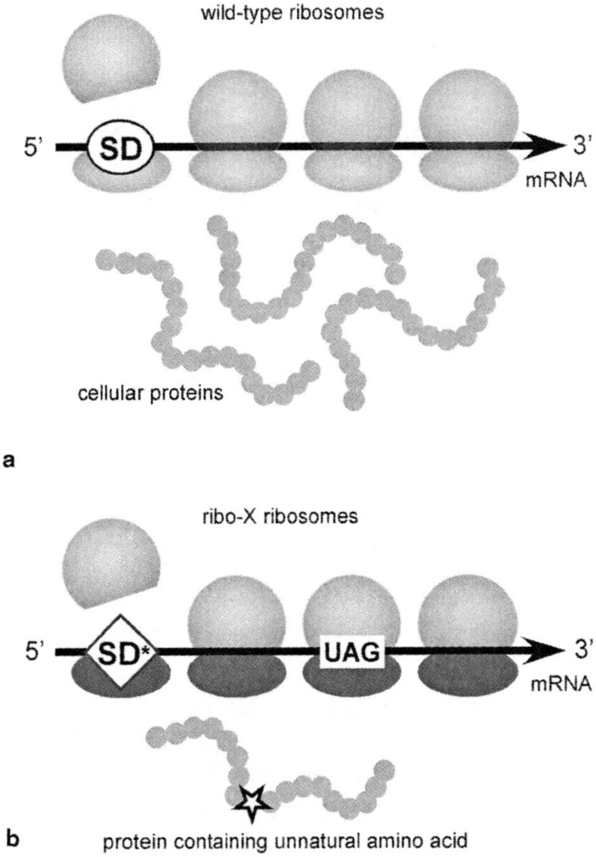

Fig. 7 Specialized ribosomes for improved suppression of amber codons: ribo-X. Using a genetic selection system, Wang et al. (2007a) isolated specialized ribo-X ribosomes with mutations in the 16S rRNA, leading to reduced affinity for release factor 1. Ribo-X ribosomes, which are consequently more efficient in suppression of amber codons, are directed exclusively to the UAG-containing messenger RNA (*mRNA*) encoding the target protein, through a modified Shine–Dalgarno/anti-Shine–Dalgarno interaction

translation of amber codons (Wang et al. 2007a). A library of mutants was designed around the well-characterized 530-loop in 16S rRNA and passed through a genetic selection system using chloramphenicol acetyltransferase mRNA with a UAG codon along with an appropriate amber suppressor tRNA. The 530-loop of 16S rRNA comprises part of the ribosomal A-site and is known to be involved in binding of RF-1 and RF-2 (Yusupov et al. 2001; Klaholz et al. 2003; Petry et al. 2005; Selmer et al. 2006). A mutant ribosome, referred to as ribo-X, carrying two point mutations (U531G and U534A) in the 530-loop was isolated. Suppression efficiency of a single UAG codon in an mRNA by a suppressor tRNA aminoacylated with the unnatural amino acid Bzp increased from 24% for wild-type ribosomes to 62% for ribo-X. Even more impressively, concomitant suppression efficiency of two UAG codons increased from less than 1 to 22%.

3.3 *Systems Independent of Release Factor Activity: Alternative Codon/Anticodon Pairs*

The expansion of the genetic code by virtue of nonsense suppression is limited to the three naturally occurring stop codons. Alternative strategies that include novel unnatural base pairs in the codon/anticodon interaction or frameshift suppressor tRNAs open up the possibility of expanding the genetic code further and facilitating the incorporation two or more different unnatural amino acids. In addition, codons translated by such alternative suppressor tRNAs are not recognized by release factors, thus ensuring high suppression yields.

3.3.1 Novel Unnatural Base Pairs

Benner, Chamberlin, and coworkers demonstrated the incorporation in vitro of 3-iodo-L-tyrosine into a peptide through the use of a novel base pair, *isoC-isoG*, based on unnatural nucleoside bases (Bain et al. 1992). An *isoC*AG codon-containing mRNA was used alongside a tRNA featuring a CU*isoG* anticodon, where *isoG* represents a nonstandard purine complementary to *isoC*, a nonstandard pyrimidine. Both, mRNA and tRNA were synthesized chemically for this purpose. In an in vitro protein synthesis system, readthrough efficiency of the *isoC*AG codon, and thereby incorporation of 3-iodo-L-tyrosine, was specific and was as high as 90%, compared with 60% readthrough efficiency of a standard UAG codon in the presence of an amber suppressor tRNA. The increased readthrough efficiency of the novel *isoC*AG codon confirms the specificity of release factors, which do not recognize *isoC*AG as a stop codon and, therefore, allow ribosomes to translate *isoC*AG with high efficiency.

To overcome the shortcomings of this approach, mainly caused by the requirement for chemical synthesis of both mRNA and tRNA, new unnatural bases, which can be incorporated into RNA through in vitro transcription by T7 RNA polymerase, have been developed. For example, Yokoyama and coworkers designed new base pairs on the basis of hydrogen-bonding pattern and shape complementarity (Ohtsuki et al. 2001; Hirao et al. 2002; Mitsui et al. 2002, 2003). They showed that the unnatural base pyridin-2-one (*y*) was inserted into mRNA in response to bases 2-amino-6-methylaminopurine (*x*) or 2-amino-6-(2-thienyl)purine (*s*) in the template DNA, but not in response to any of the standard bases. Base pairs *y·x* and *y·s*, characterized by high specificity in transcription, were used in a coupled transcription–translation system for site-specific insertion of amino acid analogues.

Similarly, the groups of Schultz and Romesberg have used the concept of hydrophobic bases as building blocks for novel base pairs (McMinn et al. 1999; Wu et al. 2000; Ogawa et al. 2000a, b). Further improvement of systems employing unnatural base pairs has involved the design of novel unnatural base-pairs that are also accepted by the replication machinery, allowing amplification of the analogue-containing DNA template.

A detailed review of design, characterization, and use of unnatural base pairs in protein engineering can be found in the chapters by Hirao et al. and Leconte and Romesberg in this volume.

3.3.2 Four- and Five-Base Codon/Anticodon Interactions

Following the discovery of Crick, Brenner, and coworkers, who established the triplet nature of the genetic code (Crick 1958; Crick et al. 1961), work by Riddle and Roth (1970) and Yourno and coworkers (Yourno 1972; Yourno and Kohno 1972) showed that a codon consisting of four bases could be translated by a mutant tRNA containing an extra nucleotide in the anticodon (frameshift suppressor tRNA). Sisido and coworkers have exploited four- and five-base codons extensively for site-specific insertion of one or two different unnatural amino acids into proteins in vitro (Hohsaka et al. 1996, 1999, 2001a, b). In their studies, a wide variety of four- and five-base codons were evaluated; those derived from codons that are rarely used in bacteria were found to be most suitable, working efficiently without serious competition from endogenous tRNAs. For example, concomitant incorporation of 2-naphthyl-L-alanine and *p*-nitro-L-phenylalanine into streptavidin was accomplished at 60% efficiency in response to four-base codons CGGG and GGGU (Hohsaka et al. 2001b). The successful incorporation of various aromatic amino acid analogues, some carrying relatively large and bulky (expanded) side groups, also demonstrated the flexibility of the translation machinery (elongation factor, ribosome) to accept amino acid analogues differing significantly from natural amino acid substrates.

Hecht and coworkers have used an amber suppressor tRNA along with a frameshift suppressor tRNA to synthesize in vitro a protein containing two different amino acid analogues to demonstrate fluorescence resonance energy transfer between two unnatural amino acids, 7-aza-L-tryptophan (fluorescence donor) and a dabcyl diaminopropionic acid derivative (fluorescence acceptor) in a structurally modified dihydrofolate reductase (Anderson et al. 2002).

In contrast to the in vitro work, Anderson et al. (2004) reported the first example of an orthogonal frameshift suppressor tRNA/aminoacyl-tRNA synthetase pair using a type I LysRS from *Pyrococcus horikoshii* in *E. coli*. The frameshift suppressor, a $tRNA^{Lys}$ derivative, incorporated the unnatural amino acid L-homoglutamine (hGln) in response to the AGGA codon. Combining this new frameshift suppressor tRNA/aminoacyl-tRNA synthetase pair with an orthogonal amber suppressor tRNA/aminoacyl-tRNA synthetase pair developed earlier (Wang et al. 2001) permitted con-

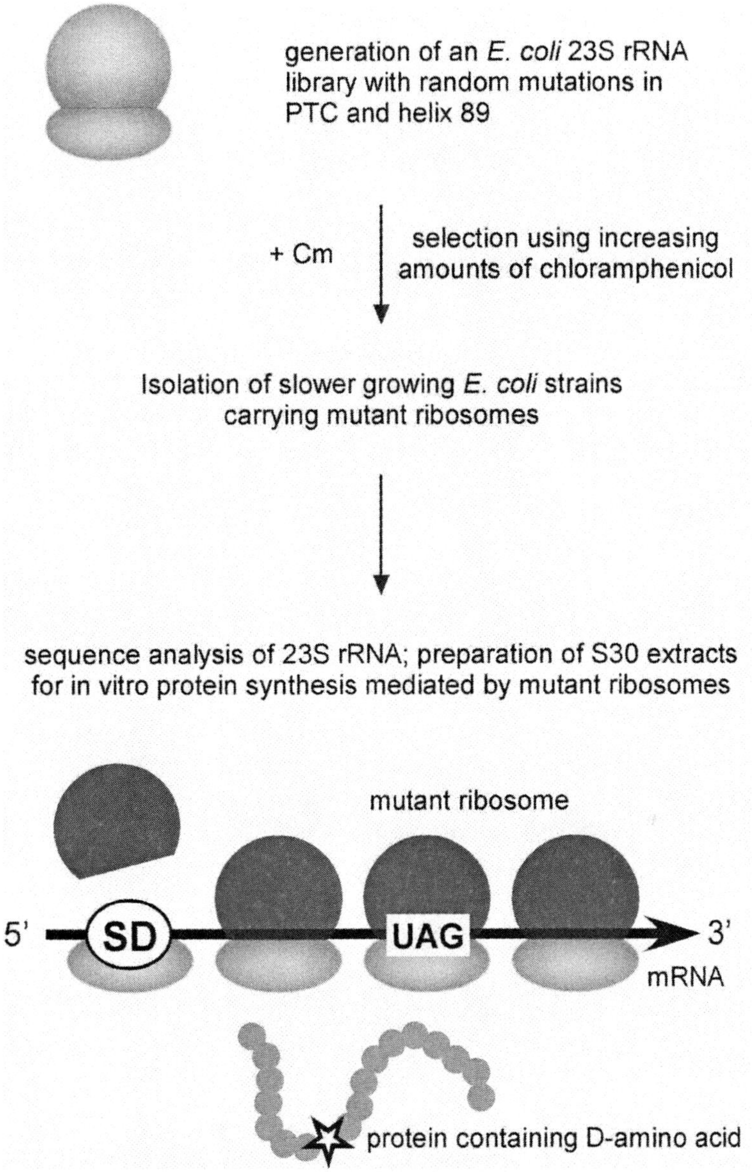

Fig. 8 Selection strategy for specialized ribosomes capable of utilizing D-amino acids in protein synthesis. *PTC* peptidyltransferase center, *rRNA* ribosomal RNA (Dedkova et al. 2003, 2006)

comitant incorporation of two unnatural amino acids, hGln and OMeTyr, in vivo (Anderson et al. 2004). Similarly, a combination of nonsense and frameshift suppressor tRNAs was used to site-specifically incorporate up to three unnatural amino acids simultaneously into a neuroreceptor in *Xenopus* oocytes (Rodriguez et al. 2006).

3.4 Incorporation of D-Amino Acids Using Evolved Ribosomes

D-Amino acids are present in all living organisms, from bacteria to higher eukaryotes, in the form of free amino acids, peptides, and proteins (Kreil 1997; Fujii 2002). The protein synthesis machinery discriminates efficiently between L-amino acids and D-amino acids, rejecting D-amino acids from being used under conditions where the corresponding L-amino acids are also present (Pingoud and Urbanke 1980; Yamane et al. 1981; Soutourina et al. 2000). For example, Calendar and Berg (1966) showed that although *E. coli* TyrRS is able to aminoacylate $tRNA^{Tyr}$ with either L-tyrosine or D-tyrosine, activation is more efficient for L-tyrosine than it is for D-tyrosine and the resulting D-Tyr-$tRNA^{Tyr}$ is hydrolyzed by D-tyrosyl-tRNA deacylase, an essential enzyme in *E. coli* (Soutourina et al. 1999). Yamane et al. (1981) estimate that the combined effects of discrimination at each stage, from aminoacylation to peptide bond formation, favor L-tyrosine by a factor greater than 10^4. Because of this, D-amino acid containing peptides and proteins are produced only through post-translational modification or nonribosomal peptide synthesis (Kreil 1997).

Hecht and colleagues used directed evolution to isolate ribosomes that allowed enhanced incorporation of D-amino acids into proteins (Dedkova et al. 2003, 2006). Two regions in 23S rRNA (2447–2450 and 2457–2462), corresponding to the peptidyltransferase center (2447, 2449) and helix 89, were randomized. The mutated 23S rRNA, which was also thiostrepton resistant, was expressed from a multicopy plasmid (Dedkova et al. 2003). To study the incorporation of D-amino acids, S30 extracts were prepared from individual *E. coli* clones having different mutant ribosomes (Fig. 8). All S30 extracts were active in the presence of thiostrepton, which inactivated chromosomally encoded wild-type ribosomes. The modified ribosomes produced D-amino acid containing proteins with increased efficiency, although the fidelity of these ribosomes was somewhat lower than that of wild-type ribosomes (Dedkova et al. 2006).

4 Conclusion

Expansion of the genetic code using a stop codon along with the corresponding suppressor tRNA/aminoacyl-tRNA synthetase pairs has provided the necessary tools to incorporate unnatural amino acids with diverse chemical, physical, and biological properties into proteins. Modifications and optimization of various components of the translational machinery have made possible the site-specific incorporation of one or more unnatural amino acids into a target protein in prokaryotic and eukaryotic expression systems. More than 200 different unnatural amino acids have been incorporated into proteins using suppressor tRNAs, both in vitro and in vivo, adding a variety of functionalities useful for studies of protein folding, stability, structure, function, protein–protein interactions, and protein localization.

Acknowledgments The work described in this chapter was supported by grants from the National Institutes of Health (GM17151, GM67741) and the US Army Office (W911NF-04-1-0353).

References

Agris PF (2004) Decoding the genome: a modified view. Nucleic Acids Res 32:223–238

Agris PF, Vendeix FAP, Graham WD (2007) tRNA's wobble decoding of the genome: 40 years of modification. J Mol Biol 366:1–13

Anderson JC, Schultz P (2003) Adaptation of an orthogonal archaeal leucyl-tRNA and synthetase pair for four-base, amber, and opal suppression. Biochemistry 42:9598–9608

Anderson RD, Zhou J, Hecht SM (2002) Fluorescence resonance energy transfer between unnatural amino acids in a structurally modified dihydrofolate reductase. J Am Chem Soc 124:9674–9675

Anderson JC, Wu N, Santoro SW, Lakshman V, King DS, Schultz PG (2004) An expanded genetic code with a functional quadruplet codon. Proc Natl Acad Sci USA 101:7566–7571

Bain JD, Switzer C, Chamberlin AR, Benner SA (1992) Ribosome-mediated incorporation of a non-standard amino acid into a peptide through expansion of the genetic code. Nature 356:537–539

Beene DL, Dougherty DA, Lester HA (2003) Unnatural amino acid mutagenesis in mapping ion channel function. Curr Opin Neurobiol 13:264–270

Björk GR (1995) Biosynthesis and function of modified nucleosides. In: Söll D, RajBhandary UL (eds) tRNA: structure, biosynthesis, and function. American Society for Microbiology, Washington DC, pp 165–205

Bossi L, Roth JR (1980) The influence of codon context on genetic code translation. Nature 286:123–127

Brick P, Bhat TN, Blow DM (1989) Structure of tyrosyl-tRNA synthetase refined at 2.3 A resolution. Interaction of the enzyme with the tyrosyl adenylate intermediate. J Mol Biol 208:83–98

Buckingham RH (1994) Codon context and protein synthesis: enhancements of the genetic code. Biochimie 76:351–354

Calendar R, Berg P (1966) The catalytic properties of tyrosyl ribonucleic acid synthetases from *Escherichia coli* and *Bacillus subtilis*. Biochemistry 5:1690–1695

Carnes J, Jacobson M, Leinwand L, Yarus M (2003) Stop codon suppression via inhibition of eRF1 expression. RNA 9:648–645

Chin JW, Cropp TA, Anderson JC, Mukherji M, Zhang Z, Schultz PG (2003) An expanded eukaryotic genetic code. Science 301:964–967

Chow CM, RajBhandary UL (1993) *Saccharomyces cerevisiae* cytoplasmic tyrosyl-tRNA synthetase gene. Isolation by complementation of a mutant *Escherichia coli* suppressor tRNA defective in aminoacylation and sequence analysis. J Biol Chem 268:12855–12863

Clark JMJ, Eyzaguirre JP (1962) Tyrosine activation and transfer to soluble ribonucleic acid. I. Purification and study of the enzyme of hog pancreas. J Biol Chem 237:3698–3702

Colby DS, Schedl P, Guthrie C (1976) A functional requirement for modification of the wobble nucleotide in the anticodon of a T4 suppressor tRNA. Cell 9:449–463

Crick FHC (1958) On protein synthesis. Symp Soc Exp Biol 12:138–163

Crick FH, Barnett L, Brenner S, Watts-Tobin RJ (1961) General nature of the genetic code for proteins. Nature 192:1227–1232

Dedkova LM, Fahmi NE, Golovine SY, Hecht SM (2003) Enhanced D-amino acid incorporation into protein by modified ribosomes. J Am Chem Soc 125:6616–6617

Dedkova LM, Fahmi NE, Golovine SY, Hecht SM (2006) Construction of modified ribosomes for incorporation of D-amino acids into proteins. Biochemistry 45:15541–15551

Drabkin HJ, Park HJ, RajBhandary UL (1996) Amber suppression in mammalian cells dependent upon expression of an *Escherichia coli* aminoacyl-tRNA synthetase gene. Mol Cell Biol 16:907–913

Drabkin HJ, Estrella M, RajBhandary UL (1998) Initiator-elongator discrimination in vertebrate tRNAs for protein synthesis. Mol Cell Biol 18:1459–1466

Drugeon G, Jean-Jean O, Frolova L, Le Goff X, Philippe M, Kisselev L, Haenni AL (1997) Eukaryotic release factor 1 (eRF1) abolishes readthrough and competes with suppressor tRNAs at all three termination codons in messenger RNA. Nucleic Acids Res 25:2254–2258

Edwards H, Schimmel P (1990) A bacterial amber suppressor in *Saccharomyces cerevisiae* is selectively recognized by a bacterial aminoacyl-tRNA synthetase. Mol Cell Biol 10:1633–1641

Ericson JU, Björk GR (1991) tRNA anticodons with the modified nucleoside 2-methylthio-N6-(4-hydroxyisopentenyl)adenosine distinguish between bases 3′ of the codon. J Mol Biol 218: 509–516

Fujii N (2002) D-Amino acids in living higher organisms. Orig Life Evol Biosph 32:103–127

Hartman MCT, Josephson K, Szostak JW (2006) Enzymatic aminoacylation of tRNA with unnatural amino acids. Proc Natl Acad Sci USA 103:4356–4361

Hartman MCT, Josephson K, Lin C-W, Szostak JW (2007) An expanded set of amino acid analogs for the ribosomal translation of unnatural peptides. PLoS ONE 2:e972

Heckler TG, Chang LH, Zama Y, Naka T, Chorghade MS, Hecht SM (1984) T4 RNA ligase mediated preparation of novel "chemically misacylated" $tRNA^{Phe}_{s}$. Biochemistry 23:1468–1473

Hendrickson TL, de Crécy-Lagard V, Schimmel P (2004) Incorporation of nonnatural amino acids into proteins. Annu Rev Biochem 73:147–176

Hino N, Okazaki Y, Kobayashi T, Hayashi A, Sakamoto K, Yokoyama S (2005) Protein photo-cross-linking in mammalian cells by site-specific incorporation of a photoreactive amino acid. Nat Methods 2:201–206

Hirao I, Kimoto M, Mitsui T, Harada Y, Fujiwara T, Sato A, Yokoyama S (2002) An unnatural base pair between imidazolin-2-one and 2-amino-6-(2-thienyl)purine in replication and transcription. Nucleic Acids Res Suppl:37–38

Hohsaka T, Ashizuka Y, Murakami H, Sisido M (1996) Incorporation of nonnatural amino acids into streptavidin through *in vitro* frame-shift suppression. J Am Chem Soc 118:9778–9779

Hohsaka T, Ashizuka Y, Sasaki H, Murakami H, Sisido M (1999) Incorporation of two different nonnatural amino acids independently into a single protein through extension of the genetic code. J Am Chem Soc 121:12194–12195

Hohsaka T, Ashizuka Y, Murakami H, Sisido M (2001a) Five-base codons for incorporation of nonnatural amino acids into proteins. Nucleic Acids Res 29:3646–3651

Hohsaka T, Ashizuka Y, Taira H, Murakami H, Sisido M (2001b) Incorporation of nonnatural amino acids into proteins by using various four-base codons in an *Escherichia coli in vitro* translation system. Biochemistry 40:11060–11064.

Hui A, de Boer HA (1987) Specialized ribosome system: Preferential translation of a single mRNA species by a subpopulation of mutated ribosomes in *Escherichia coli*. Proc Natl Acad Sci USA 84:4762–4766

Ilegems E, Pick HM, Vogel H (2002) Monitoring mis-acylated tRNA suppression efficiency in mammalian cells via EGFP fluorescence recovery. Nucleic Acids Res 30, e128:1–6

Ilegems E, Pick HM, Vogel H (2004) Downregulation of eRF1 by RNA interference increases mis-acylated tRNA suppression efficiency in human cells. Protein Eng 17:821–827

Jackson JC, Hammill JT, Mehl RA (2007) Site-specific incorporation of a ^{19}F-amino acid into proteins as an NMR probe for characterizing protein structure and reactivity. J Am Chem Soc 129:1160–1166

Jacob WF, Santer M, Dahlberg AE (1987) A single base change in the Shine–Dalgarno region of 16S rRNA of *Escherichia coli* affects translation of many proteins. Proc Natl Acad Sci USA 84:4757–4761

Jakubowski H, Fersht AR (1981) Alternative pathways for editing non-cognate amino acids by aminoacyl-tRNA synthetases. Nucleic Acids Res 9:3105–3117

Janzen DM, Geballe AP (2004) The effect of eukaryotic release factor depletion on translation termination in human cell lines. Nucleic Acids Res 32:4491–4502

Kiga D, Sakamoto K, Kodama K, Kigawa T, Matsuda T, Yabuki T, Shirouzu M et al (2002) An engineered *Escherichia coli* tyrosyl-tRNA synthetase for site-specific incorporation of an unnatural amino acid into proteins in eukaryotic translation and its application in a wheat germ cell-free system. Proc Natl Acad Sci USA 99:9715–9720

Kisselev L, Ehrenberg M, Frolova L (2003) Termination of translation: interplay of mRNA, rRNAs and release factors? EMBO J 22:175–182

Klaholz BP, Pape T, Zavialov AV, Myasnikov AG, Orlova EV, Vestergaard B, Ehrenberg M et al (2003) Structure of the *Escherichia coli* ribosomal termination complex with release factor 2. Nature 421:90–94

Kleina LG, Masson JM, Normanly J, Abelson J, Miller JH (1990) Construction of *Escherichia coli* amber suppressor tRNA genes. II. Synthesis of additional tRNA genes and improvement of suppressor efficiency. J Mol Biol 213:705–717

Kobayashi T, Sakamoto K, Takimura T, Sekine R, Kelly VP, Kamata K, Nishimura S et al (2005) Structural basis of nonnatural amino acid recognition by an engineered aminoacyl-tRNA synthetase for genetic code expansion. Proc Natl Acad Sci USA 102:1366–1371

Köhrer C, RajBhandary UL (2005) Proteins carrying one or more unnatural amino acids. In: Ibba M, Francklyn C, Cusack S (eds) Aminoacyl-tRNA synthetases. Landes Bioscience, pp 353–363

Köhrer C, Xie L, Kellerer S, Varshney U, RajBhandary UL (2001) Import of amber and ochre suppressor tRNAs into mammalian cells: a general approach to site-specific insertion of amino acid analogues into proteins. Proc Natl Acad Sci USA 98:14310–14315

Köhrer C, Yoo J, Bennett M, Schaack J, RajBhandary UL (2003) A possible approach to site-specific insertion of two different unnatural amino acids into proteins in mammalian cells via nonsense suppression. Chem Biol 10:1095–1102

Köhrer C, Sullivan EL, RajBhandary UL (2004) Complete set of orthogonal 21st aminoacyl-tRNA synthetase-amber, ochre and opal suppressor tRNA pairs: concomitant suppression of three different termination codons in an mRNA in mammalian cells. Nucleic Acids Res 32:6200–6211

Kowal AK, Köhrer C, RajBhandary UL (2001) Twenty-first aminoacyl-tRNA synthetase-suppressor tRNA pairs for possible use in site-specific incorporation of amino acid analogues into proteins in eukaryotes and in eubacteria. Proc Natl Acad Sci USA 98:2268–2273

Kreil G (1997) D-amino acids in animal peptides. Annu Rev Biochem 66:337–345

Lee CP, RajBhandary UL (1991) Mutants of *Escherichia coli* initiator tRNA that suppress amber codons in *Saccharomyces cerevisiae* and are aminoacylated with tyrosine by yeast extracts. Proc Natl Acad Sci USA 88:11378–11382

Liu W, Brock A, Chen S, Chen S, Schultz PG (2007) Genetic incorporation of unnatural amino acids into proteins in mammalian cells. Nat Methods 4:239–244

Makosky PC, Dahlberg AE (1987) Spectinomycin resistance at site 1192 in 16S ribosomal RNA of *E. coli*: An analysis of three mutants. Biochimie 69:885–889

McMinn DL, Ogawa AK, Wu Y, Liu J, Schultz PG, Romesberg FE (1999) Efforts toward expansion of the genetic alphabet: DNA polymerase recognition of a highly stable, self-pairing hydrophobic base. J Am Chem Soc 121:11585–11586

Mitsui T, Kimoto M, Harada Y, Sato A, Kitamura A, To T, Hirao I et al (2002) Enzymatic incorporation of an unnatural base pair between 4-propynyl-pyrrole-2-carbaldehyde and 9-methyl-imidazo [(4,5)-b]pyridine into nucleic acids. Nucleic Acids Res Suppl:219–220

Mitsui T, Kitamura A, Kimoto M, To T, Sato A, Hirao I, Yokoyama S (2003) An unnatural hydrophobic base pair with shape complementarity between pyrrole-2-carbaldehyde and 9-methyl-imidazo[(4,5)-b]pyridine. J Am Chem Soc 125:5298–5307

Monahan SL, Lester HA, Dougherty DA (2003) Site-specific incorporation of unnatural amino acids into receptors expressed in mammalian cells. Chem Biol 10:573–580

Motorin Y, Bec G, Tewari R, Grosjean H (1997) Transfer RNA recognition by the *Escherichia coli* delta2-isopentenyl-pyrophosphate:tRNA delta2-isopentenyl transferase: dependence on the anticodon arm structure. RNA 3:721–733

Murakami H, Bonzagni NJ, Suga H (2002) Aminoacyl-tRNA synthesis by a resin-immobilized ribozyme. J Am Chem Soc 124:6834–6835

Murakami H, Saito H, Suga H (2003) A versatile tRNA aminoacylation catalyst based on RNA. Chem Biol 10:655–662

Murakami H, Ohta M, Ashigai H, Suga H (2006) A highly flexible tRNA acylation method for non-natural polypeptide synthesis. Nat Methods 3:357–359

Myslinski E, Amé JC, Krol A, Carbon P (2001) An unusually compact external promoter for RNA polymerase III transcription of the human H1RNA gene. Nucleic Acids Res 29:2502–2509

Nakamura Y, Ito K, Isaksson LA (1996) Emerging understanding of translation termination. Cell 87:147–150

Ninomiya K (2003) Facile aminoacylation of pdCpA dinucleotide with a nonnatural amino acid in cationic micelle. Chem Commun (17):2242–2243

Ninomiya K, Minohata T, Nishimura M, Sisido M (2004) In situ chemical aminoacylation with amino acid thioesters linked to a peptide nucleic acid. J Am Chem Soc 126:15984–15989

Noren CJ, Anthony-Cahill SJ, Griffith MC, Schultz PG (1989) A general method for site-specific incorporation of unnatural amino acids into proteins. Science 244:182–188

Nowak MW, Kearney PC, Sampson JR, Saks ME, Labarca CG, Silverman SK, Zhong W et al (1995) Nicotinic receptor binding site probed with unnatural amino acid incorporation in intact cells. Science 268:439–442

Ogawa AK, Wu Y, Berger M, Schultz PG, Romesberg FE (2000a) Rational design of an unnatural base pair with increased kinetic selectivity. J Am Chem Soc 122:8803–8804

Ogawa AK, Wu Y, McMinn DL, Liu J, Schultz PG, Romesberg FE (2000b) Efforts toward the expansion of the genetic alphabet: information storage and replication with unnatural hydrophobic base pairs. J Am Chem Soc 122:3274–3287

Ohtsuki T, Kimoto M, Ishikawa M, Mitsui T, Hirao I, Yokoyama S (2001) Unnatural base pairs for specific transcription. Proc Natl Acad Sci USA 98:4922–4925

Petry S, Brodersen DE, Murphy FVt, Dunham CM, Selmer M, Tarry MJ, Kelley AC et al (2005) Crystal structures of the ribosome in complex with release factors RF1 and RF2 bound to a cognate stop codon. Cell 123:1255–1266

Pingoud A, Urbanke C (1980) Aminoacyl transfer ribonucleic acid binding site of the bacterial elongation factor Tu. Biochemistry 19:2108–2112

Rackham O, Chin JW (2005) A network of orthogonal ribosome x mRNA pairs. Nat Chem Biol 1:159–166

RajBhandary UL (1994) Initiator transfer RNAs. J Bacteriol 176:547–552

Riddle DL, Roth JR (1970) Suppressors of frameshift mutations in *Salmonella typhimurium*. J Mol Biol 54:131–144

Rodriguez EA, Lester HA, Dougherty DA (2006) In vivo incorporation of multiple unnatural amino acids through nonsense and frameshift suppression. Proc Natl Acad Sci USA 103:8650–8655

Rodriguez EA, Lester HA, Dougherty DA (2007a) Improved amber and opal suppressor tRNAs for incorporation of unnatural amino acids *in vivo*. Part 1: Minimizing misacylation. RNA 13:1703–1714

Rodriguez EA, Lester HA, Dougherty DA (2007b) Improved amber and opal suppressor tRNAs for incorporation of unnatural amino acids *in vivo*. Part 2: evaluating suppression efficiency. RNA 13:1715–1722

Saito H, Suga H (2001) A ribozyme exclusively aminoacylates the 3′-hydroxyl group of the tRNA terminal adenosine. J Am Chem Soc 123:7178–7179

Saito H, Kourouklis D, Suga H (2001) An *in vitro* evolved precursor tRNA with aminoacylation activity. EMBO J 20:1797–1806

Sakamoto K, Hayashi A, Sakamoto A, Kiga D, Nakayama H, Soma A, Kobayashi T et al (2002) Site-specific incorporation of an unnatural amino acid into proteins in mammalian cells. Nucleic Acids Res 30:4692–4699

Santoro SW, Anderson JC, Lakshman V, Schultz PG (2003) Nucleic Acids Res 31:6700–6709

Selmer M, Dunham CM, Murphy FVt, Weixlbaumer A, Petry S, Kelley AC, Weir JR et al (2006) Structure of the 70S ribosome complexed with mRNA and tRNA. Science 313:1935–1942

Seyedsayamdost MR, Xie J, Chan CT, Schultz PG, Stubbe J (2007) Site-specific insertion of 3-aminotyrosine into subunit alpha2 of *E. coli* ribonucleotide reductase: Direct evidence for involvement of Y730 and Y731 in radical propagation. J Am Chem Soc 129:15060–150671

Shimizu Y, Inoue A, Tomari Y, Suzuki T, Yokogawa T, Nishikawa K, Ueda T (2001) Cell-free translation reconstituted with purified components. Nat Biotechnol 19:751–755

Short GF, Golovine SY, Hecht SM (1999) Effects of release factor 1 on *in vitro* protein translation and the elaboration of proteins containing unnatural amino acids. Biochemistry 38:8808–8819

Soutourina J, Plateau P, Delort F, Peirotes A, Blanquet S (1999) Functional characterization of the D-Tyr-tRNATyr deacylase from *Escherichia coli*. J Biol Chem 274:19109–19114

Soutourina J, Plateau P, Blanquet S (2000) Metabolism of D-aminoacyl-tRNAs in *Escherichia coli* and *Saccharomyces cerevisiae* cells. J Biol Chem 275:32535–32542

Sprague KU (1994) Transcription of eukaryotic tRNA genes. AMS Press, Washington DC

Wang L, Brock A, Herberich B, Schultz PG (2001) Expanding the genetic code of *Escherichia coli*. Science 292:498–500

Wang L, Xie J, Schultz PG (2006) Expanding the genetic code. Annu Rev Biophys Biomol Struct 35:225–249

Wang K, Neumann H, Peak-Chew SY, Chin JW (2007a) Evolved orthogonal ribosomes enhance the efficiency of synthetic genetic code expansion. Nat Biotechnol 25:770–777

Wang W, Takimoto JK, Louie GV, Baiga TJ, Noel JP, Lee KF, Slesinger PA et al (2007b) Genetically encoding unnatural amino acids for cellular and neuronal studies. Nat Neurosci 10: 1063–1072

Wu Y, Ogawa AK, Berger M, McMinn DL, Schultz PG, Romesberg FE (2000) Efforts toward expansion of the genetic alphabet: optimization of interbase hydrophobic interactions. J Am Chem Soc 122:7621–7632

Wu N, Deiters DA, Cropp TA, King DS, Schultz PG (2004) J Am Chem Soc 126:14306–14307

Xie J, Schultz PG (2006) A chemical toolkit for proteins: An expanded genetic code. Nat Rev Mol Cell Biol 7:775–782

Xie J, Supekova L, Schultz PG (2007) A genetically encoded metabolically stable analogue of phosphotyrosine in *Escherichia coli*. ACS Chem Biol 2:474–478

Yamane T, Miller DL, Hopfield JJ (1981) Discrimination between D- and L-tyrosyl transfer ribonucleic acids in peptide chain elongation. Biochemistry 20:7059–7064

Yanofsky C, Soll L (1977) Mutations affecting $tRNA^{Trp}$ and its charging and their effect on regulation of transcription termination at the attenuator of the tryptophan operon. J Mol Biol 113:663–677

Yarus M (1982) Translational efficiency of transfer RNA's: uses of an extended anticodon. Science 218:646–652

Yarus M, Cline S, Raftery L, Wier P, Bradley D (1986) The translational efficiency of tRNA is a property of the anticodon arm. J Biol Chem 261:496–505

Ye S, Köhrer C, Huber T, Kazmi M, Sachdev P, Yan ECY, Bhagat A et al (2008) Site-specific incorporation of keto amino acids into functional G protein-coupled receptors using unnatural amino acid mutagenesis. J Biol Chem 283:1525–1533

Yourno J (1972) Externally suppressible +1 "glycine" frameshift: possible quadruplet isomers for glycine and proline. Nat New Biol 239:219–221

Yourno J, Kohno T (1972) Externally suppressible proline quadruplet ccc U. Science 175:650–652

Yusupov MM, Yusupova GZ, Baucom A, Lieberman K, Earnest TN, Cate JH, Noller HF (2001) Crystal structure of the ribosome at 5.5 A resolution. Science 292:883–896

Zhang Z, Alfonta L, Tian F, Bursulaya B, Uryu S, King DS, Schultz PG (2004) Selective incorporation of 5-hydroxytryptophan into proteins in mammalian cells. Proc Natl Acad Sci USA 101:8882–8887

In Vivo Studies of Receptors and Ion Channels with Unnatural Amino Acids

D.A. Dougherty

Contents

Abstract The combination of nonsense suppression for unnatural amino acid incorporation and heterologous expression in Xenopus oocytes provides a powerful means to probe neuroreceptors and ion channels with chemical precision. Here we describe a range of studies that illustrate the broad potential of this approach. A variety of biophysical probes and reactive moieties can be incorporated. In addition, subtle systematic variations allow detailed, physical organic chemistry studies of the complex proteins of neuroscience.

D.A. Dougherty
Division of Chemistry & Chemical Engineering, California Institute of Technology, Pasadena, CA 91125, USA, e-mail: dadougherty@caltech.edu

C. Köhrer and U.L. RajBhandary (eds.) *Protein Engineering.*
Nucleic Acids and Molecular Biology 22,

1 Introduction

The integral membrane proteins of the mammalian central nervous system (CNS) present special challenges to investigators interested in the structure and function of proteins. As with other membrane proteins, the powerful tools of structural biology – X-ray crystallography and nuclear magnetic resonance spectroscopy – are of limited applicability. However, additional issues of incompatibility with typical overexpression systems, complex and variable subunit composition, and multiple post-translational modifications make CNS proteins even more challenging. In fact, the significant advances that are being made in membrane protein structural biology include only a single mammalian channel, the $K_v1.2$ potassium channel from rat brain (Long et al. 2005).

As such, structure–function studies have played an especially prominent role in studies of CNS proteins. Fortunately, an excellent functional probe exists. Many of the key proteins of the CNS are ion channels, or can be made to act through ion channels. This enables the powerful tools of electrophysiology, and arguably no functional probe can match the dynamic and temporal range and the detailed mechanistic insights afforded to studies of ion channels by electrophysiology techniques. And while single molecule studies are, justifiably, a topic of great current interest, recall that for 30 years the patch clamp technique has provided detailed, single channel recordings of ion channels (Neher and Sakmann 1976). A number of additional tools, most notably fluorescence (Cha et al. 1999; Glauner et al. 1999), spin labeling (Hubbell et al. 2003; Perozo 2006), and chemical modification approaches such as the substituted cysteine accessibility method (Wilson and Karlin 1998; Karlin 2002) have also provided invaluable information. When considering structure–function studies of channels and receptors, the availability of a functional probe is not the major limitation.

To vary structure and thus allow a structure–function study, the standard tool is site-directed mutagenesis. Throughout the 1980s, representatives of most of the major classes of receptors/channels of the CNS were cloned, and this allowed them to be expressed under controlled conditions, enabling detailed electrophysiological characterization. Combine site-directed mutagenesis with this protocol and we have molecular neurobiology: structure–function studies of key proteins of the CNS.

While it is certainly useful, conventional site-directed mutagenesis is far from optimal because of the limited structural and functional variation among the 20 natural amino acids. As such, unnatural amino acid mutagenesis is a natural extension that could enable insightful probes of CNS proteins. Building on early work by Hecht and others (Heckler et al. 1984), Schultz reported in 1989 the successful incorporation of unnatural amino acids into a protein (Noren et al. 1989), and Chamberlin incorporated an unnatural amino acid into a small peptide (Bain et al. 1989). Subsequent work primarily from Schultz's laboratory established that the *nonsense suppression* approach could be used to incorporate a wide range of unnatural amino acids into a variety of proteins, enabling an array of structure–function studies (Cornish et al. 1995).

Despite the evident potential of the unnatural amino acid approach, it did not become a broadly adopted addition to the toolbox of chemical biology. No doubt the major factor that discouraged other groups from employing this powerful procedure was the quite limited quantities of protein that could be made. More than a decade ago, we recognized that the remarkable sensitivity of electrophysiology techniques could provide a solution. With an ultrasensitive assay, the low protein yields of unnatural amino acid mutagenesis are no longer an issue. Such reasoning, however, led to a new challenge. The unnatural amino acid method was compatible only with in vitro protein synthesis, but the channels and receptors of the CNS require a more advanced expression system of the sort associated with a vertebrate cell.

In this context, our group, in a fruitful and ongoing collaboration with Henry Lester, set out to develop the in vivo unnatural amino acid method (Nowak et al. 1995). We will refer to protein expression in a live, vertebrate cell as in vivo, recognizing that some in the biological community use this term differently. We also note that more recent advances in the unnatural amino acid method promise to overcome the quantity limitations mentioned above for some cases, although the applicability of these approaches to CNS proteins has yet to be established (Link and Tirrell 2005; Wang and Schultz 2005). The cell of choice was the *Xenopus* oocyte, a favorite vehicle of molecular neurobiology that allows expression and electrophysiological evaluation of most of the important receptors/channels of neurobiology. In the present work we summarize the results of studies of a range of receptors/channels in vertebrate cells. Over a dozen different channels and receptors have been probed with over 100 different residues, resulting in a number of insights into the structure and function of these important molecules. We will present a brief description of the method, but the major focus will be on the types of studies that can be performed and the insights that can be obtained when the proteins of the CNS are probed with unnatural amino acid mutagenesis.

2 In Vivo Nonsense Suppression Method for Unnatural Amino Acid Incorporation

The essential methodological aspects are outlined in Fig. 1, and have been described in detail elsewhere (Nowak et al. 1998). A nonsense (stop) codon, typically TAG (amber), is inserted at the site of interest, and the mutant messenger RNA (mRNA) is prepared. Separately, a designed transfer RNA (tRNA) (termed a "suppressor") that recognizes the nonsense codon is prepared and chemically aminoacylated with the desired unnatural amino acid. Note that this chemically synthesized, aminoacyl tRNA is a stoichiometric reagent in the process, and this is the factor that most limits the quantity of protein that can be produced. A key feature of the suppressor tRNA is that it must be *orthogonal* to the natural machinery of the chosen expression system, in this case the *Xenopus* oocyte. An orthogonal tRNA is one that is not

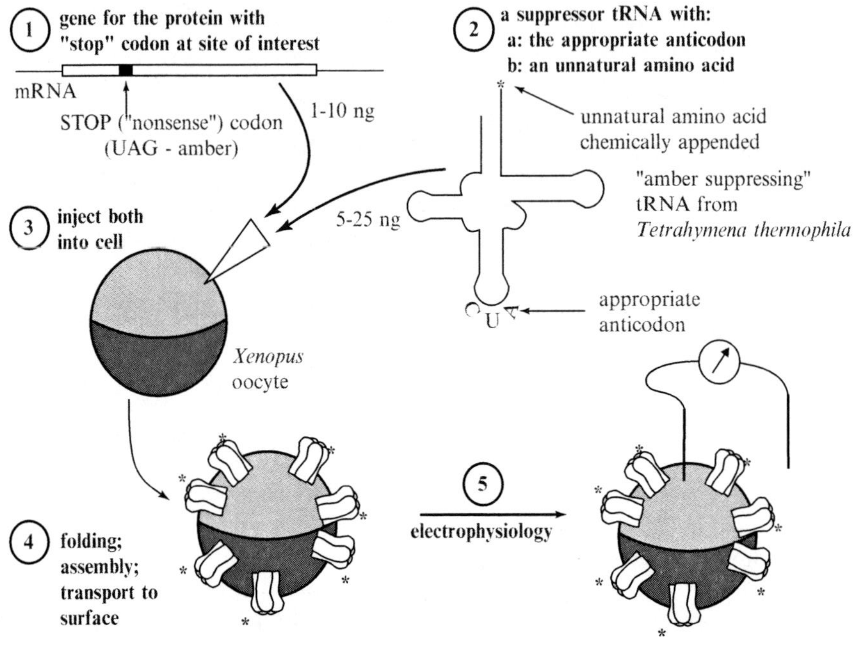

Fig. 1 In vivo nonsense suppression method for unnatural amino acid incorporation into receptors/channels expressed in a *Xenopus* oocyte

recognized by any of the endogenous aminoacyl tRNA synthetases, and so it is never charged with a natural amino acid. Were such a complication to arise, the resultant protein would be a mixture of molecules, some containing the desired unnatural amino acid, but others containing a, typically unidentified, natural amino acid at the mutation site. This would render all further study meaningless, and so the development of an orthogonal tRNA is essential.

Early studies established that the tRNA developed by Schultz for in vitro nonsense suppression, a modified yeast tRNA-Phe, was not orthogonal to the *Xenopus* oocyte expression system (Nowak et al. 1995). In designing a better tRNA, we took advantage of the fact that *Tetrahymena thermophila* uses TAG as a coding codon (for glutamine). We reasoned this might be an especially efficient suppressor, and after making one mutation to improve orthogonality, we produced THG73, an efficient, orthogonal suppressor tRNA. The overwhelming majority of our studies in *Xenopus* oocytes have used THG73 (Saks et al. 1996), and others have found this tRNA useful also. More recently we have developed two new tRNAs that employ frameshift suppression, that is, the recognition of four-base codons (Rodriguez et al. 2006), a remarkable process that was pioneered by Sisido (Hohsaka et al. 1996). These new tRNAs recognize CGGG and GGGU quadruplet codons, and are "*more orthogonal*" than THG73. Combining these with THG73 has allowed us to incorporate three different unnatural amino acids site-specifically into a single protein expressed in a *Xenopus* oocyte. Very recently, we have developed new amber suppressors and an opal suppressor that function well in *Xenopus* oocytes and further expand the collection of viable tRNAs (Rodriguez et al. 2007a, b).

The mutant mRNA and the aminoacyl suppressor tRNA are coinjected into the *Xenopus* oocyte, a process facilitated by the large size of the oocyte (approximately 1 mm across; volume of approximately 1 μl). After 24–48 h, standard whole-cell, two-electrode, voltage-clamp electrophysiology recordings reveals that the oocyte has obligingly expressed the mutant protein, assembled the typically multisubunit receptor, and transported it to the surface of the cell. Importantly, the pharmacologic properties and physiologic function of the protein expressed in the *Xenopus* oocyte are identical to what is seen in the native environment, be it neuron, muscle, kidney, etc. For the rare cases where this is not true, this method should not be applied. Substantial electrical signals can be obtained from oocytes expressing as little as 10 amol of channel; the quantity of protein expressed is rarely an issue.

While the overwhelming majority of studies of receptors and channels using unnatural amino acids have been performed in *Xenopus* oocytes, there is no fundamental limitation to moving into other cell types. Using a microelectroporation protocol, we have demonstrated that nonsense suppression and unnatural amino acid incorporation can be achieved in cultured mammalian cells such as Chinese hamster ovary cells, human embryonic kidney cells, and hippocampal neurons (Monahan et al. 2003). Others have shown that nonsense suppression in mammalian cells can be achieved through microinjection and through chemical transfection (Ilegems et al. 2002; Sakamoto et al. 2002). Thus, we can anticipate further expansion of this protocol to other cell types.

Figures 2 and 3 show a collection of residues that have been successfully incorporated into functioning channels or receptors expressed in *Xenopus* oocytes. As others have seen with other expression systems, the ribosome of the *Xenopus* oocyte is remarkably tolerant. Extraordinary variation in side chain structure can be accommodated. Even the protein backbone can be changed, in that α-hydroxy acids can be incorporated instead of α-amino acids. The most substantive limitation is that the natural stereochemistry at the α-carbon is essential for efficient incorporation; D-amino acids are not readily accepted. A byproduct of this limitation is that suppressor tRNA linked to a racemic unnatural amino acid can be employed; the ribosome will perform an efficient kinetic resolution.

The method described is applicable to a broad range of receptors and channels (Dougherty 2000; Beene et al. 2003). To date, the following systems have been successfully probed with unnatural amino acids using the nonsense suppression method: nicotinic acetylcholine receptors (both muscle-type and neuronal); 5-HT_3 and MOD-1 serotonin receptors; $GABA_A$ and $GABA_C$ receptors; G protein coupled receptors (NK2 and M2); K^+ channels (Shaker, hERG, ROMK1, $K_{ir}2.1$, GIRK1, GIRK4); Na^+ channels ($Na_v1.4$); the GABA transporter GAT-1; and the NMDA receptor. It seems safe to conclude that any system that is compatible with the *Xenopus* oocyte heterologous expression system can be probed with unnatural amino acids.

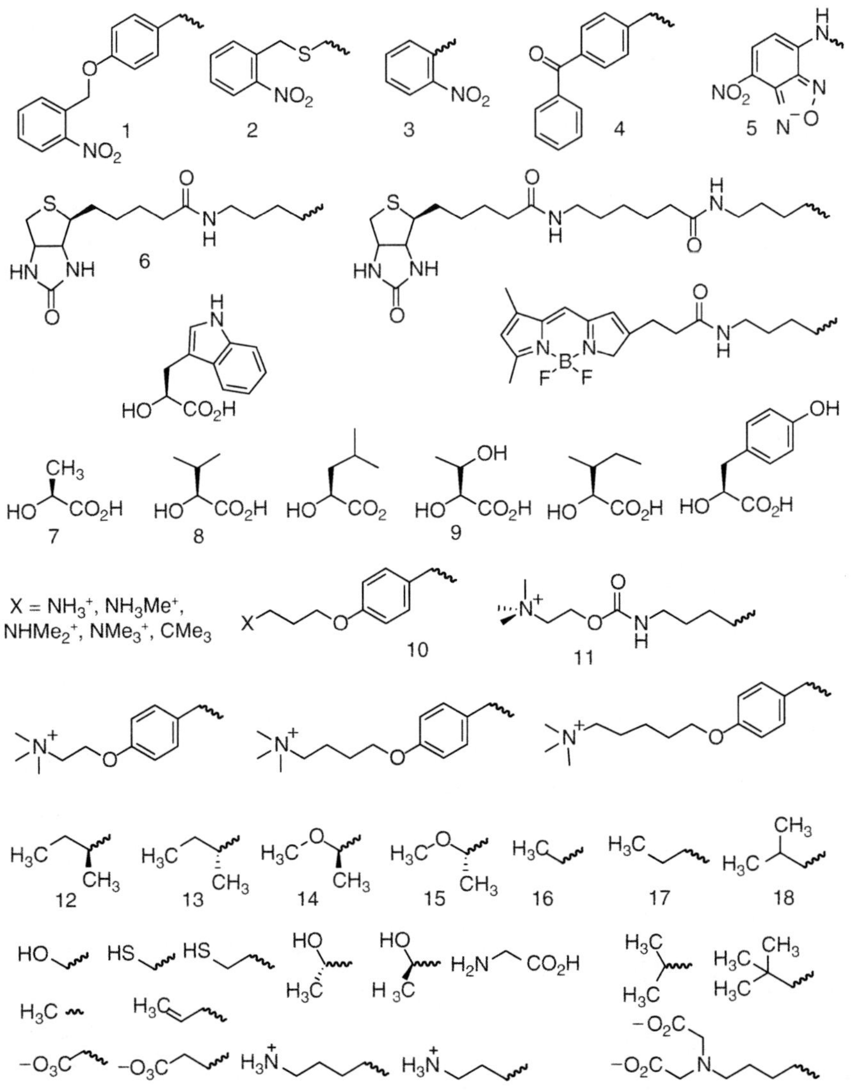

Fig. 2 Select residues that have been incorporated into functional receptors/channels expressed in *Xenopus* oocytes. Examples include photoresponsive residues, biotin-containing residues, α-hydroxy acids, tethered agonists, small nonpolar residues, and small polar residues

A majority of our studies have probed members of the so-called Cys-loop superfamily of neuroreceptors (Corringer et al. 2000; Karlin 2002; Lester et al. 2004; Unwin 2005). These comprise five identical or homologous subunits that are symmetrically or pseudo-symmetrically arrayed around the central channel. They are ligand-gated ion channels that contain both a binding site for an appropriate neurotransmitter and an ion channel that spans the membrane and is typically closed. Binding of neurotrans-

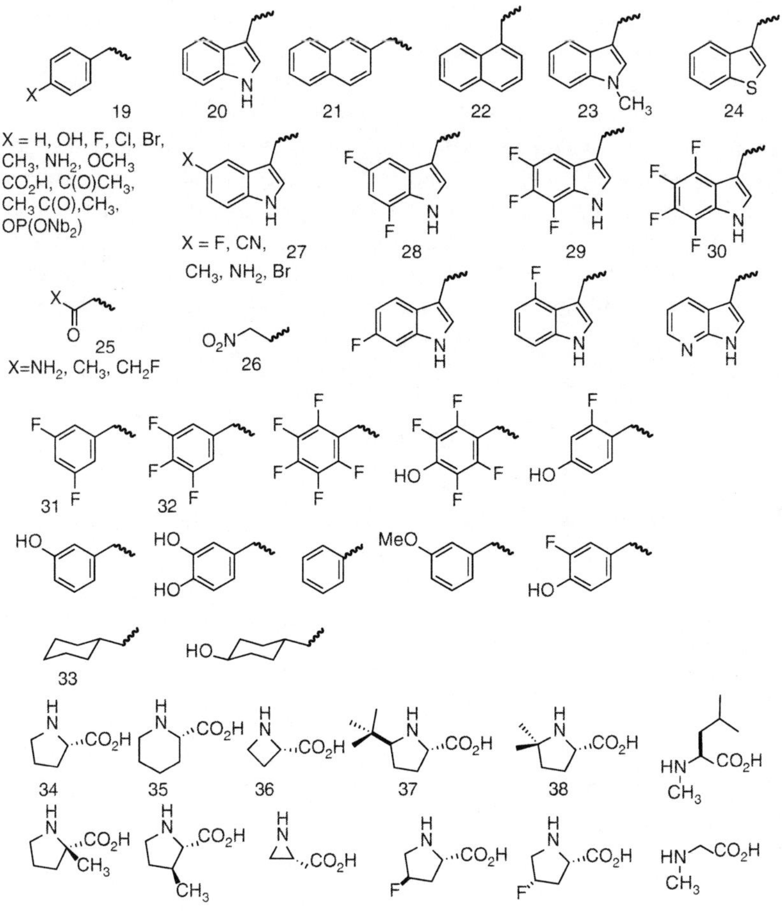

Fig. 3 Select residues that have been incorporated into functional receptors/channels expressed in *Xenopus* oocytes. Examples include phenylalanine/tyrosine analogs, tryptophan analogs, asparagine analogs, nitrohomoalanine, and proline analogs

mitter or any of a number of unnatural agonists initiates a conformational change that ultimately leads to opening (gating) of the channel. Cys-loop receptors thus enable fast synaptic transmission throughout the CNS and the peripheral nervous system. Members of the Cys-loop superfamily include the nicotinic acetylcholine receptors of both the neuromuscular junction and the CNS, the 5-HT_3 and MOD-1 serotonin receptors, the $GABA_A$ and $GABA_C$ receptors, and the glycine receptor.

The present work will emphasize studies of receptors and channels performed in living cells. However, it is important to appreciate that the breadth of successful unnatural amino acid incorporation demonstrated in Figs. 2 and 3 parallels that seen in other expression systems, both in vitro and in vivo. Therefore, useful unnatural amino acids that have been incorporated into soluble proteins with other expression systems will almost certainly be applicable to studies of receptors and channels (see the chapter by Köhrer and RajBhandary, this volume).

The wide range of experiments one can envision with such an extensive collection of residues can be conceptually grouped into two separate types. One involves unnatural amino acids that are substantially different from the natural set and that deliver a novel functionality to the protein. Examples include photoreactive amino acids, spectroscopic probes, and tethered functionalities such as agonists or antagonists. The other class involves unnatural amino acids that are only subtly different from the natural set. These allow systematic structure–function studies of the sort that have been a hallmark of small molecule chemistry. We will provide examples of both classes here, beginning with the former.

3 "Highly Unnatural" Amino Acids

3.1 Photoresponsive Amino Acids

Figure 2 shows a number of unnatural amino acids that are designed to be responsive to light. A common strategy is the so-called *caged* amino acid, a side chain that contains, in effect, a photocleavable protecting group, most typically an *o*-nitrobenzyl derivative (Petersson et al. 2003). Photolysis *decages* the side chain, revealing the

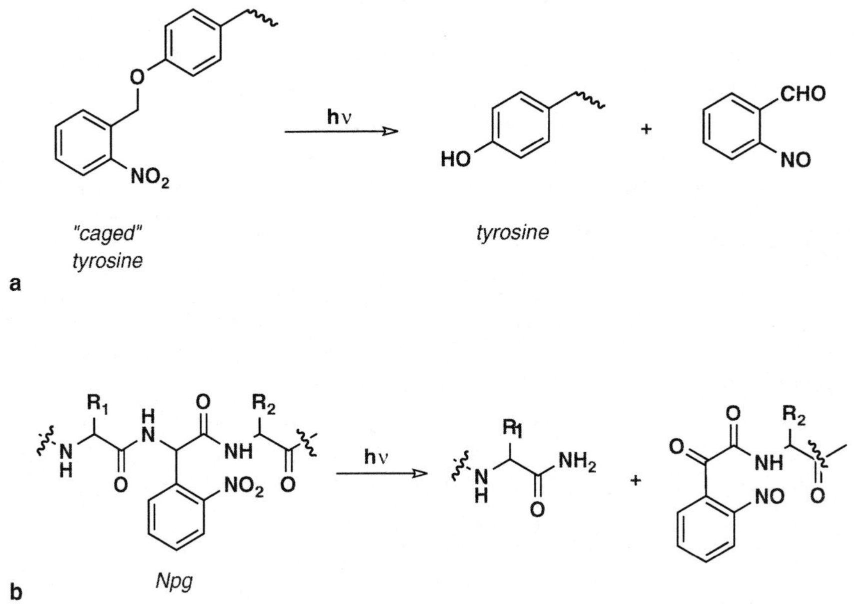

Fig. 4 Examples of photoreactive unnatural amino acids: **a** caged tyrosine; **b** *o*-nitrophenylglycine (*Npg*)

native functionality. Examples include caged tyrosine (**1**, Fig. 4a) and caged cysteine (**2**) (Philipson et al. 2001). Some time ago we demonstrated the feasibility of this approach by incorporating caged tyrosine into several sites of the nicotinic acetylcholine receptor (nAChR) (Miller et al. 1998). Using a setup that allowed delivery of millisecond pulses of light to a *Xenopus* oocyte that was voltage-clamped allowed real-time electrophysiological monitoring of the decaging process. When the caged tyrosine was incorporated at a site where it was expected to substantially distort the acetylcholine (ACh) binding site, flash decaging was followed by a slow recovery of receptor activity, allowing us to monitor in real time the conformational change associated with reforming the natural binding site.

Subsequently, we incorporated caged tyrosine into the inward rectifying potassium channel $K_{ir}2.1$ at a site that was believed to be modulated by kinases (Tong et al. 2001). When the caged protein was expressed in the presence of a kinase such as Src and a phosphatase inhibitor, photolysis launched a biochemical cascade that could be monitored by electrophysiology techniques. Interestingly, two kinase-dependent processes were revealed. One was a direct modulation of the channel, leading to a decrease in current. The other was a dynemin-dependent endocytosis, in which decaged channels were removed from the surface of the cell in a kinase-dependent manner. The caged residue thus revealed a complexity in the kinase modulation of channel behavior that conventional studies could not have detected. Since kinases are thought to play a critical role in many aspects of synaptic plasticity, the temporal and spatial control of kinase signaling afforded by caged residues could be of great value to neuroscience.

A novel caged amino acid is *o*-nitrophenylglycine (Npg, **3**, Fig. 4b) (England et al. 1997). When incorporated into a protein this residue in effect cages a backbone nitrogen. Photolysis then cleaves the peptide bond; in a process we have termed "site-specific, nitrobenzyl-induced, photochemical proteolysis" (SNIPP). We have used Npg and SNIPP to convert the Shaker K^+ channel from one that undergoes rapid (N-type) inactivation, the wild type, to one that does not inactivate. When Npg is placed near the N-terminus of the channel, photolysis snips off the N-terminal residues. This removes the inactivation domain, which is, in effect, the *ball* in a *ball and chain* mechanism for inactivating the channel by plugging the pore. The recapitulation of this well-known phenotype (termed "Shaker IR", for *inactivation removed*) established the viability of SNIPP.

We also incorporated Npg into the *Cys-loop* of the nAChR, a 15-residue disulfide loop that is the definitive structural feature of the nicotinic superfamily of ligand-gated ion channels. Earlier conventional mutagenesis studies that converted either cysteine to serine destroyed the loop and produced receptors that were unable to fold/assemble properly. This eliminated any possibility of functional characterization of a receptor with a compromised Cys-loop. Incorporating Npg into the Cys-loop, however, produced functional receptors that expressed well and functioned like the wild type. Then, photolysis completely destroyed receptor function, establishing that the Cys-loop not only had a role in receptor folding/assembly but also in proper functioning of this ligand-gated ion channel.

A valuable type of photoreactive unnatural amino acid is one that has the potential to cross-link two proteins or two regions of a single protein. The benzophenone derivative **4** in Fig. 2 is one example. Other potential cross-linking residues such as phenylazide and diazirine derivatives have been incorporated by nonsense suppression (Link and Tirrell 2005; Wang and Schultz 2005). Since the proteins of neuroscience are frequently multisubunit proteins and are often components of highly organized protein assemblies, one can envision many applications of unnatural amino acids that involve cross-linking.

A potentially very important application of the unnatural amino acid method is the incorporation of fluorescent unnatural amino acids that can serve as probes of conformational changes by either direct fluorescence changes or, with two fluorophores, by fluorescence resonance energy transfer (FRET). The potential is especially great for receptors and channels. The signaling proteins of neuroscience all exist in multiple states, and it is the interconversions among those states that are essential to understanding receptor/channel function. The potential of the method has been established by several groups (Cha et al. 1999; Glauner et al. 1999), including ours (Dahan et al. 2004), using receptors/channels that are made fluorescent by a more conventional, cysteine-labeling strategy. But it is clear that the unnatural amino acid approach could be very powerful in this regard. One of the earliest applications of the unnatural amino acid method in *Xenopus* oocytes was reported by Chollet and coworkers and involved incorporation of the fluorescent unnatural amino acid 2-amino-3-(7-nitrobenzo[1,2,5]oxadiazol-4-ylamino)propionic acid (**5**) into the NK2 receptor (Turcatti et al. 1996, 1997). When this was combined with a fluorescent antagonist, it allowed the determination of several geometrical parameters of this G protein-coupled receptor. Jan and Boxer have introduced the highly environmentally sensitive aladan into a K^+ channel expressed in a *Xenopus* oocyte (Cohen et al. 2002). The groups of Hecht (Anderson et al. 2002) and Sisido (Taki et al. 2002) have combined nonsense suppression and frameshift suppression to incorporate two fluorescent unnatural amino acids and demonstrate FRET in a soluble protein expressed in vitro, and Schultz has shown that several fluorescent unnatural amino acids are compatible with ribosomal incorporation into proteins (Summerer et al. 2006; Wang et al. 2006).

Despite these encouraging results, the most desirable fluorescence/FRET experiment – monitoring a conformational/structural change in a receptor or channel in a living cell using one or two fluorescent unnatural amino acids – has yet to be achieved. The NK2 study involved protein expression in *Xenopus* oocytes, but also subsequent harvesting of receptors from many oocytes and in vitro fluorescence measurements. Other studies have involved soluble proteins. While most workers consider fluorescence measurement to be a very sensitive method, electrophysiological monitoring is *much* more sensitive. Quantities of protein that are readily detectable by electrophysiology techniques can be invisible to fluorescence techniques. In addition, the *Xenopus* oocyte presents a significant background autofluorescence from the yolk, which comprises most of the oocyte, and that substantially complicates in situ measurements. Thus, successful exploitation of FRET in receptors/channels using two unnatural amino acids will require either very high expression levels in *Xenopus* oocytes or use of mammalian cells that are more compatible with fluorescence.

3.2 Chemically Reactive Functionality

There is increasing interest in the introduction of *bio-orthogonal* functionality into proteins (Prescher and Bertozzi 2005; van Swieten et al. 2005). These are reactive functionalities that are not found in the molecules of biology and that react specifically and efficiently with designed targets but do not react with the functional groups of biology. These bio-orthogonal functionalities are appealing because they allow selective functionalization of a protein in situ. The *reagent* can be a fluorescent group, a spin label, or another protein. The advantage over the more typical strategies such as cysteine modification is that there is essentially no undesired labeling of other sites and other proteins. The system is bio-orthogonal, and so labeling can be quite specific. One can anticipate a number of interesting applications of the bio-orthogonal strategy to studies of receptors and channels. We note that the caging strategy outlined above offers a similar approach. After naturally occurring cysteines have been passivated, photodecaging of a caged cysteine would allow selective modification.

Almost a decade ago, we showed how this strategy could be implemented in the *Xenopus* oocyte system by incorporating the amino acid biocytin, **6** (Gallivan et al. 1997). Since this residue terminates in a biotin moiety, its incorporation allows labeling of an expressed protein with streptavidin and derivatives thereof (fluorescent streptavidin, etc.). However, such labeling will only be efficient in a cellular context if the biocytin residue is exposed on the surface of the cell and thus accessible to the streptavidin. As such, biocytin incorporation can serve as a probe of surface accessibility of particular residues in a complex membrane protein. To evaluate this method, we probed the main immunogenic region (MIR) of the nAChR of the neuromuscular junction. The MIR is a ten-residue stretch of the α subunit of the nAChR that binds many antibodies, including those associated with myasthenia gravis, a human autoimmune disease. As such, we anticipated that much, if not all, of the MIR would be surface-exposed and thus amenable to the biocytin strategy. Biocytin was incorporated efficiently at five of the ten residues of the MIR. Interestingly, though, efficient streptavidin labeling was seen at only one of the five biocytin-containing sites, position Asp70. This suggested that the biocytin–streptavidin system was fairly sensitive to steric environment and is thus responsive to only highly exposed residues.

Several other bio-orthogonal strategies involving unnatural amino acids have been employed with in vitro expression and should be adaptable to the in vivo approach (Cornish et al. 1996; Prescher and Bertozzi 2005; Dieterich et al. 2006). These include ketones, which are efficiently modified by hydroxylamines, and azides, which are compatible with both click chemistry and Staudinger-type ligations. We can anticipate application of these types of strategies to many in vivo problems.

3.3 Backbone Esters

It has been known since 1970 that a remarkable aspect of ribosomal protein synthesis is that α-hydroxy acids are readily accepted by the ribosome just like α-amino acids (Fahnesto et al. 1970). Schultz showed that α-hydroxy acids could be incorporated by in vitro nonsense suppression, and evaluated the disruption of an α-helix that results from incorporating an α-hydroxy acid (Koh et al. 1997).

We have incorporated a number of α-hydroxy acids (**7–9** and others) into receptors expressed in the *Xenopus* oocyte. The frequently observed improved protein expression yields with α-hydroxy acids could reflect the fact that aminoacyl tRNAs are chemically unstable, with hydrolysis of the ester linkage between the amino acid and the tRNA being accelerated by the positive charge of the amino acid's NH_3^+ group. Hydroxyacyl tRNAs are more chemically stable, and so may have a longer lifetime in the *Xenopus* oocyte. Another valuable feature of α-hydroxy acid incorporation is that the backbone ester that results is, in a sense, bio-orthogonal. That is, treatment of proteins that have a backbone ester with ammonium hydroxide hydrolyzes the ester, thus cleaving the protein backbone. In a study of a number of sites in the nAChR, visualization of the base hydrolysis process with western blotting established that cleavage efficiency, and thus the efficiency of nonsense suppression, was very high (England et al. 1999a, b). This led to a novel approach to mapping disulfide connectivities in complex proteins (Fig. 5). In the extracellular

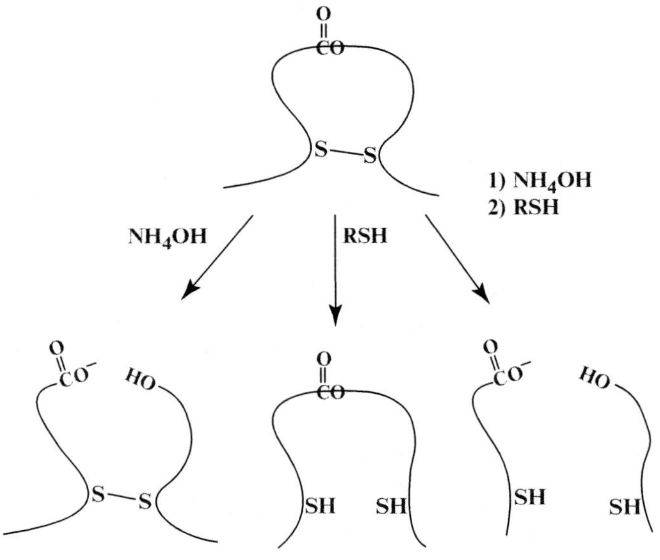

Fig. 5 The use of backbone esters obtained by incorporating α-hydroxy acids to map out disulfide connectivity

domains of many receptors/channels, there are many cysteines that are known to be cross-linked, but the exact connectivity pattern is unclear. If a backbone ester is incorporated within a disulfide loop, treatment with base alone should still produce full-length protein, as the disulfide bond keeps the chain intact. Subsequent treatment with dithiothreitol after the base hydrolysis will give two fragments. Alternatively, if the ester is outside the disulfide loop, base hydrolysis alone will give two fragments; dithiothreitol will not be necessary. Thus, backbone modification via unnatural amino acid mutagenesis and simple western blotting provides a definitive test (England et al. 1999a).

One potentially informative use of α-hydroxy acids is as surrogates for proline, in that both lack the backbone NH bond that contributes so substantially to α-helix and β-sheet formation. In other ways backbone esters are a poor mimic of proline: while prolines show an increased propensity for the s-*cis* conformer relative to all other natural amino acids (see below), esters actually prefer the s-*trans* conformer *more* strongly than amides. Also, the rotation barrier around the backbone ester bond is much smaller than for the amide of amino acids.

We have used backbone esters to evaluate a number of proline sites in receptors and channels. In the Cys-loop receptors there is a completely conserved proline residue in the middle of a particular transmembrane α-helix – the first of four such α-helices in each subunit which is termed "M1." A proline in the middle of a transmembrane helix seems likely to have an important functional or structural role, but attempting to probe that role with conventional mutagenesis simply produced nonfunctional receptors, a not overly informative result. However, we found that incorporation of essentially any α-hydroxy acid at the proline site produced a wild-type-like receptor (England et al. 1999b; Dang et al. 2000). That is, the α-hydroxy analogs of alanine, valine, or leucine (**7–9**) all function very well at this site. These results suggest that it is the absence of the backbone NH in a proline linkage that is the essential feature provided by this particular proline.

Replacing a natural amino acid with its α-hydroxy analog does not always produce such a benign result. We replaced a number of non-proline amino acids with their α-hydroxy analogs throughout the second transmembrane region, M2, which lines the ion channel (England et al. 1999b). Significant changes in gating behavior were seen throughout M2, and these results supported a gating model that proposed significant backbone conformational changes in M2 associated with gating.

A subtle aspect of backbone ester incorporation is that the backbone carbonyl becomes a measurably poorer hydrogen-bond acceptor than the amide carbonyl of the conventional peptide bond. In the context of both an α-helix and a β-sheet, this effect has a significantly destabilizing influence on secondary structural stability. We have used this aspect of α-hydroxy acid incorporation to probe drug–receptor interactions in the ACh receptor (Cashin et al. 2005). In the nAChR the natural agonist ACh is a quaternary ammonium ion, but nicotine and related potent agonists such as epibatidine are protonated ammonium ions. As such, the latter can participate in hydrogen-bonding interactions with the receptor, while ACh cannot. Model structures suggested that a particular backbone carbonyl was involved in just such a hydrogen bond. We replaced the $i + 1$ residue with the α-hydroxy analog, converting the implicated carbonyl to an ester rather than an amide. Indeed, both nicotine and epibatidine showed diminished activity at the receptor, but ACh did not, supporting the proposed hydrogen bond.

3.4 Spin Labels

Another valuable biophysical probe would be a spin label, allowing many studies of receptor/channel structure and function with relatively sensitive electron paramagnetic resonance techniques. The method has produced many interesting results on receptors/channels with the spin label introduced by cysteine modification (Hubbell et al. 2003; Perozo 2006), and it seems clear that valuable information could be obtained if spin labels could be incorporated biosynthetically. Early studies evaluated this possibility and found generally low efficiencies for incorporation of the spin label (Cornish et al. 1994). Voss and coworkers employed the *Xenopus* oocyte expression system and established incorporation into protein using a luciferase assay on homogenized oocytes (Shafer et al. 2004), but no electron paramagnetic resonance studies on proteins prepared in this way were reported. At present, this promising approach awaits validation.

3.5 Tethered Agonists

Another type of novel unnatural amino acid is the tethered agonist, a structure that mimics the natural ligand of a neuroreceptor but is covalently linked to the receptor. We have exploited this approach extensively in the nAChR, and several examples are shown in Fig. 2 (**10, 11**, and others) (Zhong et al. 1998; Li et al. 2001; Petersson et al. 2002). If the unnatural amino acid containing the agonist analog is incorporated at a position that allows the tethered agonist to occupy the natural agonist binding site, the receptor will be locked in an open state, forming a so-called constitutively active receptor. In the nAChR, a simple tethered quaternary ammonium ion can be enough to create a constitutively active receptor, *if* it is incorporated at the right position and has the proper tether length. Such experiments can provide important geometrical information. For example, in the nAChR field it was universally accepted that the α subunit of the pentameric receptor was a major contributor to the agonist binding site. However, there was a debate as to whether the agonist binding site was buried deep within the α subunit or whether it lay at the interface of two adjacent subunits. Tethered agonists incorporated into the α subunit functioned well. In addition, a tethered agonist incorporated into an adjacent, non-α subunit was able to reach the α subunit agonist binding site, establishing that the interfacial agonist binding site model was correct.

4 Structure–Function Studies

While the ability to incorporate unnatural amino acids that are qualitatively distinct from the natural set presents a powerful tool to biophysical chemists, subtler variations have proven to be equally valuable. Conventional mutagenesis often establishes that a particular residue is important to receptor/channel function, in that changing the residue significantly alters receptor/channel function. However, the precise role of the residue is often unclear. For example, if conversion of a tyrosine to a phenylalanine produces a large effect, it is often concluded that the OH of the tyrosine is involved in an important hydrogen bond. However, whether that OH is acting as a hydrogen-bond donor and/or an acceptor is not clear. Also, we will show later that such reasoning is not at all compelling, and sometimes the OH of a tyrosine is simply a steric placeholder. It may be that biochemists are uninterested in questions at this level; however, we seek *chemical-scale* insights into receptor structure and function. By chemical scale we mean, in effect, the distance scale to which chemists are accustomed: the functional group; the specific bond rotation or local conformational change; the precise noncovalent interaction. We have found that the unnatural amino acid method is very well suited to providing chemical-scale insights into receptor structure and function.

Along with providing more subtle structural variation than conventional mutagenesis, the unnatural amino acid method enables multiple, systematic changes. We will see later several examples of trends across four or five related amino acids in which receptor/channel function correlates well with the variation of a particular property of the series of residues. This provides the most compelling kind of argument. Even when the change is subtle, a single variant of the natural residue will always be open to multiple interpretations. Any change will impact steric, electronic, solvation properties, etc. and it is a challenge to sort these out. However, when a consistent trend across a systematically varying series of residues is seen, compelling conclusions can be reached. Many of the studies we will describe are very much of the form of classical physical organic chemistry. Structure–function studies that closely resemble Hammett plots have been produced. The target molecules are much more complex than the small molecules of classical physical organic chemistry, but the mindset and the types of conclusions that can be reached are strikingly similar.

4.1 Hydrophobic Effects

Hydrophobic residues contribute substantially to the transmembrane regions of receptors and channels, and often this hydrophobicity is considered to be the defining aspect of a residue. However, with the natural 20 amino acids, efforts to vary hydrophobicity inevitably involve concomitant changes in size and shape. The unnatural amino acid method provides several ways to address this ambiguity. For example, the natural amino acid isoleucine (**12**) and the unnatural *allo*-isoleucine (**13**) have identical size and hydrophobicity. If the latter is the only feature that matters, the two should give receptors with nearly identical functionality. However, if the amino acid side chain is involved in a key structural interaction, the isomeric side chains could be sensitive to this in the chiral environment of the protein. Coupled to the isoleucine/*allo*-isoleucine pair is the duo of *O*-methylthreonine (**14**) and *allo*-*O*-methylthreonine (**15**). These are isosteric to isoleucine/*allo*-isoleucine but have a single CH_2 replaced by an ether oxygen, a change that reduces hydrophobicity in a very subtle way.

In an example of the use of such strategies, we studied a very highly conserved leucine residue of the Cys-loop receptors termed "Leu 9′." It lies in the middle of a transmembrane helix (M2), and conventional mutagenesis studies established that alteration of this residue significantly affected receptor function, and that polar residues such as serine had large effects (Revah et al. 1991; Labarca et al. 1995). To a chemist, a leucine-to-serine mutation in not at all subtle, and we set out to see if hydrophobicity was really the issue at Leu 9′ (Kearney et al. 1996b). Systematic alterations of hydrophobicity can be achieved in many ways with unnatural amino acids. For example, aminobutyric acid (**16**) to norvaline (**17**) extends the chain length by one CH_2; norvaline to isoleucine adds a β branch; and norvaline to leucine (**18**) adds a γ branch. Dozens of substitutions of this sort established a consistent trend that increasing hydrophobicity made the receptor more difficult to open. However, other subtleties were revealed. At one specific site, isoleucine and *allo*-isoleucine showed a fivefold difference in response to agonist. This observation was confirmed by single channel, patch clamp studies, which showed comparable changes in channel open times. Seeing a noticeable distinction between stereoisomeric side chains indicates that hydrophobicity alone is not the whole story at the 9′ position; apparently the residue fits into a well-defined pocket.

4.2 Hydrogen Bonding

Hydrogen bonding can also be probed using unnatural amino acids. As noted already, a large effect seen for a tyrosine-to-phenylalanine mutation is often interpreted to indicate that the OH of tyrosine is involved in a hydrogen bond. Unnatural amino acids allow more insightful distinctions (Nowak et al. 1995; Kearney et al. 1996a; Price et al. 2003; Beene et al. 2004; Lummis et al. 2005a). In some instances, *O*-methyltyrosine (methoxyphenylalanine, Fig. 3, **19**, X is OCH_3) produces results comparable to tyrosine, suggesting that the tryrosine is a hydrogen-bond acceptor. In other cases, results with O-Me-Tyr resemble those with phenylalanine, suggesting tyrosine is a hydrogen-bond donor. In still other cases, however, *O*-methyltyrosine is similar to tyrosine, but so are 4-fluorophenylalanine and even 4-methylphenylalanine (**19**, X is F, CH_3). In such cases the OH of tyrosine is more likely a steric placeholder, rather than a specific hydrogen bonder. Similar kinds of studies are possible with serine and threonine.

Possible hydrogen bonding involving the NH of the indole side chain of tryptophan (**20**) can be nicely probed with naphthalene analogs (**21, 22**), *N*-methyltryptophan (**23**), and the benzothiophene analog (**24**) (Zhong et al. 1998).

Other potential hydrogen-bonding groups can also be evaluated systematically. To probe the amide groups of asparagine, the methyl ketone analogs (**25**) remove the NH_2 but keep the carbonyl oxygen. A large effect would suggest the NH_2 is a hydrogen-bond donor. The ketone carbonyl is a poorer hydrogen-bond acceptor than the amide carbonyl, and so a small impact from ketone substitution is an ambiguous result. However, progressively fluorinating the methyl group further diminishes the hydrogen bond accepting ability of the carbonyl, and so observation of an appropriate trend establishes the side chain as a hydrogen-bond acceptor. As noted before, α-hydroxy acids can be used to evaluate hydrogen-bonding involving backbone amides.

A subtle but informative unnatural amino acid substitution is available for anionic amino acids such as glutamic acid. A nitro group (NO_2) and a carboxylate (CO_2^-) are isosteric and isoelectronic. The only substantive difference between the two is the charge. As such, substituting nitrohomoalanine (**26**) for glutamic acid provides an ideal test of the importance of charge for a particular glutamic acid. We recently used this approach to evaluate a highly conserved aspartate residue in the nAChR that is in the vicinity of but apparently is not part of the ACh binding site (Cashin et al. 2007). Conventional mutagenesis concluded that the negative charge of the aspartate was essential for proper receptor function. However, near-wild-type behavior could be obtained with a receptor that contained a nitro group instead of a carboxylate, suggesting a more subtle function.

4.3 Cation–π Interaction

Substitution of aromatic amino acids is facile, and a great many derivatives of phenylalanine, tyrosine, and tryptophan have been incorporated into receptors/channels (Fig. 3). Our long-standing interest in the cation–π interaction (Dougherty and Stauffer 1990; Dougherty 1996; Ma and Dougherty 1997) was a prime motivator for developing the in vivo nonsense suppression method, and, indeed, the cation–π interaction has proven to be especially well suited to evaluation through unnatural amino acid mutagenesis. Basic studies of the cation–π interaction established the influence of many substituents on the aromatic ring (Mecozzi et al. 1996a, b). In particular, fluorine is deactivating, and multiple fluorines show additive deactivating effects. The steric perturbation from fluorine substitution is very small, and so fluorination is an ideal probe of whether an aromatic amino acid is binding to a cationic ligand or side chain. The approach is to successively incorporate monofluoro, difluoro, trifluoro, etc. derivatives [**27–30, 19** (X is F), **31, 32**] and look for a systematic change in receptor/channel function.

We have used this strategy to establish key ligand binding contacts in a number of neuroreceptors and ion channels, including the nAChR (Zhong et al. 1998), the 5-HT_3 serotonin receptor (Beene et al. 2002; Mu et al. 2003), the $GABA_A$ (Padgett

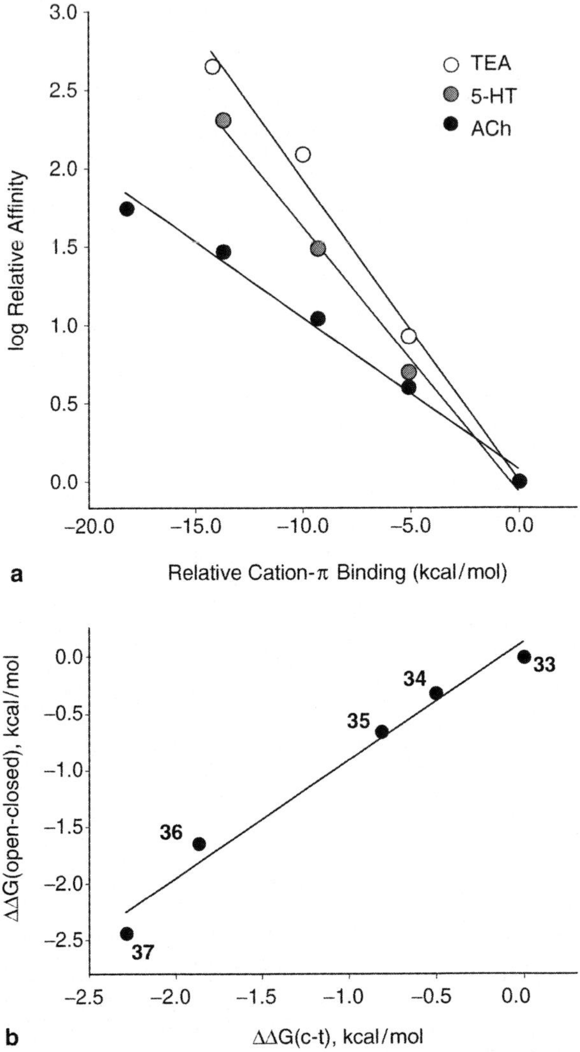

Fig. 6 Structure–function relationships obtained from unnatural amino acid mutagenesis. **a** Cation–π interactions revealed by progressive fluorination of an aromatic amino acid. The *x*-axis describes the innate cation–π binding ability of the aromatic ring, referenced to the parent system. The y-axis is log of the efficiency of inhibition (tetraethylammonium, *TEA*) or activation (acetylcholine, *ACh*; 5-hydroxytryptophan, *5-HT*), again referenced to the parent system. Moving from right to left, residues have no, one, two, three, or four fluorines. Note the point at (0,0) is shared by all three plots. Systems are as follows: TEA, extracellular blockade of the Shaker K^+ channel by TEA (Ahern et al. 2006); 5-HT, activation of the 5-HT_3 receptor by serotonin (Beene et al. 2002); ACh, activation of the nicotinic ACh receptor of the neuromuscular junction by ACh (Zhong et al. 1998). **b** Effect on receptor gating (y-axis) vs. intrinsic *cis/trans* preference for a series of proline analogs (residues **33–37**, Fig. 3) (Lummis et al. 2005b). The site probed is Pro308 of the 5-HT_3 receptor

et al. 2007) and $GABA_C$ (Lummis et al. 2005a) receptors, the Shaker K^+ channel (Ahern et al. 2006), and the sodium channel $Na_v1.4$ (Santarelli et al. 2007). Figure 6a shows several such *fluorination* plots for various ligands binding to differing receptor/channel systems. Such studies drive home the power of observing a systematic trend across a series of subtly varying series of residues. Compelling conclusions can be reached from such studies.

Not all key aromatic interactions, of course, are cation–π interactions. Often an aromatic, especially phenylalanine, is simply a hydrophobic residue. An especially useful probe in such instances is cyclohexylalanine (**33**). Phenylalanine and cyclohexylalanine are very similar in size, shape, rigidity, and hydrophobicity. If cyclohexylalanine functions well at a phenylalanine site, clearly the phenylalanine is not there for a cation–π interaction. Such a situation was seen in our study of the NMDA glutamate receptor, a critical neuroreceptor in models of learning and memory that contains several binding sites, including a Mg^{2+} binding site that occludes the ion channel. A cation–π interaction had been proposed to be important to Mg^{2+} binding on the basis of conventional mutagenesis; however, fluorination had no effect, and cyclohexylalanine performed well at the site of interest, ruling out a cation–π interaction (McMenimen et al. 2006).

4.4 Conformational Changes

As noted already, prolines often play a pivotal role in receptor/channel structure and function. Along with disrupting backbone hydrogen bonding, prolines facilitate s-*cis*/s-*trans* isomerization about the peptide bond. Using unnatural amino acid mutagenesis, we were able to directly associate the isomerization of a proline amide with the conformational change that opens (gates) the ion channel (Lummis et al. 2005b). The series proline, pipecolic acid, azetidine carboxylic acid, 5-*t*-butyl proline, 5,5-dimethylproline (**34–38**) shows a progressive increase in intrinsic s-*cis* preference, varying from 5% for proline to 71% for dimethylproline. At a particular proline of the 5-HT_3 serotonin receptor, a linear free-energy relationship was seen between the intrinsic *cis* preference and the open probability of the channel (Fig. 6b). This established a clear link between the s-*cis*/s-*trans* isomerization of the particular proline and channel opening, and thus provided the first chemical-scale model for the gating of a neuroreceptor.

5 Conclusion

To a chemist interested in protein structure and function, the ability to site-specifically incorporate unnatural amino acids into any protein of interest immediately suggests an almost limitless number of interesting and informative experiments. We felt from the outset that the area of protein science that could most benefit from the method was the field of integral membrane proteins. Membrane proteins comprise approximately 30% of the genome and approximately 60% of drug targets, but structural information is sparse. Much more so than for soluble proteins, for which crystallography has become such a broadly applicable tool, membrane proteins need sophisticated tools to tease out structure–function relationships. Happily, receptors and channels naturally provide a way to circumvent the long-standing limitation of unnatural amino acid mutagenesis – the high sensitivity of electrophysiology techniques allows very small quantities of proteins to be evaluated.

Indeed, the nonsense suppression approach to unnatural amino acid incorporation has translated very effectively to receptors and channels. The key advance was to develop a suppressor tRNA that is orthogonal to the translational machinery of the *Xenopus* oocyte. Dozens of receptors and scores of unnatural residues have been evaluated, and the approach has produced a number of valuable insights into the structure and function of many neuroreceptors and ion channels. Challenges remain, most notably routine incorporation of unnatural amino acids into receptors/channels expressed in mammalian cells, and, hopefully, the adaptation of the more recently developed nonstoichiometric approaches to unnatural amino acid incorporation to the proteins of neuroscience. We can anticipate many more exciting applications of unnatural amino acids to studies of receptors and ion channels.

Acknowledgments Our work on unnatural amino acid mutagenesis is part of an enjoyable and productive collaboration with Henry Lester, Division of Biology, California Institute of Technology. We have also benefited from fruitful collaborations with Sarah Lummis, Cambridge University, and Richard Horn, Jefferson Medical College. I am very grateful to the many exceptional students who have done all the work and provided many of the creative insights described here. Our work has been supported by the NIH (NS 34407).

References

Ahern CA, Eastwood AL, Lester HA, Dougherty DA, Horn R (2006) A cation–π interaction between extracellular TEA and an aromatic residue in potassium channels. J Gen Physiol 128:649–657

Anderson RD, Zhou J, Hecht SM (2002) Fluorescence resonance energy transfer between unnatural amino acids in a structurally modified dihydrofolate reductase. J Am Chem Soc 124: 9674–9675

Bain JD, Glabe CG, Dix TA, Chamberlin AR (1989) Biosynthetic site-specific incorporation of a non-natural amino acid into a polypeptide. J Am Chem Soc 111:8013–8014

Beene DL, Brandt GS, Zhong WG, Zacharias NM, Lester HA, Dougherty DA (2002) Cation–pi interactions in ligand recognition by serotonergic (5-HT3A) and nicotinic acetylcholine receptors: the anomalous binding properties of nicotine. Biochemistry 41:10262–10269

Beene DL, Dougherty DA, Lester HA (2003) Unnatural amino acid mutagenesis in mapping ion channel function. Curr Opin Neurobiol 13:264–270

Beene DL, Price KL, Lester HA, Dougherty DA, Lummis SCR (2004) Tyrosine residues that control binding and gating in the 5-hydroxytryptamine(3) receptor revealed by unnatural amino acid mutagenesis. J Neurosci 24:9097–9104

Cashin AL, Petersson EJ, Lester HA, Dougherty DA (2005) Using physical chemistry to differentiate nicotinic from cholinergic agonists at the nicotinic acetylcholine receptor. J Am Chem Soc 127:350–356

Cashin AL, Torrice MM, McMenimen KA, Lester HA, Dougherty DA (2007) Chemical-scale studies on the role of a conserved aspartate in preorganizing the agonist binding site of the nicotinic acetylcholine receptor. Biochemistry 46:630–639
Cha A, Snyder GE, Selvin PR, Bezanilla F (1999) Atomic scale movement of the voltage-sensing region in a potassium channel measured via spectroscopy. Nature 402:809–813
Cohen BE, McAnaney TB, Park ES, Jan YN, Boxer SG, Jan LY (2002) Probing protein electrostatics with a synthetic fluorescent amino acid. Science 296:1700–1703
Cornish VW, Benson DR, Altenbach CA, Hideg K, Hubbell WL, Schultz PG (1994) Site-specific incorporation of biophysical probes into proteins. Proc Natl Acad Sci USA 91:2910–2914
Cornish VW, Mendel D, Schultz PG (1995) Probing protein structure and function with an expanded genetic code. Angew Chem Int Ed Engl 34:621–633
Cornish VW, Hahn KM, Schultz PG (1996) Site specific protein modification using a ketone handle. J Am Chem Soc 118:8150–8151
Corringer P-J, Le Novère N, Changeux J-P (2000) Nicotinic receptors at the amino acid level. Annu Rev Pharmacol Toxicol 40:431–458
Dahan DS, Dibas MI, Petersson EJ, Auyeung VC, Chanda B, Bezanilla F, Dougherty DA, Lester HA (2004) A fluorophore attached to nicotinic acetylcholine receptor beta M2 detects productive binding of agonist to the alpha delta site. Proc Natl Acad Sci USA 101:10195–10200
Dang H, England PM, Sarah Farivar S, Dougherty DA, Lester HA (2000) Probing the role of a conserved M1 proline residue in 5-hydroxytryptamine3 receptor gating. Mol Pharm 57:1114–1122
Dieterich DC, Link AJ, Graumann J, Tirrell DA, Schuman EM (2006) Selective identification of newly synthesized proteins in mammalian cells using bioorthogonal noncanonical amino acid tagging (BONCAT). Proc Natl Acad Sci USA 103:9482–9487
Dougherty DA (1996) Cation–π interactions in chemistry and biology. a new view of benzene, Phe, Tyr, and Trp. Science 271:163–168
Dougherty DA (2000) Unnatural amino acids as probes of protein structure and function. Curr Opin Chem Biol 4:645–652
Dougherty DA, Stauffer DA (1990) Acetylcholine binding by a synthetic receptor. Implications for biological recognition. Science 250:1558–1560
England PM, Lester HA, Davidson N, Dougherty DA (1997) Site-specific, photochemical proteolysis applied to ion channels in vivo. Proc Natl Acad Sci USA 94:11025–11030
England PM, Lester HA, Dougherty DA (1999a) Mapping disulfide connectivity using backbone ester hydrolysis. Biochemistry 38:14409–14415
England PM, Zhang Y, Dougherty DA, Lester HA (1999b) Backbone mutations in transmembrane domains of a ligand-gated ion channel: implications for the mechanism of gating. Cell 96:89–98
Fahnesto S, Neumann H, Shashoua V, Rich A (1970) Ribosome-catalyzed ester formation. Biochemistry 9:2477–2478
Gallivan JP, Lester HA, Dougherty DA (1997) Site-specific incorporation of biotinylated amino acids to identify surface-exposed residues in integral membrane proteins. Chem Biol 4: 739–749
Glauner KS, Mannuzzu LM, Gandhi CS, Isacoff EY (1999) Spectroscopic mapping of voltage sensor movement in the Shaker potassium channel. Nature 402:813–817
Heckler TG, Chang L-H, Zama Y, Naka T, Chorghade MS, Hecht SM (1984) T4 RNA ligase mediated preparation of novel "chemically misacylated" $tRNA^{Phe}$s. Biochemistry 23: 1468–1473
Hohsaka T, Ashizuka Y, Murakami H, Sisido M (1996) Incorporation of nonnatural amino acids into streptavidin through in vitro frame-shift suppression. J Am Chem Soc 118:9778–9779
Hubbell WL, Altenbach C, Hubbell CM, Khorana HG (2003) Rhodopsin structure, dynamics, and activation: a perspective from crystallography, site-directed spin labeling, sulfhydryl reactivity, and disulfide cross-linking. In: Membrane proteins, vol 63: advances in protein chemistry, pp 243–290
Ilegems E, Pick HM, Vogel H (2002) Monitoring mis-acylated tRNA suppression efficiency in mammalian cells via EGFP fluorescence recovery. Nucleic Acids Res 30:e128

Karlin A (2002) Emerging structure of the nicotinic acetylcholine receptors. Nat Rev Neurosci 3:102–114

Kearney P, Nowak M, Zhong W, Silverman S, Lester H, Dougherty D (1996a) Dose-response relations for unnatural amino acids at the agonist binding site of the nicotinic acetylcholine receptor: tests with novel side chains and with several agonists. Mol Pharmacol 50:1401–1412

Kearney PC, Zhang H, Zhong W, Dougherty DA, Lester HA (1996b) Determinants of nicotinic receptor gating in natural and unnatural side chain structures at the M2 9′ position. Neuron 17:1221–1229

Koh JT, Cornish VW, Schultz PG (1997) An experimental approach to evaluating the role of backbone interactions in proteins using unnatural amino acid mutagenesis. Biochemistry 36:11314–11322

Labarca C, Nowak MW, Zhang H, Tang L, Deshpande P, Lester HA (1995) Channel gating governed symmetrically by conserved leucine residues in the M2 domain of nicotinic receptors. Nature 376:514–516

Lester HA, Dibas MI, Dahan DS, Leite JF, Dougherty DA (2004) Cys-loop receptors: new twists and turns. Trends Neurosci 27:329–336

Li LT, Zhong WG, Zacharias N, Gibbs C, Lester HA, Dougherty DA (2001) The tethered agonist approach to mapping ion channel proteins toward a structural model for the agonist binding site of the nicotinic acetylcholine receptor. Chem Biol 8:47–58

Link AJ, Tirrell DA (2005) Reassignment of sense codons in vivo. Methods 36:291–298

Long SB, Campbell EB, MacKinnon R (2005) Crystal structure of a mammalian voltage-dependent Shaker family K+ channel. Science 309:897–903

Lummis SCR, Beene DL, Harrison NJ, Lester HA, Dougherty DA (2005a) A cation–pi binding interaction with a tyrosine in the binding site of the GABA(C) receptor. Chem Biol 12:993–997

Lummis SCR, Beene DL, Lee LW, Lester HA, Broadhurst RW, Dougherty DA (2005b) *Cis–trans* isomerization at a proline opens the pore of a neurotransmitter-gated ion channel. Nature 438:248–252

Ma JC, Dougherty DA (1997) The cation–π interaction. Chem Rev 97:1303–1324

McMenimen KA, Petersson EJ, Lester HA, Dougherty DA (2006) Probing the Mg2+ blockade site of an N-methyl-D-aspartate (NMDA) receptor with unnatural amino acid mutagenesis. Acs Chem Biol 1:227–234

Mecozzi S, West Jr AP, Dougherty DA (1996a) Cation–π Interactions in aromatics of biological and medicinal interest: electrostatic potential surfaces as a useful qualitative guide. Proc Natl Acad Sci USA 93:10566–10571

Mecozzi S, West Jr AP, Dougherty DA (1996b) Cation–π interactions in simple aromatics. Electrostatics provide a predictive tool. J Am Chem Soc 118:2307–2308

Miller JC, Silverman SK, England PM, Dougherty DA, Lester HA (1998) Flash decaging of tyrosine sidechains in an ion channel. Neuron 20:619–624.

Monahan SL, Lester HA, Dougherty DA (2003) Site-specific incorporation of unnatural amino acids into receptors expressed in mammalian cells. Chem Biol 10:573–580

Mu TW, Lester HA, Dougherty DA (2003) Different binding orientations for the same agonist at homologous receptors: a lock and key or a simple wedge? J Am Chem Soc 125:6850–6851

Neher E, Sakmann B (1976) Single channels recorded from membrane of denervated frog muscle fibres. Nature 260:799–802

Noren CJ, Anthony-Cahill SJ, Griffith MC, Schultz PG (1989) A general-method for site-specific incorporation of unnatural amino-acids into proteins. Science 244:182–188

Nowak MW, Kearney PC, Sampson JR, Saks ME, Labarca CG, Silverman SK, Zhong W, Thorson J, Abelson JN, Davidson N, Schultz PG, Dougherty DA, Lester HA (1995) Nicotinic receptor binding site probed with unnatural amino acid incorporation in intact cells. Science 268: 439–442

Nowak MW, Gallivan JP, Silverman SK, Labarca CG, Dougherty DA, Lester HA (1998) In vivo incorporation of unnatural amino acids into ion channels in a Xenopus oocyte expression system. Methods Enzymol 293:504–529

Padgett CL, Hanek AP, Lester HA, Dougherty DA, Lummis SC (2007) Unnatural amino acid mutagenesis of the GABA(A) receptor binding site residues reveals a novel cation–π interaction between GABA and β2Tyr97. J Neurosci 27:886–892
Perozo E (2006) Gating prokaryotic mechanosensitive channels. Nat Rev Mol Cell Biol 7: 109–119
Petersson EJ, Choi A, Dahan DS, Lester HA, Dougherty DA (2002) A perturbed pK(a) at the binding site of the nicotinic acetylcholine receptor: implications for nicotine binding. J Am Chem Soc 124:12662–12663
Petersson EJ, Brandt GS, Zacharias NM, Dougherty DA, Lester HA (2003) Caging proteins through unnatural amino acid mutagenesis. Methods Enzymol 360:258–273
Philipson KD, Gallivan JP, Brandt GS, Dougherty DA, Lester HA (2001) Incorporation of caged cysteine and caged tyrosine into a transmembrane segment of the nicotinic ACh receptor. Am J Physiol Cell Physiol 281:195–206
Prescher JA, Bertozzi CR (2005) Chemistry in living systems. Nat Chem Biol 1:13–21
Price KL, Beene DL, Dougherty DA, Lester HA, Lummis SCR (2003) The role of tyrosine residues at the mouse 5-HT3A receptor ligand binding site investigated by unnatural amino acid mutagenesis. Brit J Pharmacol 140
Revah F, Bertrand D, Galzi JL, Devillers-Theiry A, Mulle C (1991) Mutations in the channel domain alter desensitization of a neuronal nicotinic receptor. Nature 353:846–849
Rodriguez EA, Lester HA, Dougherty DA (2006) In vivo incorporation of multiple unnatural amino acids through nonsense and frameshift suppression. Proc Natl Acad Sci USA 103: 8650–8655
Rodriguez EA, Lester HA, Dougherty DA (2007a) Improved amber and opal suppressor tRNAs for incorporation of unnatural amino acids in vivo, part 1: minimizing misacylation. RNA 13: 1703–1714
Rodriguez EA, Lester HA, Dougherty DA (2007b) Improved amber and opal suppressor tRNAs for incorporation of unnatural amino acids in vivo, part 2: evaluating suppression efficiency. RNA 13:1715–1722
Sakamoto K, Hayashi A, Sakamoto A, Kiga D, Nakayama H, Soma A, Kobayashi T, Kitabatake M, Takio K, Saito K, Shirouzu M, Hirao I, Yokoyama S (2002) Site-specific incorporation of an unnatural amino acid into proteins in mammalian cells. Nucleic Acids Res 30:4692–4699
Saks ME, Sampson JR, Nowak MW, Kearney PC, Du F, Abelson JN, Lester HA, Dougherty DA (1996) An engineered Tetrahymena tRNAGln for in vivo incorporation of unnatural amino acids into proteins by nonsense suppression. J Biol Chem 271:23169–23175
Santarelli VP, Eastwood AL, Dougherty DA, Horn R, Ahern CA (2007) A cation–π interaction discriminates among sodium channels that are either sensitive or resistant to tetrodotoxin block. J Biol Chem 282:8044–8051
Shafer AM, Kalai T, Liu SQB, Hideg K, Voss JC (2004) Site-specific insertion of spin-labeled L-amino acids in Xenopus oocytes. Biochemistry 43:8470–8482
Summerer D, Chen S, Wu N, Deiters A, Chin JW, Schultz PG (2006) A genetically encoded fluorescent amino acid. Proc Natl Acad Sci USA 103:9785–9789
Taki M, Hohsaka T, Murakami H, Taira K, Sisido M (2002) Position-specific incorporation of a fluorophore-quencher pair into a single streptavidin through orthogonal four-base codon/anticodon pairs. J Am Chem Soc 124:14586–14590
Tong Y, Brandt GS, Li M, Shapovalov G, Slimko E, Karschin A, Dougherty DA, Lester HA (2001) Tyrosine decaging leads to substantial membrane trafficking during modulation of an inward rectifier potassium channel. J Gen Physiol 117:103–118
Turcatti G, Nemeth K, Edgerton MD, Meseth U, Talabot F, Peitsch M, Knowles J, Vogel H, Chollet A (1996) Probing the structure and function of the tachykinin neurokinin-2 receptor through biosynthetic incorporation of fluorescent amino acids at specific sites. J Biol Chem 271:19991–19998
Turcatti G, Nemeth K, Edgerton MD, Knowles J, Vogel H, Chollet A (1997) Fluorescent labeling of NK2 receptor at specific sites in vivo and fluorescence energy transfer analysis of NK2 ligand-receptor complexes. Receptors Channels 5:201–207

Unwin N (2005) Refined structure of the nicotinic acetylcholin receptor at 4 A resolution. J Mol Biol 346:967–989

van Swieten PF, Leeuwenburgh MA, Kessler BM, Overkleeft HS (2005) Bioorthogonal organic chemistry in living cells: Novel strategies for labeling biomolecules. Org Biomol Chem 3: 20–27

Wang L, Schultz PG (2005) Expanding the genetic code. Angew Chem Int Ed Engl 44:34–66

Wang JY, Xie JM, Schultz PG (2006) A genetically encoded fluorescent amino acid. J Am Chem Soc 128:8738–8739

Wilson GG, Karlin A (1998) The location of the gate in the acetylcholine receptor channel. Neuron 20:1269–1281

Zhong W, Gallivan JP, Zhang Y, Li L, Lester HA, Dougherty DA (1998) From ab initio quantum mechanics to molecular neurobiology: A cation–π binding site in the nicotinic receptor. Proc Natl Acad Sci USA 95:12088–12093

Synthesis of Modified Proteins Using Misacylated tRNAs

S.M. Hecht

Contents

Abstract While ribosomally mediated protein synthesis is ordinarily limited to the amino acids utilized by nature, the ability to misacylate tRNAs with a wide variety of amino acids and amino acid analogs provides the wherewithal to dramatically expand the repertoire of amino acids that can be incorporated into proteins for study. Dissection of the mechanism of protein biosynthesis is also possible using misacylated tRNAs. This chapter summarizes efforts to create misacylated tRNAs, and the several ways in which they have been employed to prepare modified proteins of diverse structure.

S.M. Hecht
Departments of Chemistry and Biology, University of Virginia, Charlottesville, Virginia 22904, USA, e-mail: sidhecht@virginia.edu

C. Köhrer and U.L. RajBhandary (eds.) *Protein Engineering.*
Nucleic Acids and Molecular Biology 22,

1 Introduction

Since the pioneering work of Chapeville et al. (1962), the use of misacylated transfer RNAs to introduce modified amino acids into specific positions in synthesized proteins has received growing attention. The ability to engineer proteins containing specific side chain modifications of utility for studying protein function, and introducing novel structures and functionally important groups at specific sites, no doubt underpins this effort.

This chapter summarizes the status of key initiatives within the overall program designed to permit the elaboration of modified proteins. Specific topics covered include amino acid protecting group strategies for preparing misacylated transfer RNAs (tRNAs), the use of alternative codons for amino acid incorporation at specific sites, and the introduction of multiple unnatural amino acids into a single protein. Also discussed are the incorporation of amino acids that afford proteins having modified backbones and the engineering of modified bacterial ribosomes to facilitate the incorporation of such unusual amino acids into proteins. Finally, the use of bisaminoacylated tRNAs in protein synthesizing systems is described.

2 Amino Acid Protecting Groups for Misacylated tRNAs

2.1 Preparing Misacylated tRNAs

Strategies for preparing misacylated tRNAs were first defined by the Hecht laboratory (Hecht et al. 1978; Heckler et al. 1983, 1984) and almost every cell-free study reported since that time has used that approach. As shown in Fig. 1, the optimized strategy involves the use of a tRNA (transcript) lacking the dinucleotide ($pCpA_{OH}$) ordinarily present at the 3′-terminus of a tRNA. T4 RNA ligase-mediated ligation

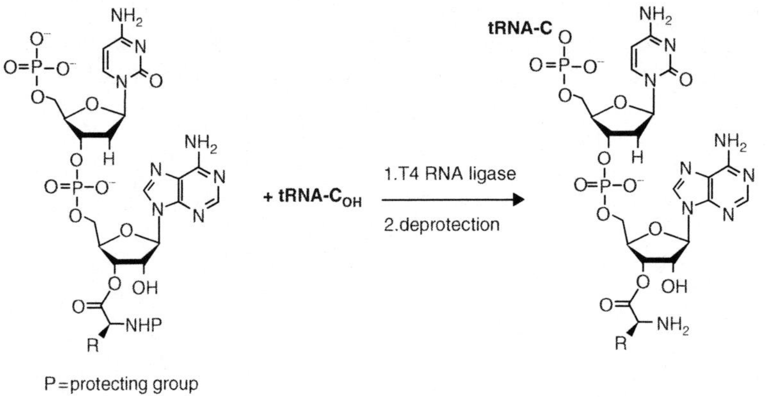

Fig. 1 Strategy employed for the preparation of misacylated transfer RNAs (tRNAs)

of this abbreviated tRNA (tRNA-C_{OH}) to a synthetic dinucleotide containing an appended amino acid on the ribose moiety of adenosine affords the requisite activated tRNA.

Relative to the initial publications, a few key technical modifications have been described that substantially facilitate the preparation of misacylated tRNAs. These include the finding that tRNA transcripts lacking post-transcriptional modifications function well in cell-free protein synthesizing systems. Thus, the abbreviated tRNA-C_{OH}s can be prepared by runoff transcription of a linearized plasmid that encodes the tRNA-C_{OH} of interest (Noren et al. 1990). This strategy has provided a convenient alternative to the removal of nucleotides from the 3′-terminus of mature tRNAs isolated and purified from natural sources (Hecht 1992).

A second innovation that has facilitated the preparation of misacylated tRNAs involves the use of an aminoacylated pCpA in which the cytidine moiety lacks the 2′-OH group on ribose, pdCpA (Robertson et al. 1989). While the finding that the OH group on the ribose moiety of the penultimate nucleotide at the 3′-terminus of tRNA is not essential to tRNA function in protein synthesis might seem to be a relatively "*minor*" technical finding, it has proven to be quite important in simplifying the preparation of misacylated tRNAs. Relative to the original scheme for the synthesis of aminoacylated pCpAs (Heckler et al. 1984), the preparation of aminoacylated pdCpAs is much simpler.

2.2 Amino Acid Protecting Groups

Perhaps the most critical and technically demanding facet of the method for preparing misacylated tRNAs has involved the choice of the protecting group for the aminoacyl moiety of the aminoacylated pdCpA derivative employed in the ligation reaction. T4 RNA ligase effects the transformation shown in Fig. 1 rather slowly. In the absence of a protecting group for N^{α} of the aminoacyl moiety of pdCpA, hydrolysis of the aminoacylated pdCpA to afford pdCpA proceeds more quickly than the ligation reaction. Thus, the ligation reaction would likely produce mostly unacylated full-length tRNA, as well as products formed from multiple additions of (deacylated) dinucleotide. While there has been one report of the use of unprotected pCpA derivatives to produce misacylated tRNAs (Baldini et al. 1988), all other reports have employed protecting groups.

The difficulty in identifying suitable N^{α} protecting groups is a direct consequence of the nature of the aminoacyl moiety of activated tRNAs. This group undergoes facile hydrolysis, especially at higher pH. The problem, then, is identifying a protecting group that is removable from N^{α} of the aminoacyl moiety of the aminoacyl-tRNA formed without effecting concomitant amino acid hydrolysis from the tRNA. Four protecting groups identified for this purpose to date are shown in Fig. 2.

The earliest protecting group described was the *o*-nitrophenylsulfenyl protecting group, removable under weakly acidic (or reductive) conditions (Hecht et al. 1978; Robertson et al. 1989). Limitations include the poor aqueous solubility properties of

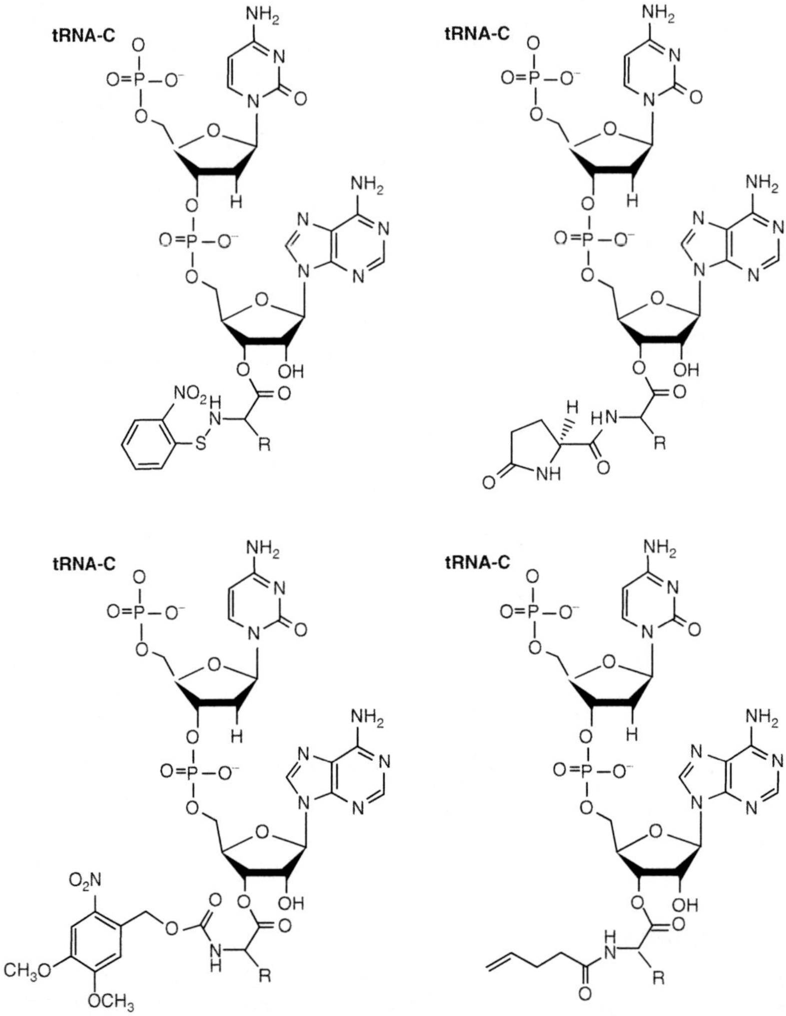

Fig. 2 Protecting groups used for N^{α} of the aminoacyl moiety in the formation of misacylated tRNAs: *o*-nitrophenylsulfenyl protecting group (*top left*); pyroglutamate protecting group (*top right*); nitroveratryloxycarbonyl protecting group (*bottom left*); pentenoyl protecting group (*bottom right*)

some of the *o*-nitrophenylsulfenyl-protected aminoacylated pCpAs, as well as relatively inefficient removal of the protecting group. A second protecting group described by the Hecht laboratory is the pyroglutamate protecting group, removable by the commercially available enzyme pyroglutamate aminopeptidase. Since the commercial preparation utilized to date lacked any detectable esterase activity, depro-tection of the pyroglutamylaminoacyl-tRNAs studied was quite specific (Roesser et al. 1989). While the enzyme has broad substrate specificity, the *Pseudomonas fluorescens* enzyme does fail to cleave the *N*-pyroglutamyl-L-proline linkage (Uliana and Doolittle 1969), and a potential limitation is that the enzyme may also fail to cleave a conjugated pyroglutamate from some amino acid analogs appended to tRNA.

The two types of protecting groups that have been employed the most extensively for amino acid protection during the misacylation of tRNAs include the nitroveratryloxycarbonyl (NVOC) protecting group, shown in Fig. 2. Originally reported by Patchornik for amino acid protection (Patchornik et al. 1970; Amit et al. 1974), the light-sensitive NVOC group was adopted for use in the preparation of misacylated tRNAs by the Schultz laboratory (Robertson et al. 1991). The NVOC group can be removed efficiently, providing that a high-intensity light source is available, and is broadly applicable to the deprotection of many amino acid analogs. One potential limitation would be for use with light-sensitive amino acid analogs.

The other protecting group that has been employed extensively for the preparation of misacylated tRNAs is the pentenoyl protecting group. This protecting group was used previously for amines in carbohydrates (Debenham et al. 1995) and other molecules (Madsen et al. 1995), and has been adapted by the Hecht laboratory for use in preparing misacylated tRNAs. *N*-Pentenoyl-L-amino acids are readily prepared (Lodder et al. 1997, 1998); the protecting group can be removed from the *N*-pentenoylaminoacyl-tRNAs by simple treatment with aqueous iodine. The mechanism of removal presumably involves an iminolactone intermediate (Debenham et al. 1995; Madsen et al. 1995). This protecting group is also quite versatile, although it clearly cannot be used with amino acids having oxidizable side chains such as cysteine, tryptophan, and tyrosine. It may be noted that an interesting variation of the pentenoyl protecting group is the amino acid allylglycine, which can be incorporated into proteins at specific sites, and then cleaved specifically at those sites by treatment with aqueous iodine (Wang et al. 2000, 2002; Baird et al. 2000).

A typical scheme for the synthesis of a protected aminoacyl-pdCpA derivative is illustrated in Fig. 3, using *N*-pentenoylphenylalanyl-pdCpA as an example

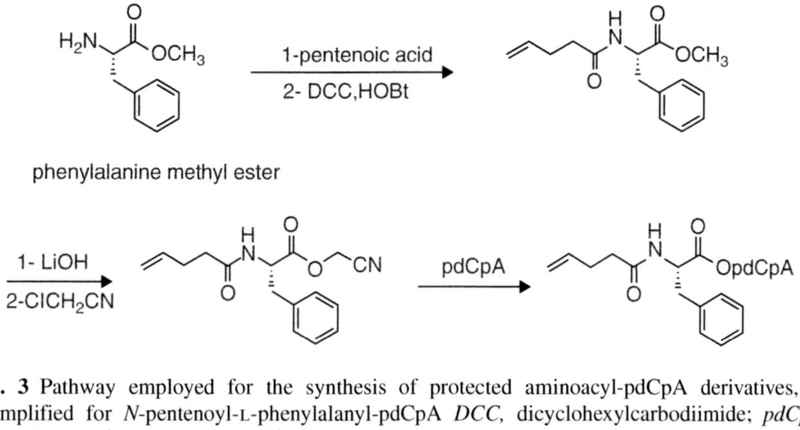

Fig. 3 Pathway employed for the synthesis of protected aminoacyl-pdCpA derivatives, as exemplified for *N*-pentenoyl-L-phenylalanyl-pdCpA *DCC,* dicyclohexylcarbodiimide; *pdCpA,* 5′-phosphoro-2′-deoxycytidylyl (3′-5′) adenosine

(Lodder et al. 1998). A transformation of particular interest in this scheme is the use of a cyanomethyl ester of *N*-pentenoylphenylalanine to effect specific derivatization of the ribose OH groups in the dinucleotide, thus obviating the need for nucleobase protection. This method of activation was first reported by Robertson et al. (1991).

3 Alternative Codons for Incorporation of Unnatural Amino Acids

While most of the misacylated tRNAs used to introduce unnatural amino acids into proteins have been used to effect readthrough of the stop codon UAG, in principle any unique codon will suffice. Thus recent studies involving mammalian cells have also utilized messenger RNAs (mRNAs) engineered to contain UAA (Kohrer et al. 2001) and UGA codons (Zhang et al. 2004), as well as various combinations of such codons (Kohrer et al. 2003, 2004).

The use of stop codons is obviously limited by the number of such codons provided by nature, but expansion of the available code has been realized by utilizing tRNAs that recognize four- and five-base codons (Hohsaka et al. 2001a, b). Clearly, this strategy can potentially be used to introduce multiple amino acids within a protein (Hohsaka et al. 2001a). It is also possible to employ a combination of stop codons and four-base codons, and an example of this strategy has been reported (Anderson et al. 2002).

Yet another strategy that has been explored to create unique codons is the creation of new base pairs, and the introduction of a novel nucleoside in the mRNA, with its complement in the corresponding position in the anticodon. Clearly this strategy requires significant effort in the elaboration of the requisite tRNA and mRNA, but has actually been realized in two studies (Bain et al. 1992; Hirao et al. 2002; for more details see the chapters by Hirao et al. and Leconte and Romesberg, this volume).

Strategies have also been defined to reassign codons intended for the incorporation of one of the 20 proteinogenic amino acids. This has been accomplished by the use of bacteria auxotrophic for a specific amino acid, which have been grown on a medium containing a structural analog of that amino acid (Hortin and Boime 1983; Wilson and Hatfield 1984; Budisa et al. 1999; Link et al. 2003). A characteristic of this procedure is that all amino acids of one type are replaced with the structural analog. While such amino acid substitutions have been used to good advantage to study the physical properties of the derived proteins as materials (Deming et al. 1996; Tang and Tirrell 2001), the use of this strategy to facilitate a study of enzyme mechanism (see the chapter by Beatty and Tirrell, this volume) is clearly more problematic. This strategy has been extended to include tRNAs activated with unnatural amino acids by "*chemical misacylation*" (Tan et al. 2004).

The use of translation systems reconstructed from purified components has the promise to permit an unprecedented level of control over the amino acid constituents incorporated into proteins in cell-free protein synthesizing systems (Forster et al. 2001; Shimizu et al. 2001) (see the chapter by Hirao et al., this volume). Forster et al. (2003) were able to reassign three sense codons for the incorporation of three unnatural amino acids. By the use of a reconstructed translation system and amino acid analogs known to be recognized by specific aminoacyl-tRNA synthetases, the Szostak laboratory has recently demonstrated the incorporation of 12 unnatural amino acid analogs in a system that involved the reassignment of 35 of the 61 sense codons (Josephson et al. 2005). Thus, in spite of the considerable experimental demands of reconstituted translation systems, the ability to incorporate multiple unnatural amino acids holds exceptional promise.

4 Incorporation of Unnatural Amino Acids Affording Proteins Having Modified Backbones

4.1 Synthesis of Proteins with Altered Backbones

While most unnatural amino acids incorporated into proteins have involved simple alteration of the amino acid side chains, there are now a number of examples involving amino acid alterations that afford proteins with altered backbones. As shown in Fig. 4, this has included the use of N-methylated amino acids (**1** and **2**) (Bain et al. 1991a, b; Ellman et al. 1992; Frankel et al. 2003; Merryman and Green 2004) and α-hydroxy acids (**3** and **4**) (Fahnestock and Rich 1971a, b; Bain et al. 1991a, b; Ellman et al. 1992; Koh et al. 1997; England et al. 1999a–c; Lu et al. 2001), as well as amino acids having the D-configuration at the α-carbon atom (Roesser et al. 1986; Bain et al. 1991a, b; Ellman et al. 1992; Starck et al. 2003). While the yields realized for incorporation of the *N*-methyl and α-hydroxy acids were adequate to

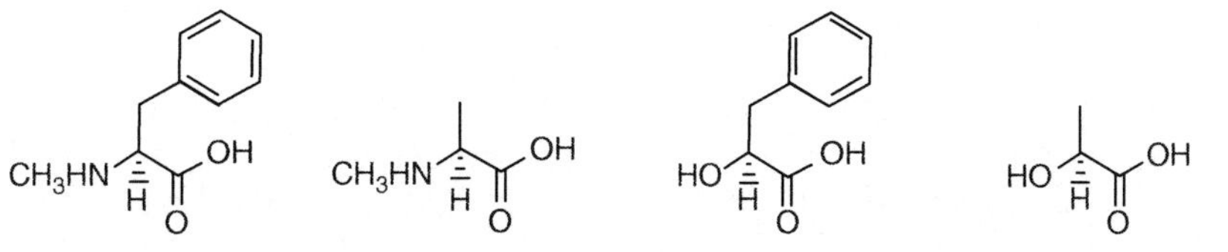

1 **2** **3** **4**

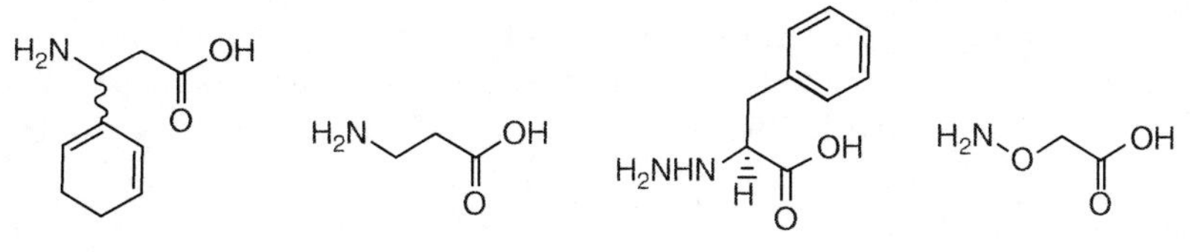

5 **6** **7** **8**

Fig. 4 Amino acid analogs whose incorporation into protein has resulted in peptides/proteins with modified backbones good, the D-amino acids were incorporated in very low yields, possibly reflecting the design of the ribosome for specific incorporation of L-amino acids.

Also studied has been the ability of the ribosome to direct the incorporation of amino acids with the formation of polypeptide analogs having altered "*connectivity*." The earliest example involved the use of a deaminated amino acid (**3**) leading to the formation of a depsipeptide (Fahnestock and Rich 1971b); this type of modified backbone has now been realized in a few different studies (Fahnestock and Rich 1971a, b; Bain et al. 1991a, b; Ellman et al. 1992; Koh et al. 1997; England et al. 1999a–c) and used to assess the contribution of individual backbone amides to protein structure (Koh et al. 1997; Shin et al. 1997). While the number of atoms in the peptide backbone is unchanged as a consequence of using an α-hydroxy acid at individual positions in a protein, the yields of incorporation of such species are generally low, perhaps reflecting the diminished nucleophilicity of the heteroatom attached to the α-carbon of the amino acid (see, however, Tan et al. 2004).

Proteins having polypeptide backbones with altered connectivity may also be accessible by the use of misacylated tRNAs containing amino acids with an NH_2 group in the β or γ rather than the α position (e.g., **5** and **6**). Studies of the incorporation of such species have been reported (Heckler et al. 1983; Roesser et al. 1989; Bain et al. 1991a, b; Ellman et al. 1992), but also reflect a diminished efficiency of utilization of such species as peptidyl acceptors in the ribosomal A-site. This may be rationalized on the basis of a suboptimal orientation of the nucleophilic nitrogen atom with respect to the carboxylate moiety of the peptidyl tRNA during the peptidyltransferase reaction. It is interesting to note that β-phenylalanine actually demonstrated an enhanced ability to function as a donor in the peptidyltransferase reaction (Heckler et al. 1988).

In this context, two additional types of amino acid analogs are worthy of note. These are the hydrazino amino acids (exemplified by hydrazinophenylalanine **7**) (Killian et al. 1998) and aminooxyglycine **8** (Eisenhauer and Hecht 2002). Both of these amino acid analogs would be expected to form proteins of altered connectivity having peptide bond analogs that might be expected to display distinct conformational preferences (Aubry et al. 1991; Lecoq et al. 1991; Dupont et al. 1993; Yang et al. 1996; Wu et al. 1999; Peter et al. 2000; Thévenet et al. 2000). Experimentally, hydrazinophenylalanine was found to be utilized for ribosome-mediated peptide bond formation, albeit with diminished efficiency. This might be considered surprising since the misacylated tRNA activated with hydrazinophenylalanine actually has two nucleophilic groups, both of which might be expected to be more nucleophilic than a single α-amino group, and one of which can assume the same nominal orientation in the peptidyltransferase reaction as the α-amino group of a proteinogenic amino acid. In spite of the fact that the two possible products of the peptidyltransferase reaction are actually formed (Fig. 5) in good yield, the efficiency of incorporation of hydrazinophenylalanine into a protein was only about 10% in a system that utilized readthrough of a UAG codon at position 10 of dihydrofolate reductase (DHFR) for incorporation of the analog (Killian et al. 1998). Plausibly, this may reflect inefficient function of the analog as a donor in the ribosomal P-site.

Aminooxyglycyl-$tRNA_{CUA}$ has also been prepared recently and employed for incorporation of aminooxyglycine into both DHFR and DNA polymerase β

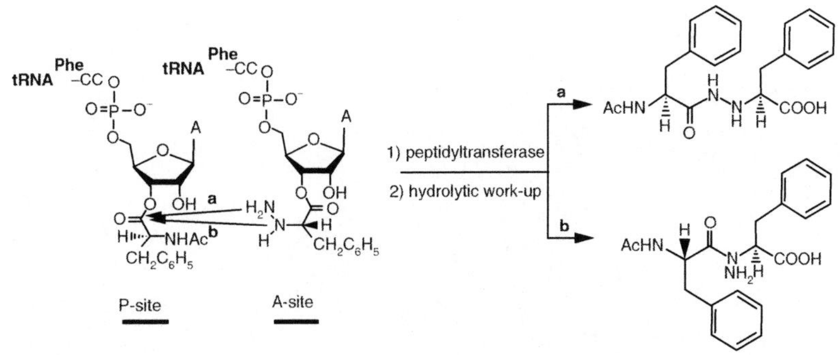

Fig. 5 Products formed in a protein synthesizing system that employed misacylated hydrazinophenylalanyl-tRNA

(Eisenhauer and Hecht 2002). In spite of the fact that this amino acid analog contains only one of the two nucleophilic groups present in hydrazino amino acid analogs (cf. **7** and **8**), the incorporation of this analog proceeded in comparable yield, and was dependent upon codon context. In addition to the formation of full-length protein in a fashion that was dependent on the presence of aminooxyglycyl-$tRNA_{CUA}$, the presence of an *N*-alkoxyamide linkage at position 72 of DNA polymerase β was verified by its ability to undergo reductive cleavage of the N–O bond at the site of aminooxyglycine incorporation (Fig. 6).

4.2 Modified Ribosomes for Incorporation of D-Amino Acids

While the ribosomally mediated formation of proteins having modified backbones is of special interest in expanding the repertoire of polypeptide structures available for study and use as novel catalysts and biomaterials, the elaboration of such species inevitably relies upon the ability of the amino acid analog building blocks to be utilized for ribosomal protein (analog) synthesis. Intrinsic limitations in the ability of both prokaryotic and eukaryotic ribosomes to utilize such unusual amino acid species are clearly reflected in the exceptionally poor incorporation yields noted for most of the analogs shown in Fig. 4.

A potential strategy for increasing the yields of proteins that incorporate unusual amino acid analogs involves reengineering the ribosome to facilitate the use of specific amino acids. The Hecht laboratory has recently reported the formation and assay of bacterial ribosomes with mutations in two regions of 23S ribosomal RNA (rRNA) (2447–2451 and 2457–2462) known to be important for the peptidyltransferase reaction (Dedkova et al. 2003). Colonies of *Escherichia coli* containing modified ribosomes were chosen on the basis of their altered growth rates in the presence of chloramphenicol.

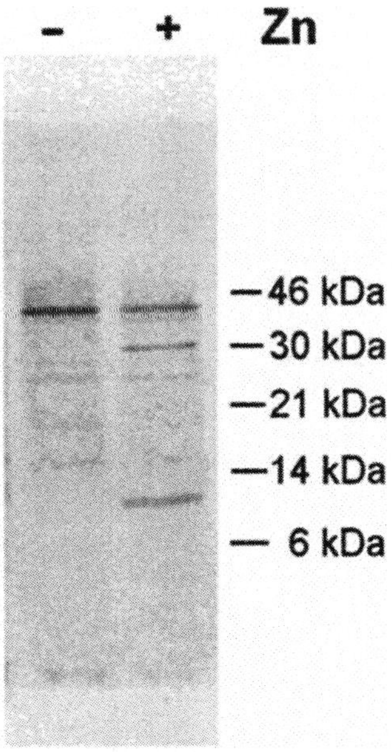

Fig. 6 Incorporation of aminooxyglycine into position 72 of DNA polymerase β and reductive cleavage of the *N*-alkoxyamide bond formed. The fragments having masses of 30.6 and 9.9 kDa correspond to the expected cleavage products

Table 1 Suppression of a UAG codon at position 22 of dihydrofolate reductase messenger RNA (mRNA) with D- aminoacyl-tRNA$_{CUA}$s and L-aminoacyl-tRNA$_{CUA}$s[a]

	Suppression (%)	
Amino acid	Wild-type ribosomes	Mutant ribosomes
L-Phenylalanine	58	54
D-Phenylalanine	3	12
L-Methionine	52	47
D-Methionine	5	23
–[b]	0.9	2

[a]Relative to the amount of protein produced using wild-type mRNA and wild-type ribosomes *tRNA*, transfer RNA

[b]No misacylated *tRNA*

The 23S rRNAs from the selected colonies were sequenced to verify that the intended sequences had been altered. Functional characterization of the ribosomes was effected using S-30 preparations from individual colonies; these were used to measure the ability of the modified ribosomes to incorporate D-phenylalanine or D-methionine from the respective aminoacyl-tRNA$_{CUA}$s by readthrough of a UAG codon. To ensure that the modified ribosomes were capable of maintaining reasonable fidelity of protein synthesis, the S-30 preparations were also used to elaborate the same wild-type proteins (DHFR and firefly luciferase). The wild-type and modified proteins were then assayed for function. As shown in Table 1, the incorporation of D-phenylalanine and D-methionine was much greater in the presence of the modified ribosomes. The modified ribosomes were also shown to produce wild-type proteins whose activity as catalysts was not dramatically different from those of the same proteins elaborated by native ribosomes.

Perhaps the most compelling evidence for the incorporation of the D-amino acids was the properties noted for the modified DHFRs and luciferases putatively containing D-amino acids at specific sites. The incorporation of D-amino acids into positions believed not to be important to protein function had little effect. For example, replacement of valine at position 10 of DHFR with L-methionine or D-methionine had little effect on protein function. In contrast, introduction of D-phenylalanine into position 22 of the same protein significantly reduced the activity of the derived protein, relative to the protein containing L-phenylalanine at the same position. Position 22, which contains tryptophan in wild-type DHFR, has been shown to be important for dihydrofolate binding by the enzyme (Bystroff et al. 1990).

The ability to create bacterial ribosomes that exhibit enhanced incorporation of D-amino acids argues that it should be possible to identify modified ribosomes that significantly facilitate the incorporation of other types of modified amino acids into proteins in cell-free protein synthesizing systems.

5 Bisaminoacylated tRNAs as Participants in Protein Synthesis

While all studies of protein synthesis reported to date have involved the use of tRNAs bearing a single amino acid at the 3′-end, Stepanov et al. (1992, 1998) have described the formation of bis(2′, 3′-*O*-phenylalanyl)-tRNA by the cognate aminoacyl-tRNA synthetase from *Thermus thermophilus*. To determine whether such tandemly activated tRNAs could function in protein synthesis, the Hecht laboratory has prepared tRNAs bisaminoacylated with alanine or methionine; the strategy employed is exemplified for bismethionyl-tRNA in Fig. 7. Following activation of pdCpA, the monomethionyl-pdCpA and the bismethionyl-pdCpA were separated and the latter was ligated to tRNA-C_{OH}. Treatment with aqueous iodine then afforded bismethionyl-tRNA.

The use of bisalanyl-$tRNA_{CUA}$ and bismethionyl-$tRNA_{CUA}$ in protein synthesis was studied in both prokaryotic and eukaryotic protein synthesizing systems. As shown in Table 2, both bisaminoacylated tRNAs effected the incorporation of the activated amino acids into three different positions of DHFR (positions –1, 10, and 27) by suppression of UAG codons at those positions. Under conditions limiting for aminoacyl-$tRNA_{CUA}$s, the yields of full-length protein were approximately twice as great those obtained in analogous experiments that employed the respective monoaminoacylated tRNAs (Wang et al. 2006).

The bisaminoacylated species were studied for stability at the level of the bisaminoacylated pdCpA derivatives. While monoaminoacylated pdCpA derivatives were found to undergo hydrolysis readily to afford pdCpA and the free amino

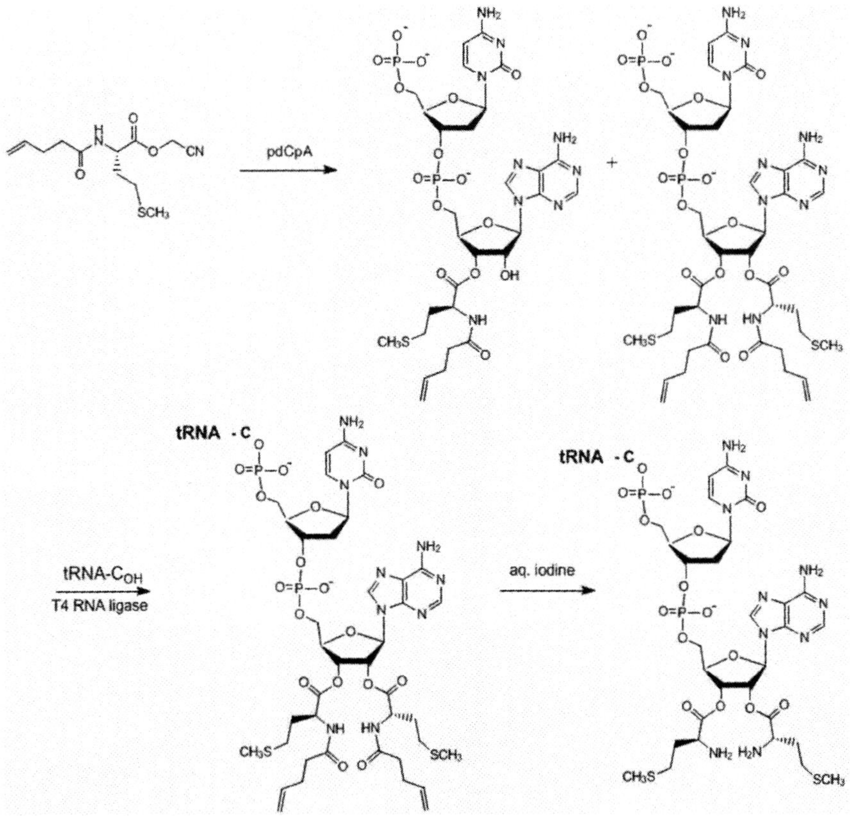

Fig. 7 Route employed for the preparation of bismethionyl-$tRNA_{CUA}$

Table 2 Relative suppression efficiencies of mono- and bisaminoacylated tRNAs in the synthesis of dihydrofolate reductase[a]

	Suppression efficiencies (%)		
Misacylated suppressor $tRNA_{CUA}$	Position 1	Position 10	Position 27
Monoalanyl	41	61	34
Bisalanyl	39	63	29
Monomethionyl	18	29	19
Bismethionyl	21	41	23

[a]Relative to the amount of protein produced using wild-type mRNA

acids, the bisaminoacylated species were found to be completely stable under a variety of conditions, including incubation at pH 7.0 for 6 h at 25 °C, incubation at pH 4.0 for 12 h at 25 °C, and incubation at pH 7.0 for 2 h at 65 °C. As the yields of proteins in cell-free protein synthesizing systems incorporating misacylated tRNAs

Table 3 Relative stabilities of mono- and bisaminoacylated tRNAs in a cell-free protein synthesizing system

	Suppression efficiency (%)[a]	
Misacylated suppressor $tRNA_{CUA}$	No preincubation	30-min preincubation
–[b]	100	48
Monoallylglycyl	72	15
Bisallylglycyl	73	41

[a]Relative to the amount of protein produced using wild-type mRNA

[b]No misacylated *tRNA*

are often limited by the stability of the misacylated tRNAs, it seemed possible that the bisaminoacylated species might therefore afford enhanced yields of proteins.

In fact, as shown in Table 3, when $tRNA_{CUA}$s activated with one or two allylglycines were incubated for 30 min in a rabbit reticulocyte protein synthesizing system prior to the addition of plasmid encoding DHFR mRNA, it was found that the yield of DHFR was much greater in comparison with the yield for wild-type DHFR synthesis when the protein synthesizing system contained bisallyglycyl-$tRNA_{CUA}$ (85%) than when it contained the monoallyglycyl-tRNA (31%). Thus, the use of bisaminoacylated tRNAs may provide an important advantage in the elaboration of proteins in cell-free systems, e.g., where continuous protein synthesizing systems are employed.

A detailed study of the mechanism of protein synthesis in the presence of bisaminoacylated $tRNA_{CUA}$s has been performed, and it demonstrated that each of the amino acids from the tandemly activated species is incorporated into different protein molecules (Wang et al. 2006).

Acknowledgments The studies in the Hecht laboratory described in this chapter were supported by NIH research grant CA77359, awarded by the National Cancer Institute.

References

Amit B, Sehavi V, Patchornik A (1974) Photosensitive protecting groups of amino sugars and their use in glycoside synthesis. 2-Nitrobenzyloxycarbonylamino and 6-nitroveratryloxycarbonylamino derivatives. J Org Chem 39:192–196

Anderson RD, Zhou J, Hecht SM (2002) Fluorescence resonance energy transfer between unnatural amino acids in a structurally modified dihydrofolate reductase. J Am Chem Soc 124:9674–9675

Aubry A, Bayeul D, Mangeot JP, Vidal J, Sterin S, Collet A, Lecoq A, Marraud M (1991) X-ray conformational study of hydrazino peptide analogs. Biopolymers 31:793–801

Bain JD, Diala ES, Glabe CG, Wacker DA, Lyttle MH, Dix TA, Chamberlin AR (1991a) Site-specific incorporation of nonnatural residues during in vitro protein biosynthesis with semi-synthetic aminoacyl-tRNAs. Biochemistry 30:5411–5421

Bain JD, Wacker DA, Kuo EE, Chamberlin AR (1991b) Site-specific incorporation of nonnatural residues into peptides – effect of residue structure on suppression and translation efficiencies. Tetrahedron 47:2389–2400

Bain JD, Switzer C, Chamberlin AR, Benner SA (1992) Ribosome-mediated incorporation of a nonstandard amino acid into a peptide through expansion of the genetic code. Nature 356:537–539

Baird TT Jr, Wang B, Lodder M, Hecht SM, Craik CS (2000) Generation of active trypsin by chemical cleavage. Tetrahedron 56:9477–9485

Baldini G, Martoglio B, Schachenmann A, Zugliani C, Brunner J (1988) Mischarging *Escherichia coli* $tRNA_{Phe}$ with L-4′-[3-(trifluoromethyl)-3H-diazirin-3-yl]phenylalanine, a photoactivatable analog of phenylalanine. Biochemistry 27:7951–7959

Budisa N, Minks C, Alefelder S, Wenger W, Dong F, Moroder L, Huber R (1999) Toward the experimental codon reassignment in vivo: protein building with an expanded amino acid repertoire. FASEB J 13:41–51

Bystroff C, Oatley SJ, Kraut J (1990) Crystal structures of *Escherichia coli* dihydrofolate reductase: the $NADP_+$ holoenzyme and the folate.$NADP_+$ ternary complex. Substrate binding and a model for the transition state. Biochemistry 29:3263–3277

Chapeville F, Lipmann F, von Ehrestein G, Weisblum B, Ray WJ, Benzer S (1962) On the role of soluble ribonucleic acid in coding for amino acids. Proc Natl Acad Sci USA 48:1086–1092

Debenham JS, Madsen R, Roberts C, Fraser-Reid B (1995) Two new orthogonal amine-protecting groups that can be cleaved under mild or neutral conditions. J Am Chem Soc 117:3302–3303

Dedkova LM, Fahmi NE, Golovine SY, Hecht SM (2003) Enhanced D-amino acid incorporation into protein by modified ribosomes. J Am Chem Soc 125:6616–6617

Deming TJ, Fournier MJ, Mason TL, Tirrell DA (1996) Structural modification of a periodic polypeptide through biosynthetic replacement of proline with azetidine-2-carboxylic acid. Macromolecules 29:1442–1444

Dupont V, Lecoq A, Mangeot JP, Aubry A, Boussard G, Marraud M (1993) Conformational perturbations induced by N-amination and N-hydroxylation of peptides. J Am Chem Soc 115:8898–8906

Eisenhauer BM, Hecht SM (2002) Site-specific incorporation of (aminooxy)acetic acid into proteins. Biochemistry 41:11472–11478

Ellman JA, Mendel D, Schultz PG (1992) Site-specific incorporation of novel backbone structures into proteins. Science 255:197–200

England PM, Lester HA, Dougherty DA (1999a) Mapping disulfide connectivity using backbone ester hydrolysis. Biochemistry 38:14409–14415

England PM, Lester HA, Dougherty DA (1999b) Incorporation of esters into proteins: Improved synthesis of hydroxyacyl tRNAs. Tetrahedron Lett 40:6189–6192

England PM, Zhang Y, Dougherty DA, Lester HA (1999c) Backbone mutations in transmembrane domains of a ligand-gated ion channel: implication for the mechanism of gating. Cell 96:89–98

Fahnestock S, Rich A (1971a) Synthesis by ribosomes of viral coat protein containing ester linkages. Nat New Biol 229:8–10

Fahnestock S, Rich A (1971b) Ribosome-catalyzed polyester formation. Science 173:340–343

Forster AC, Weissbach H, Blacklow SC (2001) A simplified reconstitution of mRNA-directed peptide synthesis: activity of the epsilon enhancer and an unnatural amino acid. Anal Biochem 297:60–70

Forster AC, Tan Z, Nalam MN, Lin H, Qu H, Cornish VW (2003) Programming peptidomimetic syntheses by translating genetic codes designed de novo. Proc Natl Acad Sci USA 100: 6353–6357

Frankel A, Millward SW, Roberts RW (2003) Encodamers: unnatural peptide oligomers encoded in RNA. Chem Biol 10:1043–1050

Hecht SM (1992) Probing the synthetic capabilities of a center of biochemical catalysis. Acc Chem Res 25:545–552

Hecht SM, Alford BL, Kuroda Y, Kitano S (1978) Chemical aminoacylation of tRNA's. J Biol Chem 253:4517–4520

Heckler TG, Zama Y, Naka T, Hecht SM (1983) Dipeptide formation with misacylated $tRNA_{Phe}$s. J Biol Chem 258:4492–4495

Heckler TG, Chang LH, Zama Y, Naka T, Hecht SM (1984) Preparation of 2′-(3′)-O-Acyl-pCpA derivatives as substrates for T4 RNA ligase-mediated chemical aminoacylation. Tetrahedron 40:87–94

Heckler TG, Roesser JR, Cheng X, Chang PI, Hecht SM (1988) Ribosomal binding and dipeptide formation by misacylated $tRNA_{Phe}$'s. Biochemistry 27:7254–7262

Hirao I, Ohtsuki T, Fujiwara T, Mitsui T, Yokogawa T, Okuni T, Nakayama H, Takio K, Yabuki T, Kigawa T, Kodama K, Nishikawa K, Yokoyama S (2002) An unnatural base pair for incorporating amino acid analogs into proteins. Nat Biotechnol 20:177–182

Hohsaka T, Ashizuka Y, Murakami H, Sisido M (2001a) Five-base codons for incorporation of nonnatural amino acids into proteins. Nucleic Acids Res 29:3646–3651

Hohsaka T, Ashizuka Y, Taira H, Murakami H, Sisido M (2001b) Incorporation of nonnatural amino acids into proteins by using various four-base codons in an *Escherichia coli* in vitro translation system. Biochemistry 40:11060–11064

Hortin G, Boime I (1983) Applications of amino acid analogs for studying co- and posttranslational modifications of proteins. Methods Enzymol 96:777–784

Josephson K, Hartman MCT, Szostak JW (2005) Ribosomal synthesis of unnatural peptides. J Am Chem Soc 127:11727–11735

Killian JA, Van Cleve MD, Shayo YF, Hecht SM (1998) Ribosome-mediated incorporation of hydrazinophenylalanine into modified peptide and protein analogues. J Am Chem Soc 120:3032–3042

Koh JT, Cornish VW, Schultz PG (1997) An experimental approach to evaluating the role of backbone interactions in proteins using unnatural amino acid mutagenesis. Biochemistry 36:11314–11322

Kohrer C, Xie L, Kellerer S, Varshney U, RajBhandary UL (2001) Import of amber and ochre suppressor tRNAs into mammalian cells: a general approach to site-specific insertion of amino acid analogues into proteins. Proc Natl Acad Sci USA 98:14310–14315

Kohrer C, Yoo JH, Bennett M, Schaack J, RajBhandary UL (2003) A possible approach to two different unnatural site-specific insertion of amino acids into proteins in mammalian cells via nonsense suppression. Chem Biol 10:1095–1102

Kohrer C, Sullivan EL, RajBhandary UL (2004) Complete set of orthogonal 21st aminoacyl-tRNA synthetase-amber, ochre and opal suppressor tRNA pairs: Concomitant suppression of three different termination codons in an mRNA in mammalian cells. Nucleic Acids Res 32: 6200–6211

Lecoq A, Marraud M, Aubry A (1991) Hydrazino and N-amino peptides – chemical and structural aspects. Tetrahedron Lett 32:2765–2768

Link AJ, Mock ML, Tirrell DA (2003) Non-canonical amino acids in protein engineering. Curr Opin Biotechnol 14:603–609

Lodder M, Golovine S, Hecht SM (1997) A chemical deprotection strategy for the elaboration of misacylated transfer RNA's. J Org Chem 62:778–779

Lodder M, Golovine S, Laikhter AL, Karginov VA, Hecht SM (1998) Misacylated transfer RNAs having a chemically removable protecting group. J Org Chem 63:794–803

Lu T, Ting AY, Mainland J, Jan IY, Schultz PG, Yang J (2001) Probing ion permeation and gating in a K+ channel with backbone mutations in the selectivity filter. Nat Neurosci 4:239–246

Madsen R, Roberts C, Fraser-Reid B (1995) The pent-4-enoyl group: a novel amine-protecting group that is readily cleaved under mild conditions. J Org Chem 60:7920–7926

Merryman C, Green R (2004) Transformation of aminoacyl tRNAs for the in vitro selection of "drug-like" molecules. Chem Biol 11:575–582

Noren CJ, Anthony-Cahill SJ, Suich DJ, Noren KA, Griffith MC, Schultz PG (1990) In vitro suppression of an amber mutation by a chemically aminoacylated transfer RNA prepared by run-off transcription. Nucleic Acids Res 18:83–88

Patchornik A, Amit B, Woodward RB (1970) Photosensitive protecting groups. J Am Chem Soc 92:6333–6335

Peter C, Daura X, van Gunsteren WF (2000) Peptides of aminoxy acids: A molecular dynamics simulation study of conformational equilibria under various conditions. J Am Chem Soc 122:7461–7466

Robertson SA, Noren CJ, Anthony-Cahill SJ, Griffith MC, Schultz PG (1989) The use of 5′-phospho-2-deoxyribocytidylylriboadenosine as a facile route to chemical aminoacylation of tRNA. Nucleic Acids Res 17:9649–9660

Robertson SA, Ellman JA, Schultz PG (1991) A general and efficient route for chemical aminoacylation of transfer RNAs. J Am Chem Soc 113:2722–2729

Roesser JR, Chorghade MS, Hecht SM (1986) Ribosome-catalyzed formation of an abnormal peptide analog. Biochemistry 25:6361–6365

Roesser JR, Xu C, Payne RC, Surratt CK, Hecht SM (1989) Preparation of misacylated aminoacyl-$tRNA_{Phe}$'s useful as probes of the ribosomal acceptor site. Biochemistry 28:5185–5195

Shimizu Y, Inoue A, Tomari Y, Suzuki T, Yokogawa T, Nishikawa K, Ueda T (2001) Cell-free translation reconstituted with purified components. Nat Biotechnol 19:751–755

Shin I, Ting AY, Schultz PG (1997) Analysis of backbone hydrogen bonding in a β-turn of staphylococcal nuclease. J Am Chem Soc 119:12667–12668

Starck SR, Qi X, Olsen BN, Roberts RW (2003) The puromycin route to assess stereo- and regiochemical constraints on peptide bond formation in eukaryotic ribosomes. J Am Chem Soc 125:8090–8091

Stepanov VG, Moor NA, Ankilova VN, Lavrik OI (1992) Phenylalanyl-transfer RNA-synthetase from *Thermus thermophilus* can attach 2 molecules of phenylalanine to transfer RNA(Phe). FEBS Lett 311:192–194

Stepanov VG, Moor NA, Anikova VN, Vasil'eva IA, Sukhanova MV, Lavrik OI (1998) A peculiarity of the reaction of tRNA aminoacylation catalyzed by phenylalanyl-tRNA synthetase from the extreme thermophile *Thermus thermophilus*. Biochim Biophys Acta 1386:1–15

Tan Z, Forster AC, Blacklow SC, Cornish VW (2004) Amino acid backbone specificity of the *E. coli* translation machinery. J Am Chem Soc 126:12752–12753

Tang Y, Tirrell DA (2001) Biosynthesis of a highly stable coiled-coil protein containing hexafluoroleucine in an engineered bacterial host. J Am Chem Soc 123:11089–11090

Thévenet L, Vanderesse R, Marraud M, Didierjean C, Aubry A (2000) Pseudopeptide fragments and local structures induced by an alpha-aminoxy acid in a dipeptide. Tetrahedron Lett 41:2361–2364

Uliana JA, Doolittle RF (1969) Pyrrolidonecarboxylyl peptidase: Studies on the specificity of the enzyme. Arch Biochem Biophys 131:561–565

Wang B, Lodder M, Zhou J, Baird TT, Brown KC, Craik CS, Hecht SM (2000) Chemically mediated site-specific cleavage of proteins. J Am Chem Soc 122:7402–7403

Wang B, Brown KC, Lodder M, Craik CS, Hecht SM (2002) Chemically mediated site-specific proteolysis. Alteration of protein-protein interaction. Biochemistry 41:2805–2813

Wang B, Zhou J, Lodder M, Anderson RD, III, Hecht SM (2006) Tandemly activated tRNAs as participants in protein synthesis. J Biol Chem 281:13865–13868

Wilson MJ, Hatfield DL (1984) Incorporation of modified amino acids into proteins in vivo. Biochim Biophys Acta 781:205–215

Wu YD, Wang DP, Chan KWK, Yang D (1999) Theoretical study of peptides formed by aminoxy acids. J Am Chem Soc 121:11189–11196

Yang D, Ng FF, Li ZJ, Wu Y, Chan KWK, Wang DP (1996) An unusual turn structure in peptides containing alpha-aminoxy acids. J Am Chem Soc 118:9794–9795

Zhang ZW, Alfonta L, Tian F, Busulaya B, Uryu S, King DS, Schultz PG (2004) Selective incorporation of 5-hydroxytryptophan into proteins in mammalian cells. Proc Natl Acad Sci USA 101:8882–8887

Cell-Free Synthesis of Proteins with Unnatural Amino Acids. The PURE System and Expansion of the Genetic Code

I. Hirao, T. Kanamori, and T. Ueda(✉)

Contents

Abstract The site-specific incorporation of unnatural amino acids into proteins enables us not only to elucidate the function of amino acid residues at particular positions but also to endow natural proteins with new features. Protein expression using living cells, however, imposes limit on efficiency and site-specificity of the introduction of amino acid analogs irregular to translation system or cellular metabolism. Cell-free translation system can contend against such difficulties by its high controllability such as removal and/or supplement of components in the

---

T. Ueda
Department of Medical Genome Sciences, Graduate School of Frontier Sciences, University of Tokyo, FSB401, 5-1-5 Kashiwanoha, Kashiwa, Chiba 277-8562, Japan, e-mail: ueda@k.u-tokyo.ac.jp

C. Köhrer and U.L. RajBhandary (eds.) *Protein Engineering.*
Nucleic Acids and Molecular Biology 22,

system. In this chapter, the advantage of the reconstituted translation system, named PURE system, for the efficient incorporation of an unnatural amino acid at amber codon is demonstrated. Moreover, introduction of nucleic acids containing unnatural base into a cell-free transcription-translation system can expand the genetic code to create novel codon-anticodon correspondence for non-standard amino acids.

1 Introduction

Natural proteins are generally composed of the 20 standard amino acids. One of the most powerful protein engineering tools available today is site-directed mutagenesis, in which a mutant protein with an amino acid substitution at a desired position can be easily obtained by expressing it *in vivo*. However, since this method relies on the use of living cells, it is typically limited to substitutions involving the 20 standard amino acids and their closely related derivatives. In contrast, an *in vitro* approach, which employs a cell-free translation system, allows for the site-specific incorporation of a wide variety of unnatural amino acids. We describe here the improvement of a reconstituted cell-free translation system for the efficient incorporation of unnatural amino acids, as well as the expansion of the genetic code using artificial nucleic acids.

2 Conventional Cell-Free Translation

2.1 Overview of Conventional Cell-Free Translation

Conventional cell-free translation systems are based on cell extracts prepared from *Escherichia coli*, wheat germ, or rabbit reticulocytes. Cell extracts comprise most of the soluble cell components, with the exception of macromolecules, such as cell-surface membrane components and genomic DNA. As the cell extract fraction contains all protein components necessary for protein synthesis, it is possible to synthesize a desired protein simply by supplying the template nucleic acids and the necessary low molecular weight compounds, including energy sources and amino acids. This system is convenient compared with *in vivo* methods using living cells because protein production is completed within a few hours. Moreover, reaction conditions are highly controllable. Optimization of conditions, such as temperature, or the inclusion of additional factors can be easily achieved. Therefore, techniques based on cell-free translation have been developed for the site-specific incorporation of unnatural amino acids into proteins.

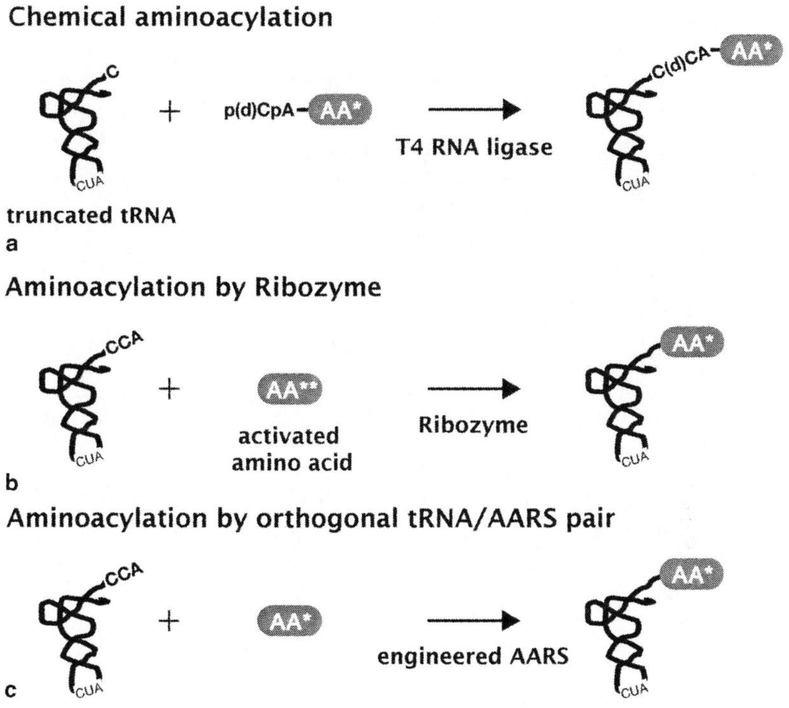

Fig. 1 Preparation of misaminoacylated transfer RNA *(tRNA)*. *AA** desired unnatural amino acid, *AARS* aminoacyl-tRNA synthetase

2.2 *Amber Suppression*

In 1976, Johnson et al. (1976) demonstrated synthesis of a protein containing a modified lysine. Lysyl transfer RNA (tRNA), which had previously been modified at its ε-amino group, was added to a rabbit reticulocyte lysate, resulting in the incorporation of modified lysine into the protein product. However, in their work, the modified lysine was uniformly introduced at lysine residues at all positions. Subsequently, Hecht and coworkers developed the chemical aminoacylation method (Heckler et al. 1984). For this, a dinucleotide pCpA was chemically aminoacylated with the desired unnatural amino acid, following (5′-phospho-2′-deoxycytidylyl-3′,5′-adenosine) which the aminoacylated pCpA was ligated to the truncated tRNA lacking the 3′-terminal dinucleotide, pCpA, by using T4 RNA ligase, as shown in Fig. 1a.

In 1989, Schultz's and Chamberlin's groups developed a method which allows for incorporation of unnatural amino acids into proteins at an amber codon (Noren et al. 1989; Bain et al. 1989). Schultz's group also modified the chemical aminoacylation method; they used pdCpA (5′-phosphocytidylyl-3′,5′-adenosine) instead of pCpA (Robertson et al. 1989) and a truncated amber suppressor tRNA generated by runoff transcription of the corresponding DNA template (Noren et al. 1990). The resultant aminoacylated suppressor tRNA and a gene containing an amber nonsense codon, TAG, at the site of interest in the open reading frame were added to an *in vitro* transcription/translation system using *E. coli* S30 extracts. This effectively allowed the incorporation of an unnatural amino acid at the position corresponding to the amber codon. They showed that more than 50 unnatural amino acids were compatible with the *E. coli* translation system (Ellman et al. 1991) and that even proteins with modified polypeptide backbone structures could be synthesized (Ellman et al. 1992).

2.3 Genetic Code and Misaminoacylated tRNA

Codon and tRNA selection are key factors in the site-specific incorporation of unnatural amino acids into a polypeptide. For example, a nonsense codon, such as the amber codon, can be used for this purpose, as described in the previous section. But, because there are only three nonsense codons, this strategy cannot be used for the incorporation of multiple unnatural amino acids into a single polypeptide. Sisido and coworkers found that by using a four-base codon instead of the amber codon, two or more unnatural amino acids could be introduced into a single polypeptide (Hohsaka et al. 2001; Kajihara et al. 2006). The frameshift caused by the artificial tRNA with a four-base anticodon results in efficient production of a polypeptide into which unnatural amino acids have been incorporated, whereas intrinsic tRNAs with canonical three-base anticodons fail to synthesize a full-length polypeptide. As described below, an artificial base pair system can also be used for the site-specific incorporation of amino acid analogs into proteins (Benner 1994; Service 2000).

Another requirement for incorporation of amino acid analogs is the aminoacylation of tRNA with an unnatural amino acid. However, while the chemical aminoacylation method is logistically difficult, several groups have reported other methods by which tRNA can be aminoacylated with the required unnatural amino acid. Suga's group developed a ribozyme, termed "Flexizyme," which has aminoacylating activity (Kourouklis et al. 2005; Murakami et al. 2006). Flexizyme facilitates tRNA acylation by recognizing the 3′ end of tRNA and the activated amino acid (Fig. 1b). Originally, Flexizyme could not recognize amino acids that lack an aromatic side chain (Kourouklis et al. 2005), but a recent modification has overcome this problem such that an almost unlimited selection of amino acids can now be used to generate aminoacylated tRNA with the Flexizyme system (Murakami et al. 2006).

Another method of utilizing the orthogonality of an aminoacyl-tRNA synthetase (AARS)/tRNA pair has been reported by several groups (Wang et al. 2001; Ohno et al. 2001; Kiga et al. 2002). In this method, a pair of an orthogonal AARSs, which does not aminoacylate any tRNA in the host translation system, and an orthogonal suppressor tRNA, which is not aminoacylated by any AARS in the host system, is used

(Fig. 1c). By using an engineered AARS, which more efficiently recognizes a specific unnatural amino acid than a canonical amino acid, one can site-specifically introduce the unnatural amino acid into the polypeptide. For example, Yokoyama's group used an *E. coli* AARS/tRNA pair specific for iodotyrosine in a wheat germ extract system (Kiga et al. 2002).

3 PURE System

3.1 Overview of the PURE System

Recently, we developed a novel cell-free translation system based on a new concept that relied solely on *E. coli* translation components. We termed it the "PURE" system, for protein synthesis using recombinant elements (Shimizu et al. 2001, 2005). The PURE system contains His-tagged versions of all protein factors considered by Weissbach and coworkers over 25 years ago (Kung et al. 1977) to be sufficient for protein synthesis. Messenger RNA (mRNA) is translated into the relevant polypeptide using aminoacyl-tRNA intermediates and a ribosome, which, in prokaryotes, consists of three ribosomal RNAs and dozens of proteins. To complete translation of one open reading frame encoded in the mRNA sequence, three reaction steps proceed on the ribosome: initiation, elongation, and termination. This is followed by a ribosome recycling step to reinitiate translation. In *E. coli*, several translation factors take part in each translation step: three initiation factors (IF1, IF2, IF3), three elongation factors (EF-G, EF-Tu, EF-Ts), three release (termination) factors (RF1, RF2, RF3), and ribosome recycling factor. In addition to this main translation reaction, three other reactions are necessary to facilitate protein synthesis (Fig. 2): transcription to synthesize mRNA from the template DNA; aminoacylation of tRNAs (including the formylation of the initiator tRNA); and energy source regeneration.

Thus, besides the ten translation factors described above, T7 RNA polymerase, 20 AARSs, methionyl-tRNA transformylase, pyrophosphatase, creatine kinase, myokinase, and nucleoside-diphosphate kinase were also prepared and integrated into the PURE system. All these protein factors were overexpressed and purified in their His-tagged forms without any loss in their activities. Ribosomes were prepared from *E. coli* (Ohashi et al. 2007), and the PURE system was reconstituted with buffer, tRNA mixture, and various substrates, such as the 20 amino acids and four nucleoside triphosphates. The present composition of the PURE system is shown as follows:

- His-tagged transcription/translation factors
  - Three initiation factors (IF1, IF2, IF3)
  - Three elongation factors (EF-Tu, EF-Ts, EF-G)
  - Three release factors (RF1, RF2, RF3)
  - Ribosome recycling factor
  - 20 AARSs

  - Methionyl-tRNA transformylase
  - T7 RNA polymerase
- *E. coli* ribosomes
- *E. coli* tRNA
- Energy regeneration system
- Four nucleoside triphosphates (ATP, GTP, CTP, UTP)
- 20 amino acids
- Salts
- Buffer

The PURE system is completely different from other conventional cell-free translation systems in that it consists only of those factors necessary for transcription, translation, and energy regeneration. One of the remarkable features of the PURE system is the complete elimination of ingredients that are irrelevant to the protein

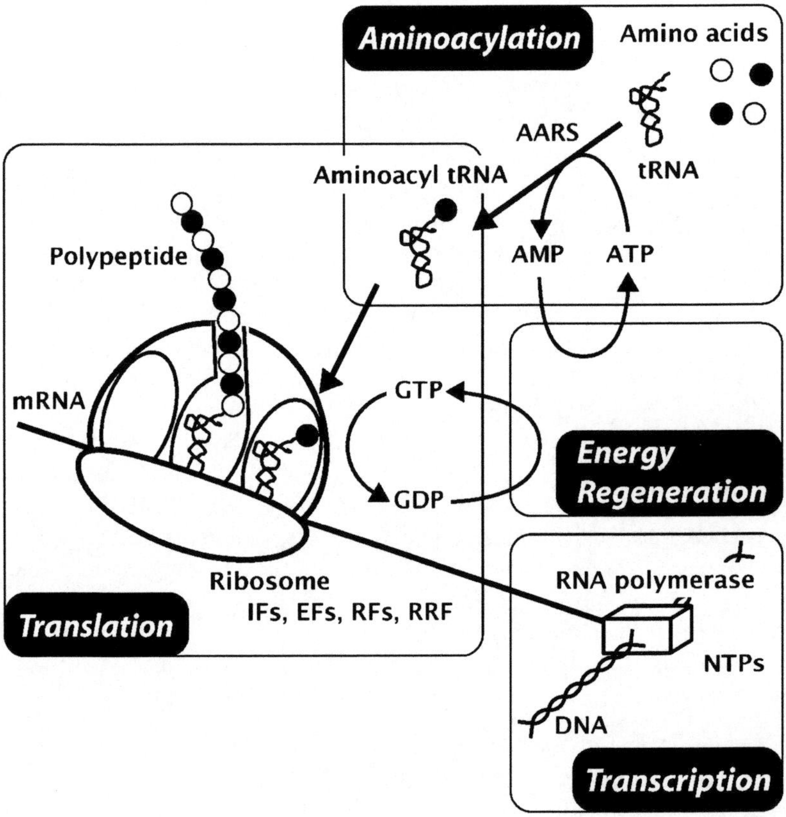

Fig. 2 The four main reactions in the PURE system. *mRNA* messenger RNA, *IFs* initiation factors, *EFs* elongation factors, *RFs* release factors, *RRF* ribosome recycling factor, *NTPs* nucleoside triphosphates

synthesis process, such as amino acid metabolic enzymes, energy-consuming factors, and degradation enzymes (e.g., nucleases and proteases). This feature enables us to circumvent problems encountered during transcription and translation, thereby resulting in a reproducible and reliable system. Another advantage is the possibility of arbitrarily altering the composition of the system depending on the purpose. For instance, we can freely omit, enhance, and/or reduce either single or multiple ingredient(s) within the reaction mixture. Overall, therefore, we believe that the PURE system is an excellent substitute for conventional cell-extract-based, cell-free translation systems.

3.2 Amber Suppression in the PURE System

There are practical difficulties in achieving efficient incorporation of unnatural amino acids using cell extracts in the amber suppression method. Release factors in the cell extract, such as RF1, inevitably compete with the aminoacylated suppressor tRNA and can result in the generation of by-products caused by translation stops at the introduced amber codon (Fig. 3a). To circumvent this competition, attempts

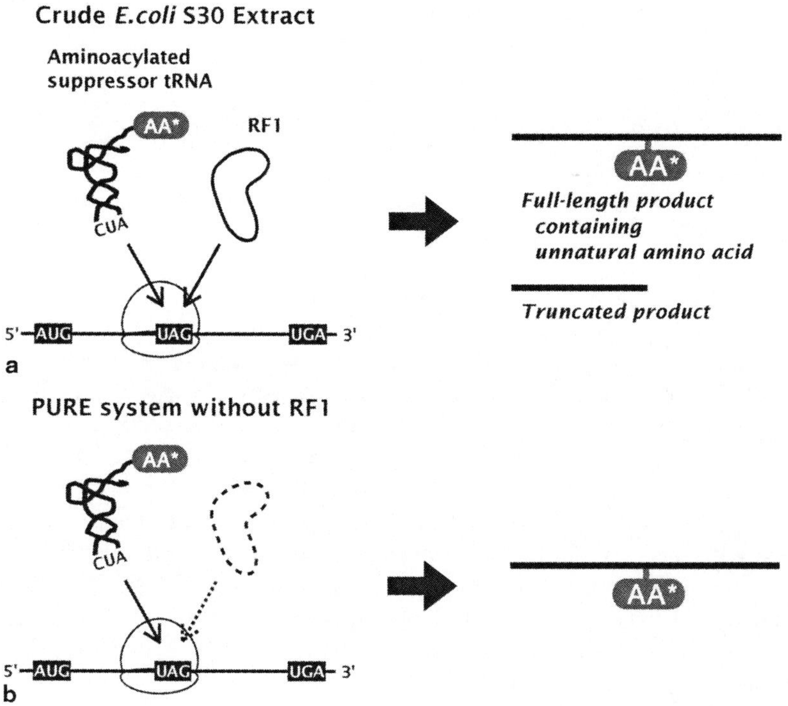

Fig. 3 Amber suppression using the PURE system without release factor 1 *(RF1)*. *AA** desired unnatural amino acid

have been made to inactivate RF1, including the use of an RNA aptamer or an antibody (Szkaradkiewicz et al. 2002; Agafonov et al. 2005). However, using the "designed" PURE system, in which RF1 is not provided, one can easily prevent this type of premature peptide termination (Fig. 3b). In fact, highly efficient suppression of almost 100% was achieved using the designed PURE system (Shimizu et al. 2001). In the designed PURE system, a near-cognate tRNA may decode the stop codon, albeit only at low efficiency (Agafonov et al. 2005). However, we found that such low-fidelity decoding can be prevented by optimizing certain factors of the PURE system, such as the magnesium ion concentration and the tRNA concentration (unpublished observations).

In the PURE system, unnatural amino acids can be easily introduced not only at nonsense codons but also at sense codons. This is because the decoding of mRNA on the ribosome can be completely separated from the aminoacylation step of tRNA by supplying the PURE system with pre-aminoacylated tRNAs (Fig. 2). Forster et al. (2003) succeeded in synthesizing a peptide that comprised only unnatural amino acids by adding misaminoacylated tRNAs to a synthetic system that included ribosomes and a subset of translation factors. Suga's group synthesized unnatural peptides in the presence of misaminoacylated tRNAs prepared by their Flexizyme system (Murakami et al. 2006). On the other hand, Szostak and coworkers synthesized an unnatural amino acid peptide by adding amino acid analogs recognized by native AARSs to the reaction mixture in place of natural amino acids (Josephson et al. 2005).

3.3 N-Terminal Labeling and the PURE System

In eubacteria, formylmethionine is usually used as the first amino acid (Laursen et al. 2005). Using *E. coli* cell extracts, Hardesty's and Rothschild's groups described the synthesis of a polypeptide that starts with a modified methionine residue (Kudlicki et al. 1994; Gite et al. 2000) and the successful use of such a system to create a polypeptide containing a fluorescent probe at its N-terminus. However, because of the competing reaction with endogenous initiator tRNA in the *E. coli* extract, the polypeptide labeling efficiency achieved using this method was very low (1–2%; Olejnik et al. 2005). To avoid this competing reaction and to improve the efficiency of unnatural amino acid incorporation at the start codon, template DNA (or mRNA) with an amber codon instead of an initiation codon, as well as an initiator tRNA with an anticodon CUA corresponding to the amber codon were employed (Varshney and RajBhandary 1990; Olejnik et al. 2005). Alternatively, a conventional translation system from which endogenous initiator tRNA is depleted could be used, but it is very difficult to remove endogenous initiator tRNA from cell extracts. In contrast, because, as described earlier, the PURE system is a completely reconstituted system, it is relatively easy to prepare a reaction mixture without endogenous initiator tRNA. Indeed, Forster et al. (2004) were able to efficiently synthesize a polypeptide in which the N-terminal methionine was biotinylated using a translation system that lacked endogenous initiator tRNA.

4 Expansion of the Genetic Code

4.1 Overview of the Expansion of the Genetic Code by Unnatural Base Pair Systems

Within the present codon table, codons that could potentially be exploited for the production of unnatural amino acids are very restricted. Although a limited number of codons, such as termination codons, can, in principle, be used for this purpose, this provides little scope for the incorporation of multiple unnatural amino acids into a single polypeptide. Expansion of the genetic code using an unnatural base pair system would provide a means to overcome this limitation and may prove to be an efficient way of achieving site-specific, multiple incorporations of amino acid analogs into proteins (Fig. 4) (Benner 1994; Service 2000). The present codon table consists of 64 different combinations of triplet-base sequences, composed of the four natural bases, encoding the standard 20 amino acids. The addition of an extra base pair to the natural A-T and G-C pairs would expand the codon table and increase the number of triplet sequences to 216 (i.e., 6^3). Of these 216 codons, 152 would be novel codons containing unnatural bases, and these could be used to produce various amino acid analogs (Fig. 5). For this expanded system to be workable, the unnatural base pairs need to function in translation, as well as in replication and transcription. For efficient translation, the novel codon–anticodon interactions of unnatural base pairs must be highly selective and thermally stable. In addition, recognition of an unnatural base pair by DNA and RNA polymerases would facilitate the generation and amplification of (1) DNA templates via replication and (2) mRNA and tRNA molecules by transcription.

4.2 Creation of Unnatural Base Pairs

Early progress in the establishment of an unnatural base pair system was made by Benner and colleagues (Switzer et al. 1989; Piccirilli et al. 1990), who developed a series of unnatural base pairs, such as isoguanine (**isoG**)–isocytosine (**isoC**) and xanthosine (**X**)–diaminopyrimidine (**K**), which have unique hydrogen-bonding patterns between pairing bases (Fig. 6b). The hydrogen-donor and hydrogen-acceptor pattern in the A-T pair is different from that in the G-C pair (Fig. 6a), thus conferring the highly orthogonal selectivity of each base pair. Benner and colleagues noticed that unnatural base pairs, such as **isoG-isoC** and **X-K**, could be designed by using combinations of the hydrogen-bonding patterns different from those of the natural base pairs. Therefore, they chemically synthesized substrates (nucleoside 5′-triphosphates) and DNA templates containing the unnatural bases. These base pairs exhibited high selectivity in replication, transcription, and translation. For example, the **isoG** substrate was site-specifically incorporated opposite **isoC** into a complementary DNA strand in DNA templates

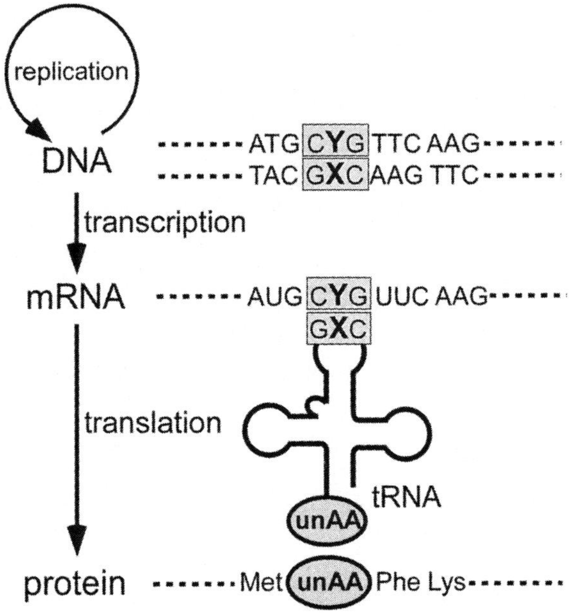

Fig. 4 The expansion of the genetic code by the incorporation of an unnatural base pair. The unnatural base pair (*X-Y*) that functions in replication, transcription, and translation enables the site-specific incorporation of an unnatural amino acid into proteins *unAA* unnatural amino acid

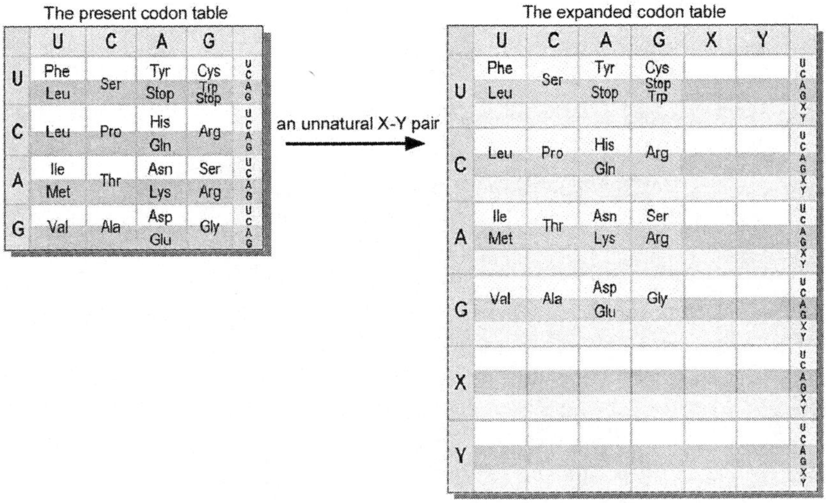

Fig. 5 Expansion of the genetic code by using an unnatural X-Y pair

using the Klenow fragment of *E. coli* DNA polymerase I. The **isoG-isoC** pair was also used in a cell-free translation system to synthesize a peptide containing an amino acid analog, 3-iodotyrosine (Bain et al. 1992). Synthetic RNA molecules, including an mRNA fragment (54-mer) with an **isoC**AG codon and an iodotyrosine-charged tRNA with a CU**isoG** anticodon, were used in an *E. coli* cell-free translation system, resulting in the synthesis of an iodotyrosine-containing peptide (16 amino acids).

Recently, replication systems that include **isoG-isoC** and **X-K** pairs were improved, enabling the use of these base pairs in PCR amplification (Johnson et al. 2004; Sismour et al. 2004; Sismour and Benner 2005). It is known that in solution, the **isoG** base forms both 2-keto and 2-hydroxy forms through tautomerization in positions 1 and 2. The enol tautomer of **isoG** preferentially pairs with T, rather than with **isoC**, because of the complementary hydrogen-bonding patterns between these two bases (Fig. 6c). To address this problem, Benner and colleagues employed 2-thiothymidine (**T**^{S}) in place of T (Fig. 6d) (Sismour and Benner 2005). The 2-thio group of **T**^{S} sterically clashes with the 2-hydroxyl group of the enol tautomer of **isoG**. Thus, the combination of **isoG-isoC**, A-**T**^{S}, and G-C pairs improved the selectivity of each base pair, thereby facilitating PCR amplification at high fidelity.

Unfortunately, the 2-amino groups of **isoC** and **K** are poorly recognized by some polymerases (Switzer et al. 1993; Horlacher et al. 1995). For example, ribonucleotide substrates of **isoC** and **K** cannot be incorporated into RNA by T7 RNA polymerase, whereas the **isoG** and **K** substrates are incorporated into RNA opposite their pairing partners. Ternary crystal structures of polymerases in complex with DNA templates and incoming substrates revealed hydrogen-bonding interactions between the amino acid residues in the polymerases and the 2-keto groups of pyrimidines or 3-nitrogens of purines (Doublié et al. 1998; Kiefer et al. 1998; Cheetham et al. 1999). This observation suggests that the presence of proton-acceptor groups is essential for position 2 of pyrimidines, and thus, that **isoC** and **K** 2-amino groups are unfavorable for polymerase recognition.

Another unnatural base pair system which combines the concepts of hydrogen-bonding patterns (Switzer et al. 1989; Piccirilli et al. 1990) and shape complementarity (Rappaport 1988; Morales and Kool 1998; Delaney et al. 2003) has been developed by Hirao and colleagues (Ohtsuki et al. 2001; Hirao et al. 2002; Mitsui et al. 2005). One example of this system is the unnatural base pairing between 2-amino-6-thienylpurine (**s**) and 2-oxopyridine (**y**) (Fig. 7a) (Hirao et al. 2002). The bulky 6-thienyl group of **s** sterically clashes with the 4-keto group of T and the 4-amino group of C, and thus prevents the formation of noncognate **s**-T and **s**-C pairs (Fig. 7b). In contrast, the relatively small hydrogen at position 6 of **y** maintains the shape complementarity of the **s**-**y** pairing. The thermal stability of the double-stranded DNA fragment (12-mer) containing the **s**-**y** pair (T_m = 43.4 °C) was higher than that observed in fragments containing the **s**-T pair (T_m = 40.3°C) or the **s**-C pair (T_m = 41.1°C) (Fujiwara et al. 2001). In transcription, the ribonucleotide substrate of **y** (**y**TP) can be site-specifically incorporated into RNA (opposite **s** in DNA templates) by T7 RNA polymerase. The incorporation selectivity of y opposite s exceeds 97%.

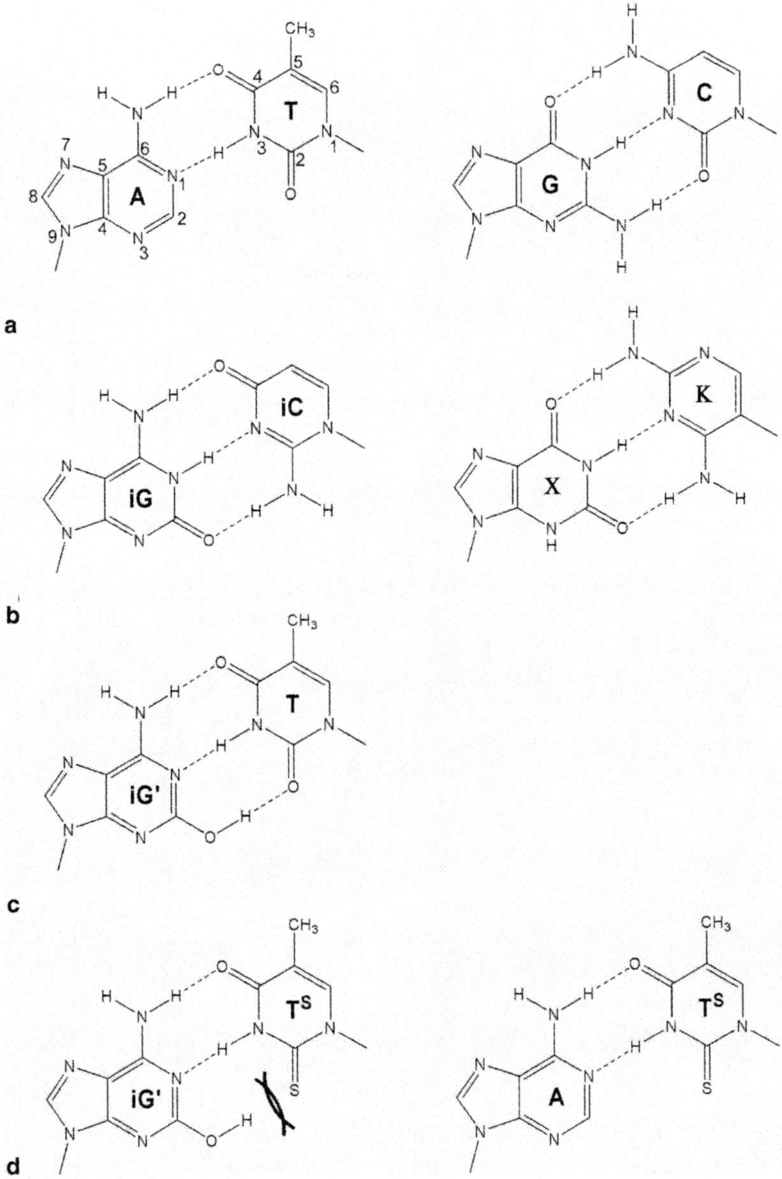

Fig. 6 The natural A-T and G-C pairs and Benner's base pairs. The unnatural isoG-isoC and xanthosine–diaminopyrimidine pairs (**b**) have different hydrogen-bonding patterns from the A-T and G-C pairs (**a**). The enol tautomer of isoG (***iG′***) pairs with T (**c**), but not with the modified T base (***T***^{S}), which can pair with A (**d**)

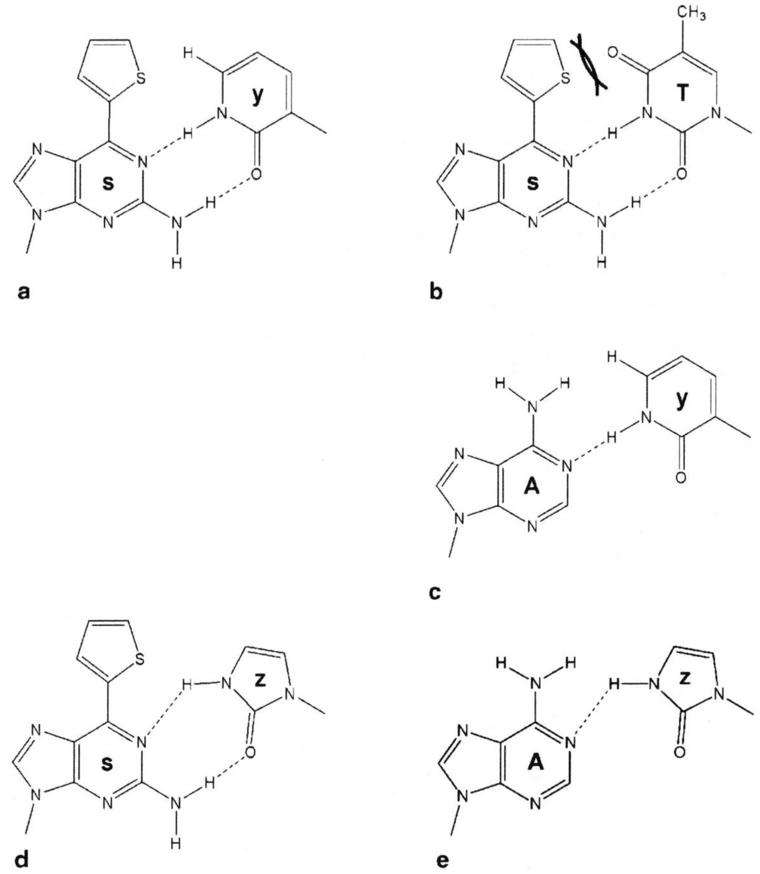

Fig. 7 The unnatural 2-amino-6-thienylpurine (*s*)–2-oxopyridine (*y*) and **s**–imidazolin-2-one (*z*) pairs. The **s** base containing the bulky 6-thienyl group preferentially pairs with **y** (**a**) rather than with T (**b**). However, the shape of **y** sterically fits that of A (**c**). The five-membered ring of z fits **s** (**d**) and reduces the fitting with A (**e**), relative to the six-membered ring of **y**

4.3 Unnatural Base Pair Systems for Unnatural Amino Acid Incorporation

To incorporate unnatural amino acids into proteins, specific transcription of nucleic acids containing unnatural bases was combined with a cell-free translation system from *E. coli* (Fig. 8) (Hirao et al. 2002). Using the **s**-**y** pair, Hirao and colleagues created an extra codon (**y**AG)–anticodon (CUs) interaction and prepared $tRNA_{CUs}$ by combination of chemical synthesis and enzymatic ligation. For the $tRNA_{CUs}$, the *Saccharomyces cerevisiae* $tRNA^{Tyr}$ sequence was employed, because it can be aminoacylated with 3-substituted tyrosine analogs by *S. cerevisiae* tyrosyl-tRNA synthetase, and is poorly recognized by

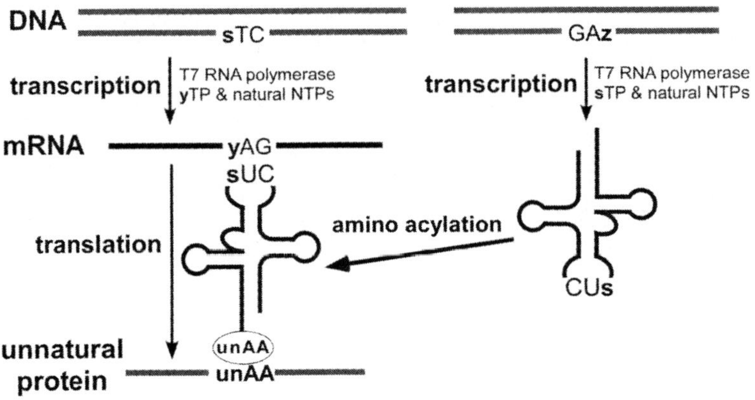

Fig. 8 System containing two unnatural base pairs for incorporating unnatural amino acids into proteins. *unAA* desired unnatural amino acid

E. coli AARSs (Ohno et al. 1998), making it orthogonal in an *E. coli* system. In the coupled transcription–translation system, the use of a 3-chlorotyrosine-charged $tRNA_{CUs}$, a DNA template containing **s**, and **y**TP led to the selective generation of human Ha-Ras protein (185 amino acids) containing 3-chlorotyrosine at position 32.

Although this coupled transcription–translation system using the **s**-**y** pair has the potential for site-specifically incorporating amino acid analogs into proteins, there is certainly room for further methodological improvement. The preparation of $tRNA_{CUs}$ is laborious, as it combines chemical and enzymatic procedures. Transcription is more convenient for the preparation of the tRNA containing **s**, but the incorporation of **s** into RNA opposite **y** is less selective than that of **y** opposite **s** in T7 transcription. The bulky thienyl group of **s** prevents pairing with natural bases, whereas **y** has no such functional group. As a result, natural purine substrates, especially A, were partially incorporated into RNA opposite **y** in DNA templates (Fig. 7c), and the incorporation selectivity of **s** opposite **y** was reduced by 70–80%. This low selectivity of **s** incorporation hinders the transcriptional preparation of RNA molecules containing **s** at specific positions.

The unidirectional complementarity of the **s**-**y** pair was overcome by using a five-membered-ring base, imidazolin-2-one (**z**) (Fig. 7d) (Hirao et al. 2004). Shape complementarity between the five-membered ring of **z** and **s** is better than that between the six-membered ring of **y** and **s**. In contrast, the fit between **z** and the natural purines is inferior to that of **y** (Fig. 7e). As a result, the **s** substrate was selectively incorporated into RNA opposite **z** in DNA templates by T7 RNA polymerase. Unexpectedly, the **z** base was less effective as a substrate than **y**, and the selectivity of **z** incorporation opposite **s** was lower than that of **y** incorporation opposite **s**. Therefore, in combination with the **s**-**y** pair, the **s**-**z** pair can be used in a coupled transcription–translation system, in which the extra codon–anticodon interaction is mediated by the **s**-**y** pair. The **y**-containing mRNA is prepared by transcription using the **s-y** pair, and the **s**-containing tRNA is prepared by transcription using the **s-z** pair (Fig. 8). This system, with two unnatural base-pairs, will be a powerful tool when combined with an orthogonal AARS/tRNA set for the site-specific incorporation of amino acid analogs into proteins (Liu and Schultz 1999; Döring et al. 2001; Wang et al. 2001; Kowal et al. 2001; Kiga et al. 2002) or with ribozymes that aminoacylate tRNA molecules (Kourouklis et al. 2005; Murakami et al. 2006).

4.4 Further Development of Unnatural Base Pair Systems

Another approach to the creation of extra base pairs is to use hydrophobic base analogs (Henry and Romesberg 2003). Kool and colleagues generated non-hydrogen-bonded base pairs between 4-methylbenzimidazole (**Z**) and difluorotoluene (**F**), and

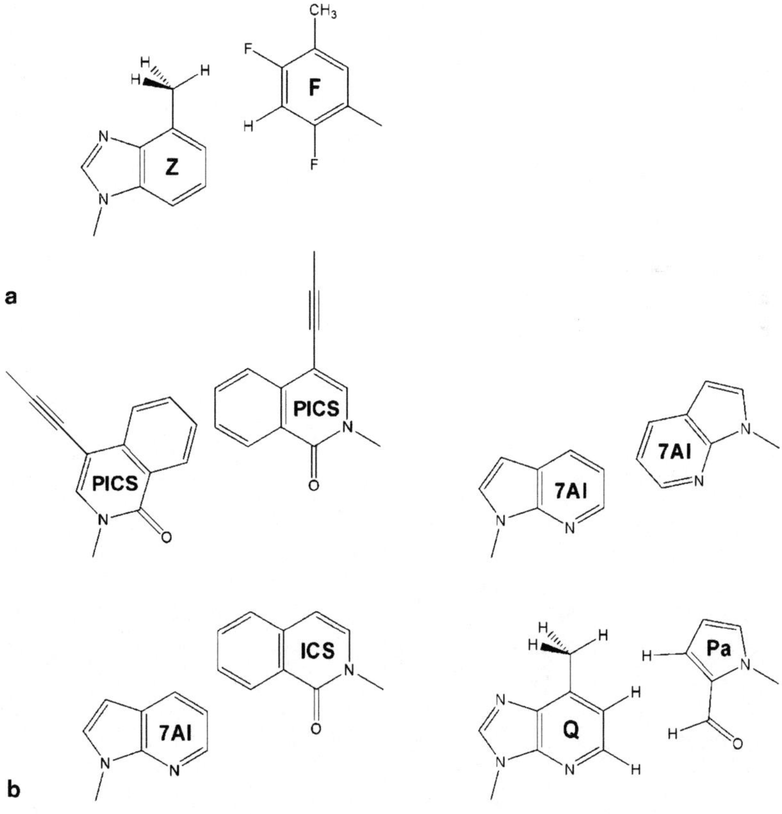

Fig. 9 Hydrophobic, non-hydrogen-bonding base pairs. The Z-F pair, an isostere of the A-T pair (**a**) and several hydrophobic base pairs (**b**)

found that each of the base analogs was complementarily incorporated into DNA by the Klenow fragment of DNA polymerase I (Fig. 9a) (Morales and Kool 1998). Their results indicated that hydrogen bonds in a base pair are not absolutely required in replication, and they suggested that shape complementarity plays an important role in faithful base pairing. Although the **Z-F** pair, which mimics the shape of the A-T pair, cannot be used as an extra base pair, the results showed that hydrophobic base pairs are good candidates for creating extra base pairs. The use of hydrophobic base analogs can avoid the tautomerization problem observed with hydrogen-bonded base analogs, and as a consequence several hydrophobic base pairs with high selectivity in replication have been developed (Fig. 9b) (McMinn et al. 1999; Ogawa et al. 2000; Wu et al. 2000; Mitsui et al. 2003; Henry et al. 2004). Recently, another hydrophobic base pair has been reported exhibiting high selectivity both in PCR amplification and transcription (Hirao et al. 2006). However, there have been no reports describing the use of hydrophobic base pairs during translation in protein synthesis. This raises the question of whether non-hydrogen-bonded base pairs can be used for the codon–anticodon interaction in ribosome-mediated translation.

At present, no unnatural base pairs that complementarily function in replication, transcription, and translation with high selectivity have been created. The selectivity of the **s-y** pair in replication is not as high as it is in transcription. Accordingly, protein synthesis using unnatural base pair systems is presently restricted to the test tube. However, studies of unnatural base pairs have rapidly advanced, and these studies have generated valuable information about the mechanisms of base pairing in DNA and RNA biosyntheses. It is expected that this in turn will facilitate the development of more sophisticated unnatural base pair systems. In addition, the development of modified and evolved polymerases that can selectively and efficiently form unnatural base pairs is in progress (Chelliserrykattil and Ellington 2004; Fa et al. 2004; Leconte et al. 2005, for more details see Chap. 11 by Leconte and Romesberg). Thus, the practical use of unnatural base pair systems for codon-expanded translation may soon become a reality.

5 Conclusion

Four billion years ago, living systems employed 20 amino acids as basic elements for the synthesis of polypeptides and four bases as the basic elements of nucleic acids. Based upon these elements, translation systems have integrated various components, such as translation factors and the ribosome, and have evolved to become extremely complicated systems to guarantee high fidelity and efficiency. Simultaneously, cells have evolved to support this fundamental system by building up metabolic pathways, including those involving amino acid and nucleotide biosynthesis. Such convoluted cellular systems seriously impede the development of new production systems for artificial proteins or nucleic acids. Thus, a cell-free approach may be a powerful means of facilitating the incorporation of an artificial amino acid at a desired site. In particular, the reconstituted PURE system increases the feasibility of producing proteins with unnatural amino acids because of its high degree of functionality. Here, we have described how the omission of RF1 resulted in the highly efficient suppression of an amber codon. If we succeed in reconstructing individual tRNA species, manipulation of the genetic code will become achievable, and this will enable us to incorporate various unnatural amino acids into one polypeptide. Moreover, with the addition of a set of unnatural bases to the genetic code, more than 100 unnatural amino acids could be harnessed for use in such a cell-free translation system.

Acknowledgments This work was supported by a Grant-in-Aid for Scientific Research (KAKENHI 15350097 and 19201046) from the Ministry of Education, Culture, Sports, Science, and Technology and by the RIKEN Structural Genomics/Proteomics Initiative (RSGI), the National Project on Protein Structural and Functional Analyses, Ministry of Education, Culture, Sports, Science and Technology of Japan.

References

Agafonov DE, Huang Y, Grote M, Sprinzl M (2005) Efficient suppression of the amber codon in *E. coli* in vitro translation system. FEBS Lett 579:2156–2160

Bain JD, Glabbe CG, Dix TA, Chamberlin AR (1989) Biosynthetic site-specific incorporation of a non-natural amino acid into a polypeptide. J Am Chem Soc 111:8013–8014

Bain JD, Switzer C, Chamberlin AR, Benner SA (1992) Ribosome-mediated incorporation of a non-standard amino acid into a peptide through expansion of the genetic code. Nature 356: 537–539

Benner SA (1994) Expanding the genetic lexicon: incorporating non-standard amino acids into proteins by ribosome-based synthesis. Trends Biotechnol 12:158–163

Cheetham GM, Jeruzalmi D, Steitz TA (1999) Structural basis for initiation of transcription from an RNA polymerase-promoter complex. Nature 399:80–83

Chelliserrykattil J, Ellington AD (2004) Evolution of a T7 RNA polymerase variant that transcribes 2′-O-methyl RNA. Nat Biotechnol 22:1155–1160

Delaney JC, Henderson PT, Helquist SA, Morales JC, Essigmann JM, Kool ET (2003) High-fidelity in vivo replication of DNA base shape mimics without Watson–Crick hydrogen bonds. Proc Natl Acad Sci USA 100:4469–4473

Döring V, Mootz HD, Nangle LA, Hendrickson TL, Crécy-Lagard V, Schimmel P, Marlière P (2001) Enlarging the amino acid set of *Escherichia coli* by infiltration of the valine coding pathway. Science 292:501–504

Doublié S, Tabor S, Long AM, Richardson CC, Ellenberger T (1998) Crystal structure of a bacteriophage T7 DNA replication complex at 2.2 Å resolution. Nature 391:251–258

Ellman JA, Mendel D, Anthony-Cahill SJ, Noren CJ, Schultz PG (1991) Biosynthetic method for introducing unnatural amino acids site-specifically into proteins. Methods Enzymol 202:301–336

Ellman JA, Mendel D, Schultz PG (1992) Site-specific incorporation of novel backbone structures into proteins. Science 255:197–200

Fa M, Radeghieri A, Henry AA, Romesberg FE (2004) Expanding the substrate repertoire of a DNA polymerase by directed evolution. J Am Chem Soc 126:1748–1754

Forster AC, Tan Z, Nalam MN, Lin H, Qu H, Cornish VW, Blacklow SC (2003) Programming peptidomimetic syntheses by translating genetic codes designed de novo. Proc Natl Acad Sci USA 100:6353–6357

Forster AC, Cornish VW, Blacklow SC (2004) Pure translation display. Anal Biochem 333:358–364

Fujiwara T, Kimoto M, Sugiyama H, Hirao I, Yokoyama S (2001) Synthesis of 6-(2-thienyl)purine nucleoside derivatives that form unnatural base pairs with pyridine-2-one nucleosides. Bioorg Med Chem Lett 11:2221–2223

Gite S, Mamaev S, Olejnik J, Rothschild K (2000) Ultrasensitive fluorescence-based detection of nascent proteins in gels. Anal Biochem 279:218–225

Heckler TG, Chang LH, Zama Y, Naka T, Chorghade MS, Hecht SM (1984) T4 RNA ligase mediated preparation of novel "chemically misacylated" tRNAPheS. Biochemistry 23:1468–1473

Henry AA, Romesberg FE (2003) Beyond A, C, G and T: Augmenting nature's alphabet. Curr Opin Chem Biol 7:727–733

Henry AA, Olsen AG, Matsuda S, Yu C, Geierstanger BH, Romesberg FE (2004) Efforts to expand the genetic alphabet: Identification of a replicable unnatural DNA self-pair. J Am Chem Soc 126:6923–6931

Hirao I, Ohtsuki T, Fujiwara T, Mitsui T, Yokogawa T, Okuni T, Nakayama H, Takio K, Yabuki T, Kigawa T, Kodama K, Yokogawa T, Nishikawa K, Yokoyama S (2002) An unnatural base pair for incorporating amino acid analogs into proteins. Nat Biotechnol 20:177–182

Hirao I, Harada Y, Kimoto M, Mitsui T, Fujiwara T, Yokoyama S (2004) A two-unnatural-base-pair system toward the expansion of the genetic code. J Am Chem Soc 126:13298–13305

Hirao I, Kimoto M, Mitsui T, Fujiwara T, Kawai R, Sato A, Harada Y, Yokoyama S (2006) An unnatural hydrophobic base pair system: site-specific incorporation of nucleotide analogs into DNA and RNA. Nat Methods 3:729–735

Hohsaka T, Ashizuka Y, Taira H, Murakami H, Sisido M (2001) Incorporation of nonnatural amino acids into proteins by using various four-base codons in an *Escherichia coli* in vitro translation system. Biochemistry 40:11060–11064

Horlacher J, Hottiger M, Podust VN, Hübscher U, Benner SA (1995) Recognition by viral and cellular DNA polymerases of nucleosides hearing bases with nonstandard hydrogen bonding patterns. Proc Natl Acad Sci USA 92:6329–6333

Johnson AE, Woodward WR, Herbert E, Menninger JR (1976) Nepsilon-acetyllysine transfer ribonucleic acid: a biologically active analogue of aminoacyl transfer ribonucleic acids. Biochemistry 15:569–575

Johnson SC, Sherrill CB, Marshall DJ, Moser MJ, Prudent JR (2004) A third base pair for the polymerase chain reaction: inserting isoC and isoG. Nucleic Acids Res 32:1937–1941

Josephson K, Hartman MC, Szostak JW (2005) Ribosomal synthesis of unnatural peptides. J Am Chem Soc 127:11727–11735

Kajihara D, Abe R, Iijima I, Komiyama C, Sisido M, Hohsaka T (2006) FRET analysis of protein conformational change through position-specific incorporation of fluorescent amino acids. Nat Methods 3:923–929

Kiefer JR, Mao C, Braman JC, Beese LS (1998) Visualizing DNA replication in a catalytically active Bacillus DNA polymerase crystal. Nature 391:304–307

Kiga D, Sakamoto K, Kodama K, Kigawa T, Matsuda T, Yabuki T, Shirouzu M, Harada Y, Nakayama H, Takio K, Hasegawa Y, Endo Y, Hirao I, Yokoyama S (2002) An engineered *Escherichia coli* tyrosyl-tRNA synthetase for site-specific incorporation of an unnatural amino acid into proteins in eukaryotic translation and its application in a wheat germ cell-free system. Proc Natl Acad Sci USA 99:9715–9720

Kourouklis D, Murakami H, Suga H (2005) Programmable ribozymes for mischarging tRNA with nonnatural amino acids and their applications to translation. Methods 36:239–244

Kowal AK, Köhrer C, RajBahandary UL (2001) Twenty-first aminoacyl-tRNA synthetase-suppressor tRNA pairs for possible use in site-specific incorporation of amino acid analogues into proteins in eukaryotes and in eubacteria. Proc Natl Acad Sci USA 98:2268–2273

Kudlicki W, Odom OW, Kramer G, Hardesty B (1994) Chaperone-dependent folding and activation of ribosome-bound nascent rhodanese. J Mol Biol 244:319–331

Kung HF, Treadwell BV, Spears C, Tai PC, Weissbach H (1977) DNA-directed synthesis in vitro of beta-galactosidase: Requirement for a ribosome release factor. Proc Natl Acad Sci USA 74:3217–3221

Laursen BS, Sorensen HP, Mortensen KK, Sperling-Petersen HU (2005) Initiation of protein synthesis in bacteria. Microbiol Mol Biol Rev 69:101–123

Leconte AM, Chen L, Romesberg FE (2005) Polymerase evolution: Efforts toward expansion of the genetic code. J Am Chem Soc 127:12470–12471

Liu DR, Schultz PG (1999) Progress toward the evolution of an organism with an expanded genetic code. Proc Natl Acad Sci USA 96:4780–4785

McMinn DL, Ogawa AK, Wu Y, Liu J, Schultz PG, Romesberg FE (1999) Efforts toward expansion of the genetic alphabet: DNA polymerase recognition of a highly stable, self-pairing hydrophobic base. J Am Chem Soc 121:11585–11586

Mitsui T, Kitamura A, Kimoto M, To T, Sato A, Hirao I, Yokoyama S (2003) An unnatural hydrophobic base pair with shape complementarity between pyrrole-2-carbaldehyde and 9-methylimidazo[(4,5)-b]pyridine. J Am Chem Soc 125:5298–5307

Mitsui T, Kimoto M, Harada Y, Yokoyama S, Hirao I (2005) An efficient unnatural base pair for a base-pair-expanded transcription system. J Am Chem Soc 127:8652–8658

Morales JC, Kool ET (1998) Efficient replication between non-hydrogen-bonded nucleoside shape analogs. Nat Struct Biol 5:950–954

Murakami H, Ohta A, Ashigai H, Suga H (2006) A highly flexible tRNA acylation method for non-natural polypeptide synthesis. Nat Methods 3:357–359

Noren CJ, Anthony-Cahill SJ, Griffith MC, Schultz PG (1989) A general method for site-specific incorporation of unnatural amino acids into proteins. Science 244:182–188

Noren CJ, Anthony-Cahill SJ, Suich DJ, Noren KA, Griffith MC, Schultz PG (1990) In vitro suppression of an amber mutation by a chemically aminoacylated transfer RNA prepared by run-off transcription. Nucleic Acids Res 18:83–88

Ogawa AK, Wu Y, McMinn DL, Liu J, Schultz PG, Romesberg FE (2000) Efforts toward the expansion of the genetic alphabet: information storage and replication with unnatural hydrophobic base pairs. J Am Chem Soc 122:3274–3287

Ohashi H, Shimizu Y, Ying BW, Ueda T (2007) Efficient protein selection based on ribosome display system with purified components. Biochem Biophys Res Commun 352:270–276

Ohno S, Yokogawa T, Fujii I, Asahara H, Inokuchi H, Nishikawa K (1998) Co-expression of yeast amber suppressor $tRNA_{Tyr}$ and tyrosyl-tRNA synthetase in *Escherichia coli*: Possibility to expand the genetic code. J Biochem 124:1065–1068

Ohno S, Yokogawa T, Nishikawa K (2001) Changing the amino acid specificity of yeast tyrosyl-tRNA synthetase by genetic engineering. J Biochem 130:417–423

Ohtsuki T, Kimoto M, Ishikawa M, Hirao I, Yokoyama S (2001) Unnatural base pairs for specific transcription. Proc Natl Acad Sci USA 98:4922–4925

Olejnik J, Gite S, Mamaev S, Rothschild K (2005) N-terminal labeling of proteins using initiator tRNA. Methods 36:252–260

Piccirilli JA, Krauch T, Moroney SE, Benner SA (1990) Enzymatic incorporation of a new base pair into DNA and RNA extends the genetic alphabet. Nature 343:33–37

Rappaport HP (1988) The 6-thioguanine/5-methyl-2-pyrimidinone base pair. Nucleic Acids Res 16:7253–7267

Robertson SA, Noren CJ, Anthony-Cahill SJ, Griffith MC, Schultz PG (1989) The use of 5′-phospho-2 deoxyribocytidylylriboadenosine as a facile route to chemical aminoacylation of tRNA. Nucleic Acids Res 17:9649–9660

Service RF (2000) Creation's seventh day. Science 289:232–235

Shimizu Y, Inoue A, Tomari Y, Suzuki T, Yokogawa T, Nishikawa K, Ueda T (2001) Cell-free translation reconstituted with purified components. Nat Biotechnol 19:751–755

Shimizu Y, Kanamori T, Ueda T (2005) Protein synthesis by pure translation systems. Methods 36:299–304

Sismour AM, Benner SA (2005) The use of thymidine analogs to improve the replication of an extra DNA base pair: a synthetic biological system. Nucleic Acids Res 33:5640–5646

Sismour AM, Lutz S, Park J-H, Lutz MJ, Boyer PL, Hughes SH, Benner SA (2004) PCR amplification of DNA containing non-standard base pairs by variants of reverse transcriptase from Human Immunodeficiency Virus-1. Nucleic Acids Res 32:728–735

Switzer C, Moreney SE, Benner SA (1989) Enzymatic incorporation of a new base pair into DNA and RNA. J Am Chem Soc 111:8322–8323

Switzer CY, Moreney SE, Benner SA (1993) Enzymatic recognition of the base pair between isocytidine and isoguanosine. Biochemistry 32:10489–10496

Szkaradkiewicz K, Nanninga M, Nesper-Brock M, Gerrits M, Erdmann VA, Sprinzl M (2002) RNA aptamers directed against release factor 1 from Thermus thermophilus. FEBS Lett 514:90–95

Varshney U, RajBhandary UL (1990) Initiation of protein synthesis from a termination codon. Proc Natl Acad Sci USA 87:1586–1590

Wang L, Brock A, Herberich B, Schultz PG (2001) Expanding the genetic code of *Escherichia coli*. Science 292:498–500

Wu Y, Ogawa AK, Berger M, McMinn DL, Schultz PG, Romesberg FE (2000) Efforts toward expansion of the genetic alphabet: Optimization of interbase hydrophobic interactions. J Am Chem Soc 122:7621–763

Engineering Nucleobases and Polymerases for an Expanded Genetic Alphabet

A.M. Leconte and F.E. Romesberg(✉)

Contents

Abstract The four natural nucleotides of DNA form base pairs capable of encoding the complex genetic information necessary for all life; additionally, the sequence specific hybridization and enzymatic synthesis of DNA has revolutionized biotechnology. Expansion of the genetic alphabet to include additional, orthogonal nucleotides to work within the context of natural DNA has the potential to greatly expand this essential biopolymer's utility. Here, we detail the three general approaches to the design of unnatural DNA base pairs: alternative hydrogen bonding, shape complementarity, and hydrophobic forces. All of these approaches have been implemented with notable success, but are still limited by DNA polymerase recognition of the unnatural pairs. Thus, we also consider the role of the DNA polymerase and highlight efforts to use alternative DNA polymerases, either natural or engineered.

F.E. Romesberg
Department of Chemistry, The Scripps Research Institute, 10550 North Torrey Pines Road, La Jolla, CA 92037, USA, e-mail: floyd@scripps.edu

C. Köhrer and U.L. RajBhandary (eds.) *Protein Engineering.*
Nucleic Acids and Molecular Biology 22,

1 Introduction

DNA is the repository for all biological information within a cell; its replication and transcription into RNA underlie information storage and retrieval, respectively. The ability to manipulate DNA sequences, and therefore genetic information, has revolutionized modern biology and is one of the most significant scientific advancements of the past century, enabling now-common techniques such as molecular cloning and protein expression. Additionally, since DNA and RNA are the only molecules with a natural link between phenotype and genotype, chemists have successfully used the power of selection to isolate and amplify individual oligonucleotides that possess a desired property (i.e., recognition of another molecule or the ability to catalyze a desired reaction) from a pool of random oligonucleotides (Joyce 2004; Klussman 2006). In vitro oligonucleotide selections promise to be, among other valuable biotechnology applications (Clark and Remcho 2002; Tombelli et al. 2005), an important and novel route to drug discovery (Nimjee et al. 2005; Lee et al. 2006), as evidenced by the recent approval of the first nucleic acid therapeutic which selectively binds a protein target (Ng et al. 2006).

While the manipulation of DNA and RNA is undoubtedly a significant advance, the potential of natural oligonucleotides is restricted by the four nucleotides that pair to form the two natural base pairs. These four nucleotides lead to a finite set of three-base codons, which limits the coding capacity of DNA. Additionally, these nucleobases are similar structurally and chemically, limiting the functional diversity of oligonucleotides. Thus, the expansion of the genetic alphabet, and the subsequent expansion of the genetic code, has captivated the imaginations of scientists for years. On an academic level, efforts towards expanding the genetic alphabet complement a large body of work which evaluates the properties of modified natural DNA (Spratt 2001; Chiaramonte et al. 2003; Adelfinskaya et al. 2005; Zhang et al. 2005) and DNA polymerase analogs (Astatke et al. 1998; Spratt 2001), and asks basic questions about the composition and characteristics of the most fundamental biopolymers and the enzymes that synthesize them. Now, with the rapidly expanding roles of nucleic acid (Clark and Remcho 2002; Tombelli et al. 2005) and protein (Pavlou and Reichert 2004; Hultschig et al. 2006) biotechnologies, expansion of the genetic alphabet and code would have broad industrial applicability, from directed evolution to drug discovery.

Generally, a candidate unnatural base pair must satisfy several basic criteria. First, the base pair must be thermally stable in duplex DNA and must be more stable than mispairs formed between unnatural and natural nucleotides. Second, the base pair must be enzymatically synthesized efficiently and selectively by DNA polymerases. The use of most unnatural base pairs is limited by inadequate DNA-polymerase-mediated synthesis, which is commonly evaluated using the exonuclease-deficient Klenow fragment of *Escherichia coli* DNA polymerase I (Kf; Kf with 3′→5′ exonuclease activity will be denoted as "Kf exo^{+}") (Jacobsen et al. 1974; Derbyshire et al. 1988). Several studies have evaluated other polymerases for their ability to accept the unnatural substrates and have found that their activities differ

Table 1 Promising candidate base pairs

Base pair	Structure	Design strategy	Notes
disoC:disoG	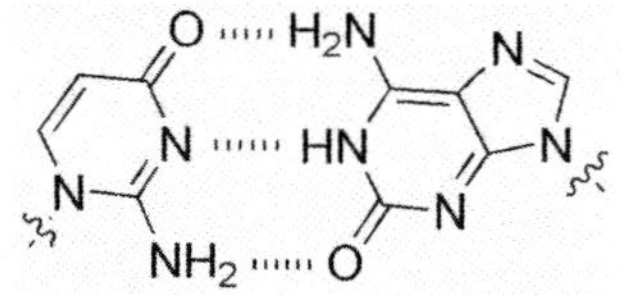	Alternative hydrogen bonding	Stable and selective in duplex DNA Full-length synthesis reported in either context dTTP inserted against isoG relatively efficiently
dκ:d**X**	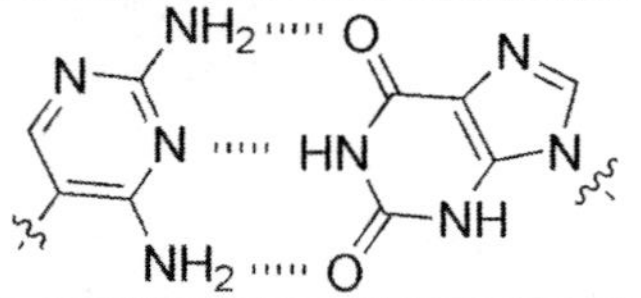	Alternative hydrogen bonding	Stable and selective in duplex DNA Full-length synthesis only in one context Mispairing fairly inefficient
d**z**:d**s**	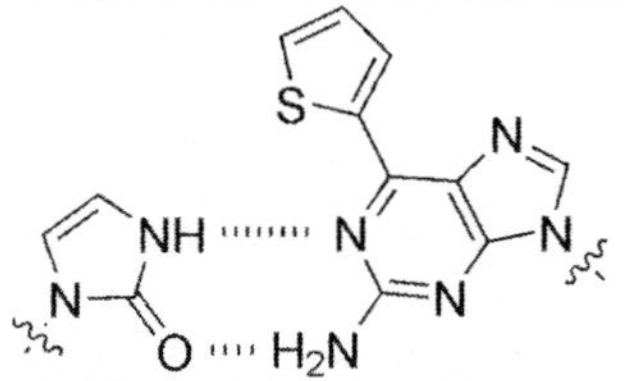	Shape complementary with hydrogen bonds	Stable and selective in duplex DNA d**s**TP incorporation efficient and selective d**z**TP incorporation only moderately efficient and selective
d**Pa**:d**Q**	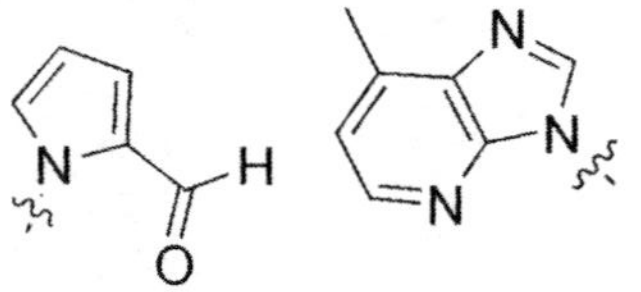	Shape complementary without hydrogen bonds	Stable in duplex DNA Full-length synthesis in either context Moderate efficiency and selectivity

(*continued*)

Table 1 (continued)

Base pair	Structure	Design strategy	Notes
dPICS:dPICS	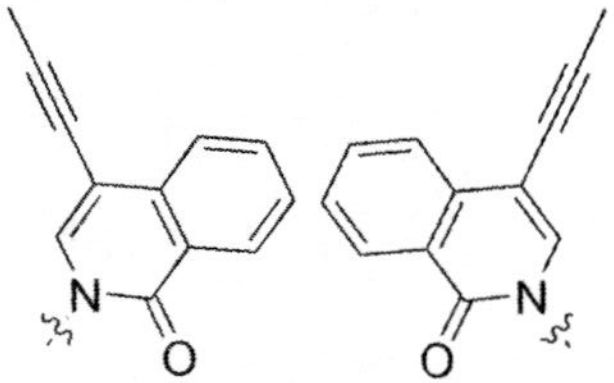	Predominantly hydrophobic forces	Highly stable and selective in duplex DNA Efficiently synthesized with moderate selectivity Inefficient extension limits utility
d3MN:d3MN	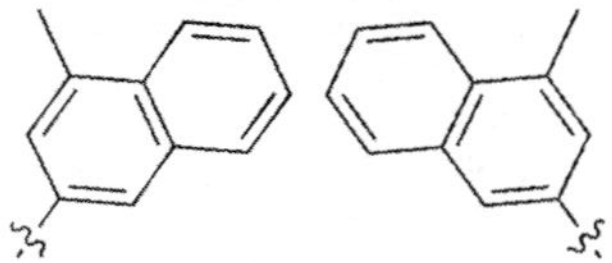	Predominantly hydrophobic forces	Stable and selective in duplex DNA Highly efficient synthesis with relatively high selectivity Inefficient extension limits utility
d3FB:d3FB	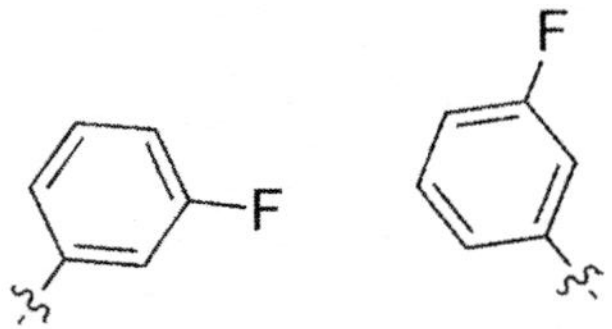	Predominantly hydrophobic forces	Moderately stable and selective in duplex DNA Efficiently synthesized and extended Modest selectivity over dA

Table 2 Applications of alternate DNA polymerases

Base pair	Structure	Polymerase	Notes
dκ:dX	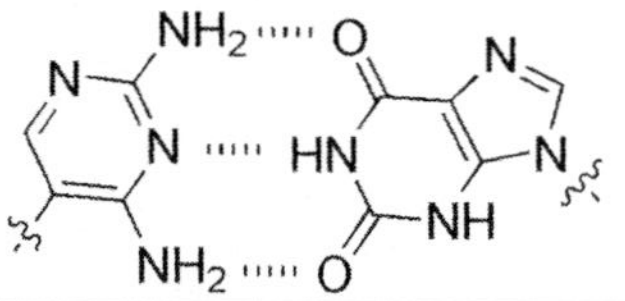	HIV-RT	First use of an alternative polymerase Only DNA polymerase capable of incorporating dκTP against dX and extending the dκ:d**X** pair
disoC:disoG	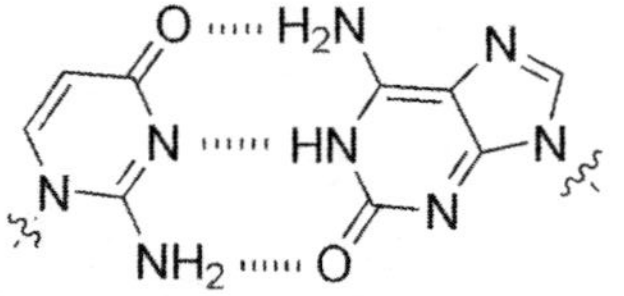	Taq	First practical PCR amplification of an unnatural pair Using 2-thio-T, achieved 98% fidelity per round
d7AI:d7AI	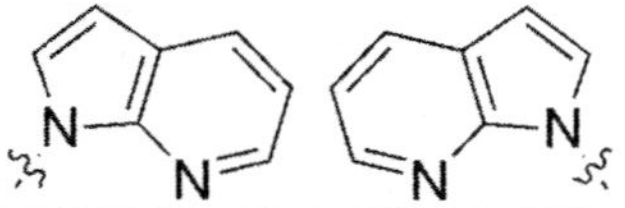	Pol β	Kf and pol β used together as first bipolymerase system able to synthesize full-length product 400-fold estimated fidelity
dPICS:dPICS	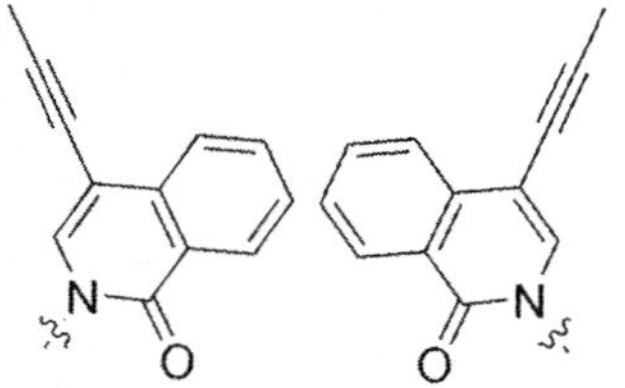	Sf P2	First polymerase mutant selected for improved recognition of an unnatural base pair Increased self-pair synthesis and extension over wild type enzyme

HIV-RT, reverse transcriptase from HIV-1; *Taq*, DNA polymerase of *Thermus aquaticus*; *Kf*, exonuclease-deficient Klenow fragment of *Escherichia coli* DNA polymerase I; *pol* β, rat polymerase β; *Sf*, Stoffel fragment of Taq significantly from that of Kf. More recently, directed evolution methods have been used to produce a DNA polymerase that more efficiently synthesizes DNA containing unnatural base pairs. Here, we review the candidate unnatural base pairs identified to date (summarized in Table 1), as well as efforts to identify polymerases which better synthesize DNA containing unnatural base pairs (summarized in Table 2).

2 Unnatural DNA Base Pairs

The earliest unnatural nucleobase design strategies were, quite reasonably, based on the presumed physical properties that make DNA stable and efficiently replicated by DNA polymerases (Watson and Crick 1953). Benner and coworkers examined pairs based on isostructural nucleobases with complementary hydrogen-bonding patterns that were different from those used by the natural pairs (Fig. 1b). More recently, strategies have also been pursued that are based on the observation of Kool and coworkers that hydrogen bonds are not absolute requirements for efficient DNA synthesis (Moran et al. 1997). Yokoyama, Hirao, and coworkers have designed unnatural base pairs which use natural scaffolds with altered shape to guide selective synthesis by shape complementarity (Fig. 1c). Contemporaneously, we have developed unnatural base pairs which use predominantly hydrophobic forces to drive selective pairing and synthesis (Fig. 1d). These efforts have resulted in the identification of pairs that are replicated efficiently by DNA polymerases despite, in some cases, possessing no conventional shape complementarity and little to no hydrogen-bonding potential. The broad success of non-purine:pyrimidine pairs has dramatically increased the number of potential nucleobase analogs for expansion of the genetic alphabet.

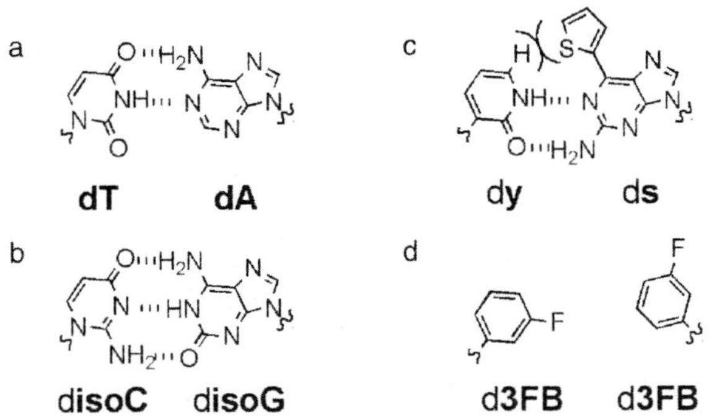

Fig. 1 Representative natural and unnatural base pairs. **a** T:A, a natural base pair, uses hydrogen bonding to drive selective pairing. **b isoC:isoG** uses nonnatural hydrogen-bonding patterns. **c y:s** uses hydrophobic packing in the major groove. **d 3FB:3FB** uses predominantly hydrophobic forces

2.1 *Alternative Hydrogen-Bonding Nucleobase Pairs*

Selective hydrogen bonding drives natural DNA synthesis; however, the natural pairs represent only two of the possible 16 purine:pyrimidine pairs which can form on the basis of three complementary hydrogen bonds. By designing pyrimidine and purine nucleobases that pair via hydrogen-bonding schemes which are orthogonal to the natural nucleobases, Benner and coworkers were the first to design and characterize unnatural base pairs (Switzer et al. 1989; Piccirilli et al. 1990). Alternative hydrogen-bonding pairs are generally thermally stable and selective against natural mispairs in duplex DNA (Geyer et al. 2003). However, these pairs are limited by DNA-polymerase-mediated DNA synthesis. It should be noted that these pairs were evaluated solely on the basis of the ability of DNA polymerases to synthesize full-length product containing the unnatural pair. Thus, the qualitative nature of these experiments makes comparison somewhat difficult, especially with later candidate base pairs, which are typically evaluated using steady-state kinetic methods. The alternative hydrogen-bonding design strategy has produced two interesting candidate base pairs, which are discussed further next.

2.1.1 disoC:disoG

The **disoC:ddisoG** pair (Fig. 1b) was the first unnatural nucleobase pair to be synthesized and characterized (Switzer et al. 1989; Switzer et al. 1993). It is more stable than, and as thermally selective as, a dA:dT pair in the same context (Geyer et al. 2003). Kf, with **disoC** in the template, inserts **disoG**TP preferentially over the natural nucleotides; however, complete full-length synthesis requires a small amount of dATP (Switzer et al. 1989, 1993). This was attributed to the deamination of **disoC**, possibly during oligonucleotide synthesis, which results in dU. Thus, the chemical instability of **disoC**, and resultant mispairing with dA, largely limits this sequence context. With **disoG** in the template, **disoC**TP is inserted, although less efficiently than dTTP. The facile insertion of dTTP is presumably due to the phenolic tautomer of **disoG**, which also results in efficient insertion of **disoG**TP against dT (Switzer

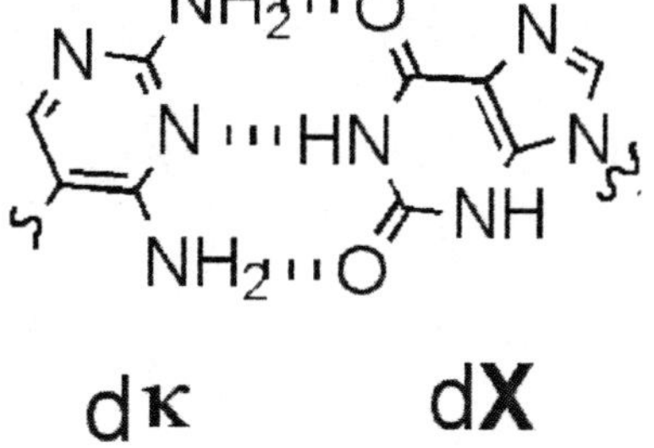

Fig. 2 dκ:dX

et al. 1993). While the d**isoC**:d**isoG** pair has been used in some biotechnological applications (Elbeik et al. 2004a, b), its utility is generally limited by poor fidelity during enzyme-mediated synthesis, especially during d**isoC**TP incorporation.

2.1.2 dκ:dX

The d**κ**:d**X** pair (Fig. 2) is also thermostable and selective in DNA, although not as stable or as selective as the d**isoC**:d**isoG** pair (Geyer et al. 2003). Considering that the riboside of d**X** has a pK_a of 5.7, and duplex DNA containing the pair is more stable at pH 5.4 than at pH 7.9, the relative destabilization of the d**κ**:d**X** pair appears to be caused by the anionic form of d**X** (Geyer et al. 2003). Despite the destabilization, this pair is thermally stable and selective in DNA; however, like d**isoC**:d**isoG**, the d**κ**:d**X** pair is predominantly limited by polymerase-mediated replication (Piccirilli et al. 1990). With d**κ** in the template, d**X**TP is inserted and extended relatively efficiently and selectively; full-length product was only observed in the presence of d**X**TP. Unfortunately, Kf does not incorporate d**κ**TP against d**X** (Horlacher et al. 1995), limiting the practical use of the base pair, and making the d**isoC**:d**isoG** pair the most viable of the alternative hydrogen-bonding nucleobase pairs.

2.2 *Shape Complementary Nucleobase Pairs*

The remarkable discovery that hydrogen bonds are not necessary for efficient DNA synthesis (Moran et al. 1997) spurred a new era of nucleobase design. Yokoyama, Hirao, and coworkers have designed base pairs using natural scaffolds which are modified to occlude mispairing with the natural nucleobases via steric interactions rather than noncomplementary hydrogen bonding. By the addition of a bulky substituent to the purine nucleobase on the major groove of the Watson–Crick face, mispairs formed between the unnatural purine and dC or dT are minimized. In a complementary fashion, the pyrimidine-like nucleobase has no functional group in the major groove, and is more shape-complementary to the unnatural purine nucleobase than to the natural purines. In a second approach, hydrophobic isosteres were derivatized to optimize their polymerase-mediated synthesis. These strategies have yielded several interesting base pair candidates, and the progression of the strategy is detailed next.

2.2.1 dx:dy and Derivatives

The first attempt at using a bulky major-groove substituent to occlude mispairs through sterics was d**x**:d**y** (Fig. 3a); d**x** features a dimethyl amine oriented in the major groove, while d**y** is a modified pyrimidine with no functional group oriented into the major groove (Ishikawa et al. 2000). With d**x** in the template, d**y**TP was inserted by

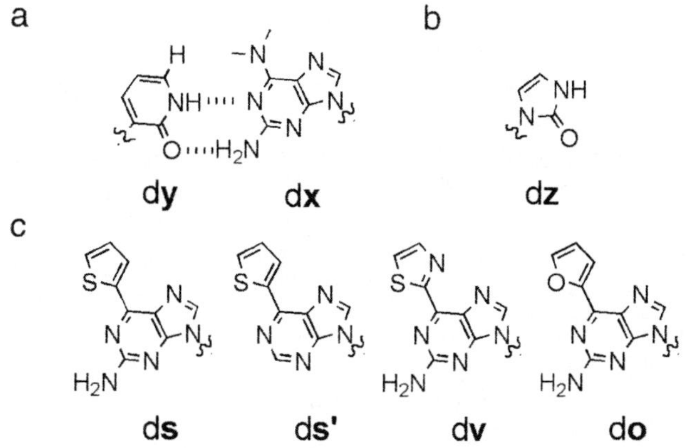

Fig. 3 Shape complementary nucleobase pairs with hydrogen bonding. (**a**) The parent pair, d**y**:d**x**. (**b**) d**y** derivative. (**c**) d**x** derivatives

Kf exo$^+$ with moderate efficiency and some selectivity. Later studies showed that the exonuclease domain of Kf was largely responsible for the fidelity seen in these experiments, as steady-state kinetics experiments revealed that Kf incorporates dTTP twofold more efficiently than d**x** (Fujiwara et al. 2001). The characterization of the reverse context, with d**x**TP incorporated against d**y**, has not been reported.

Yokoyama, Hirao, and coworkers obtained their first selective nucleobase pair by increasing the size of the purine major-groove substituent, to a five-membered ring, resulting in d**s** (Fig. 3c) (Fujiwara et al. 2001). This reduces dTTP incorporation 2.5-fold, and simultaneously increases the efficiency of d**y**TP incorporation threefold, presumably through hydrophobic packing interactions. Unfortunately, d**s** templates dCTP incorporation approximately sevenfold more efficiently than d**x** templates dCTP incorporation, perhaps owing to either weak hydrogen bonding or reduced desolvation of the amine. Thus, dCTP is the most efficiently inserted dNTP opposite d**s**, at a rate approximately twofold below that of d**y**TP. In the reverse context, with d**y** in the template, d**s**TP was inserted with moderate efficiency and sevenfold selectivity over the most efficiently inserted natural triphosphate, dATP, making it the more efficient and selective context (Hirao et al. 2004b).

The encouraging characteristics of the d**s**:d**y** pair led to later studies where the five-membered ring, oriented into the major groove, was derivatized with heteroatoms, creating d**o** (Hirao et al. 2004a) and d**v** (Mitsui et al. 2005) (Fig. 3c). The effects of these substitutions on polymerase-mediated DNA synthesis were typically minor, and resulted in less than fourfold increases in the second-order rate constant of unnatural base pair synthesis. Notably, these changes did not occur selectively, as the derivatizations also caused increased rates of mispair formation, resulting in only small gains in fidelity. The **s** nucleobase was also modified by removing the minor-groove amine, resulting in d**s**′ (Fig. 3c), which had a more pronounced effect on polymerase-mediated synthesis (Hirao et al. 2004a). While the rate of d**y**TP incorporation decreased threefold, misincorporation of dCTP, the most efficiently incorporated natural mispair with **ds**, decreased 20-fold. This large effect indicates the dCTP mispair requires the minor-groove hydrogen-bond donor, while the unnatural pair does not, demonstrating that hydrophobic interactions contribute substantially to the synthesis of the **ds:dy** pair. Changing the pyridone nucleobase, **dy**, to an imidazolinone yielded **dz** (Fig. 3b), which had a more pronounced and selective effect (Hirao et al. 2004b). The rate of misincorporation of dATP against **dz** decreased tenfold relative to the misincorporation of dATP against **dy**, while only having a minor effect on **ds**TP incorporation, giving a substantial increase in fidelity. Curiously, despite the reduction in aromatic surface area and the change in shape, **dz**TP was inserted at approximately the same rate as **dy**TP against **ds**.

Unfortunately, in all cases, incorporation of the pyrimidine analog triphosphate against the template purine analog is less efficient and less selective than the reverse context, and consistently limits the utility of these unnatural base pairs during DNA replication. Despite this slow step, the moderately efficient and selective synthesis of these pairs shows promise, and they are interesting scaffolds for further derivatization. Considering that these base pairs are also generally incorporated into RNA by T7 RNA polymerase (Kawai et al. 2005; Moriyama et al. 2005), such derivatives could prove to be strong candidates for expansion of the genetic alphabet.

2.2.2 dQ:dPa

The polymerase-mediated synthesis of base pairs formed between the hydrophobic isosteres of dA and dT, **dQ** and **dF** (Fig. 4a), respectively, is moderately efficient (Morales and Kool 1998, 1999), and **dQ**:**dF** is stable in duplex DNA (Mitsui et al. 2003b). While these nucleobases make interesting probes of polymerase activity, they make poor base pairs since **dF** templates dATP incorporation more efficiently than **dQ**TP incorporation, and extension of the **dF** primer is inefficient (Morales and Kool 1999). To circumvent both of these deficiencies, Yokoyama, Hirao, and coworkers designed **dPa** as a pairing partner for **dQ** (Fig. 4b) (Mitsui et al. 2003b). Relative to dF, **dPa** templates **dQ**TP more selectively and, unlike **dF**, does not inhibit full-length synthesis. In the reverse context, **dPa**TP is incorporated somewhat less efficiently, but also with tenfold selectivity over natural mispairs. Kf was able to synthesize full-length DNA containing the unnatural pair, confirming the selectivity and efficiency found using detailed steady-state kinetics. Propynyl derivatization of **dPa** had only nonselective effects (Mitsui et al. 2003a). Although the incorporation of **dPa**TP is somewhat inefficient, the relatively high selectivity with which the

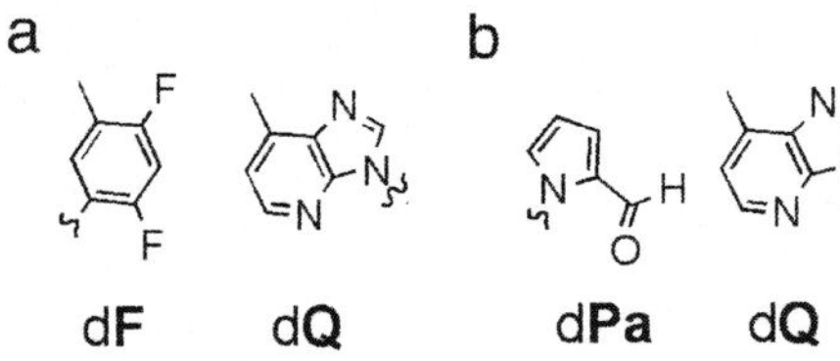

Fig. 4 Shape complementary nucleobase pairs with no hydrogen bonding. (**a**) **dF:dQ** pair. (**b**) **dPa:dQ** pair

pair is synthesized makes the **dPa**:d**Q** pair an interesting target for further derivatization, and also makes it a strong candidate for expansion of the genetic alphabet.

2.3 Predominantly Hydrophobic Nucleobase Pairs

The discovery that hydrogen bonds are not necessary for DNA-polymerase-mediated synthesis suggested that hydrophobic forces may be used to guide polymerase-mediated DNA synthesis. Rather than modify natural analogs, like Yokoyama, Hirao, and coworkers did, we have designed nucleobases which have little to no similarity to their natural counterparts and typically little to no traditional shape complementarity. We have systematically examined nucleobases of various shapes, sizes, and functional groups, and found that hydrophobic forces are capable of synthesizing unnatural base pairs at rates similar to those for natural base pairs. It should be noted that these efforts were generally focused on self-pairs, which are base pairs formed between two of the same nucleobase. Self-pairs do not practically limit the additional coding capacity of a third base pair; even the addition of a single self-pair would yield an additional 61 codons, practically doubling the coding capacity of natural DNA. More importantly, the use of a self-pair is advantageous as it limits complications due to mispairing. Some of the successful self-pairs are highlighted next. However, while not discussed here, more recent efforts have also identified promising heteropairs.

2.3.1 ICS and Derivatives

One of the earliest self-pairs we developed was that formed by d**PICS** (Fig. 5), which was the first unnatural base pair shown to be synthesized by Kf, based only on hydrophobic interactions (McMinn et al. 1999). The self-pair is as stable and nearly as selective as a natural dG:dC pair in duplex DNA. More impressively, despite the fact that it bears little similarity to a natural nucleobase, Kf synthesizes the d**PICS** self-pair with a second-order rate constant of 2.4×10^5 M^{-1} min^{-1}, which is only approximately 200-fold less than the rate of natural synthesis in the same sequence context. Importantly, the self-pair synthesis is relatively selective, as the rates of misincorporation of natural nucleoside triphosphates by Kf are all

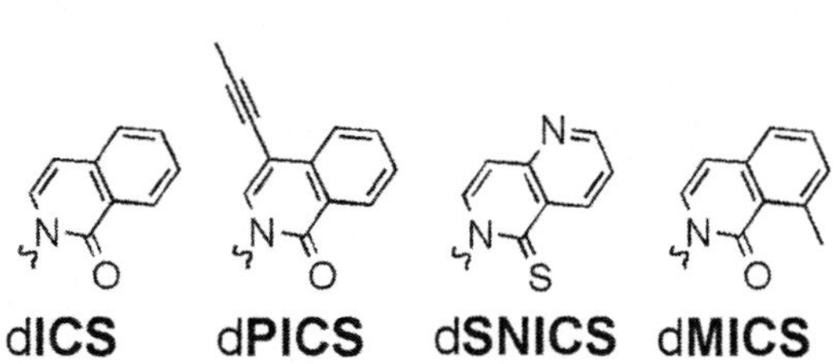

Fig. 5 d**ICS** and derivatives

less than $1.2 \times 10^4\ M^{-1}\ min^{-1}$. Unfortunately, extension, i.e., synthesis immediately following the unnatural pair, is inefficient, and limits the utility of this unnatural base pair.

To improve extension, later studies focused on derivatizing the parent **dICS** scaffold (Fig. 5) (Ogawa et al. 2000b) with methyl groups and heteroatoms as well as examining the effect of 2-position thio substitution (Wu et al. 2000; Yu et al. 2002). The **dSNICS** self-pair (Fig. 5) was the most promising of these candidates and is synthesized at a rate of $1 \times 10^6\ M^{-1}\ min^{-1}$, which is 15-fold more efficient than the most efficiently synthesized mispair, formed with dA, and 50-fold more efficient than the synthesis of the self-pair of the parent scaffold, **dICS**. While the rate of extension was 12-fold greater than the rate of **dICS** self-pair extension, the rate is still several orders of magnitude less than the rate of natural synthesis. Further improvements in the rate of extension may actually be limited by the nature of the scaffold itself.

2.3.2 d7AI and Derivatives

Another promising scaffold which is synthesized with moderate efficiency and selectivity based predominantly on hydrophobic interactions is **d7AI** (Fig. 6) (Ogawa et al. 2000b). The **d7AI** self-pair is synthesized at a rate of $2 \times 10^5\ M^{-1}\ min^{-1}$, which is 37-fold faster than the synthesis of the most efficiently formed mispair, **dA:d7AI**. Derivatization of the scaffold with heteroatoms, methyl groups, or propynyl groups (Wu et al. 2000) resulted in some improvements; for example, the **dP7AI** self-pair (Fig. 6) is synthesized with a rate of $1 \times 10^6\ M^{-1}\ min^{-1}$, fivefold faster than the parent scaffold. However, while these derivatizations increased the rate of self-pair synthesis, they did so nonselectively, as they also increased the rate of mispair synthesis. In fact, these derivatizations typically decreased the overall selectivity of the self-pair synthesis. While later studies successfully used a second DNA polymerase to extend the **d7AI** pair (see Sect. 3.3), Kf-mediated extension is highly inefficient and, as observed with the **dICS** scaffold, currently limits these base pairs.

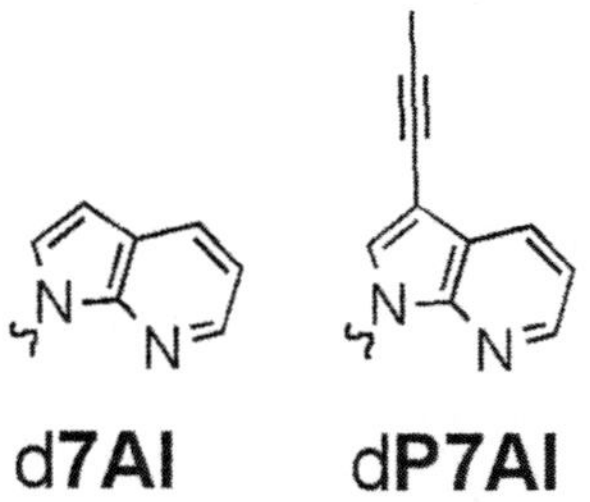

Fig. 6 d7**AI** and derivatives

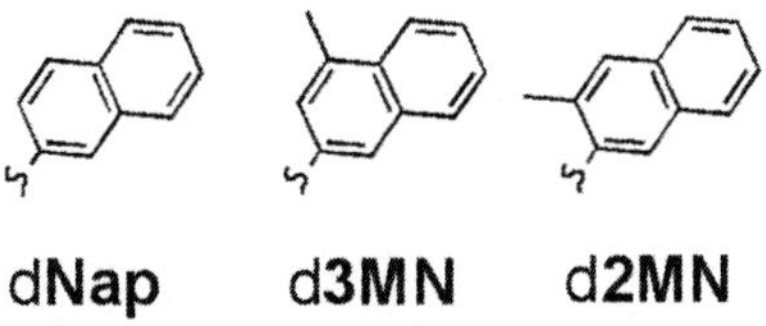

Fig. 7 dNap and derivatives

2.3.3 dNap and Derivatives

Another class of unnatural base pairs which bear large, hydrophobic nucleobases is based on the naphthyl scaffold (Ogawa et al. 2000a). These self-pairs are among the most efficiently synthesized unnatural base pairs reported to date, despite their lack of hydrogen-bonding functionalities and their nonnatural shape. The self-pair formed by the parent scaffold, d**Nap** (Fig. 7), is synthesized with a rate constant of $3 \times 10^6\ M^{-1}\ min^{-1}$, placing it among the most efficiently synthesized unnatural base pairs. Notably, methyl substitution has a significant effect on base pair synthesis. Substitution at the 2-position (d**2MN**, Fig. 7) resulted in a self-pair which is synthesized with a second-order rate constant of $4.4 \times 10^7\ M^{-1}\ min^{-1}$ (Ogawa et al. 2000b). This rate is approximately equal to the rate of natural synthesis in the same sequence context ($k_{cat}/K_M = 6.3 \times 10^7\ M^{-1}\ min^{-1}$). The most efficiently synthesized mispair is formed with dA and is synthesized approximately tenfold less efficiently that the d**2MN** self-pair. Methyl substitution at the 3-position (d**3MN**, Fig. 7) decreases the rate of dATP misincorporation, relative to d**2MN**, approximately 46-fold (Ogawa et al. 2000a). The d**3MN** self-pair is still efficiently synthesized and, considering the relatively high fidelity (62-fold selective over the fastest incorporated natural triphosphate, dTTP), it is an attractive candidate base pair. Unfortunately, as observed with other large scaffolds, synthesis beyond the primer terminus is inefficient.

2.3.4 dBEN and Derivatives

The aforementioned studies with large, aromatic nucleobases clearly demonstrate that the determinants of efficient synthesis and extension of unnatural pairs are independent. Since extension consistently limited the utility of large base pairs, we hypothesized that these pairs were interbase intercalating, resulting in geometric distortion of the unnatural primer terminus, which may prevent further synthesis (Brotschi et al. 2001; Reha et al. 2006). Geometric distortion has been suggested to be a significant cause of inefficient extension of natural mispairs (Johnson and Beese 2004; Meyer et al. 2004), and it is plausible that a mechanism which prevents the extension of purine mispairs would also prevent the extension of bicyclic hydrophobic analogs. To prevent this type of unfavorable interaction, we have synthesized and characterized self-pairs formed between smaller aromatic nucleobases, which are expected to interact in a more natural-like, edge-on

Fig. 8 Small, predominantly hydrophobic, self-pairing nucleobases; d**BEN** and derivatives

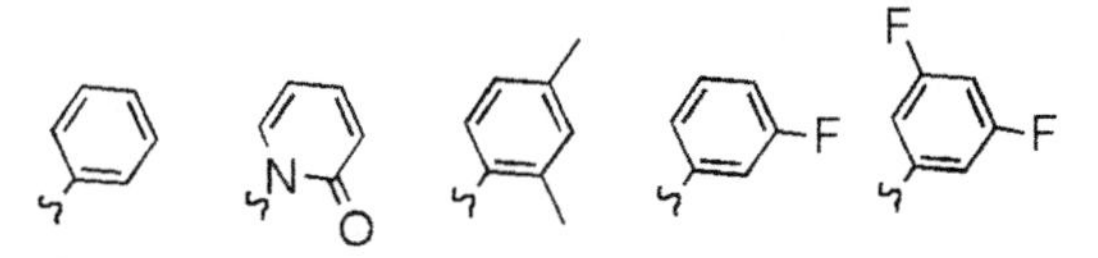

dBEN dPYR dDM5 d3FB d3,5DFB

manner. We have evaluated nucleotides bearing nucleobases consisting of simple phenyl rings derivatized with fluoro (Henry et al. 2004), methyl (Matsuda and Romesberg 2004; Matsuda et al. 2006), bromo (Hwang and Romesberg 2006), cyano (Hwang and Romesberg 2006), and pyridone (Leconte et al. 2006) functional groups. Each of these substitutions was made on a common scaffold in an attempt to deconvolute their individual effects. While the parent scaffold, d**BEN** (Fig. 8), is neither synthesized nor extended efficiently (Matsuda et al. 2006), large effects were observed with even simple modifications to the scaffold. For instance, pyridone substitution (d**PYR**, Fig. 8) increased the rate of extension approximately 50-fold when the carbonyl was directed into the developing minor groove (Leconte et al. 2006). Additionally, methyl substitution had large effects on stability, synthesis, and extension in a manner that was highly position dependent (Matsuda and Romesberg 2004, 2006). Notably, while it might appear that reduced aromatic surface area would lead to less stable base pairs, all of the self-pairs examined are stable in DNA, and several analogs, such as d**DM5** (Fig. 8), are virtually as stable and selective as a dA:dT pair in the same sequence context (Matsuda and Romesberg 2004).

The most successful of the small analogs is also one of the simplest. The **3FB** (Fig. 8) self-pair is synthesized at a rate of $2 \times 10^6\ M^{-1}\ min^{-1}$ (Henry et al. 2004). Importantly, the unnatural self-pair is extended relatively efficiently, at a rate of $3 \times 10^5\ M^{-1}\ min^{-1}$, which, at the time, was the fastest known extension of an unnatural terminus. Both the synthesis and the extension steps are selective over the most efficiently incorporated natural nucleotide, dATP, resulting in 20-fold selectivity over the two steps. Other analogs with 3-fluoro derivatization, such as **3,5DFB** (Fig. 8) are also efficiently synthesized and extended, although with diminished rates (Henry et al. 2004). Curiously, analogs with single bromo or cyano substitutions showed very different activities, both in terms of extension as well as in terms of mispair formation, even when substituted at the 3-position (Hwang and Romesberg 2006). The physical forces which drive the uniquely efficient synthesis and extension of the **3FB** pair are currently unclear. Regardless of its physical origins, the pair is one of the strongest base pair candidates to date and a promising target for further derivatization and optimization. This work illustrates that neither large aromatic surface area nor hydrogen bonds are necessary for efficient synthesis and extension.

3 Recognition of Unnatural DNA Base Pairs by DNA Polymerases: Alternatives to Kf

Although the majority of work on unnatural base pairs has focused on the development of the base pairs themselves, the DNA polymerase that replicates the unnatural base pair is equally important. Most studies, including all of those detailed so far herein, employ Kf to synthesize unnatural DNA. Kf is among the most thoroughly characterized DNA polymerases and is commercially available (Kuchta et al. 1987; Eger and Benkovic 1992; Beese et al. 1993); however, it is unclear whether other DNA polymerases are better suited to replicate unnatural DNA. Early work indicates that, at the very least, even fairly structurally similar polymerases have widely varying activities with unnatural base pairs.

3.1 Alternative DNA Polymerases

Benner and coworkers characterized the full-length synthesis of DNA containing the d**κ**:d**X** base pair (see Sect. 2.1) by several DNA polymerases. Since Kf was unable to incorporate d**κ**TP against d**X**, alternative polymerases were examined for their ability to synthesize full-length DNA in this context. Polymerases α and ε, from calf thymus, are able to incorporate d**κ**TP opposite d**X**, but neither is able to extend the unnatural pair (Horlacher et al. 1995). Interestingly, although polymerase β has been shown to efficiently synthesize a wide range of damaged templates (Kunkel et al. 1983; Huang et al. 1993; Matsuda et al. 2000) and would later be shown to be effective in a bipolymerase system using unnatural base pairs (see Sect. 3.3), it did not incorporate either unnatural or natural triphosphates against d**X** (Horlacher et al. 1995). Four thermostable DNA polymerases were also screened for their ability to synthesize DNA containing the unnatural pair; however, these reactions resulted in either inefficient synthesis or poor fidelity (Lutz et al. 1998).

The reverse transcriptase from HIV-1 (HIV-RT), which is the lowest fidelity polymerase of those tested, was the most adept at synthesizing the d**κ**:d**X** pair (Horlacher et al. 1995). With d**X** in the template and all four dNTPs, full-length DNA was the major product in the presence of high concentrations of d**κ**TP. The opposite context was also selective, as misincorporation of dNTPs against d**κ** was not observed. However, extension of the unnatural terminus appeared to be limiting, as the unextended unnatural pair was the major product. This is particularly interesting, considering that Kf, which is a higher-fidelity polymerase than HIV-RT, was shown to incorporate and extend d**X**TP against d**κ**, but not d**κ**TP against d**X**. In an analogous study, Benner and coworkers substituted the 7-deaza analog c7d**X** for d**X** and examined the recognition of the base by polymerase α, polymerase β, polymerase ε, HIV-RT, and Kf (Lutz et al. 1996). Interestingly, this small perturbation, far from the Watson–Crick face, had drastic effects on polymerase recognition. Unlike **dκ:dX**, the full-length product observed, regardless of the polymerase, was essentially the same in the presence or absence of the unnatural nucleoside triphosphates, indicating that any full-length product observed was a result of the misincorporation of a natural triphosphate.

3.2 Thermostable DNA Polymerases

The practical use of an unnatural base pair, either in vivo or in vitro, will likely require the most common of molecular biology techniques, the polymerase chain reaction (PCR). Unfortunately, the DNA polymerase most often used in kinetic studies of unnatural base pairs, Kf, is not thermostable, and is thus impractical for thermocycled PCR. While PCR amplification of DNA will likely gain more consideration once there is a truly viable unnatural pair, it is noteworthy that early efforts have met with some success. The **disoC:disoG** pair (see Sect. 2.1) was the first unnatural pair amplified in multiple rounds by PCR. However, the reaction was still not practically viable, as it utilized HIV-RT, which is not thermostable, and required long time periods for each round of amplification (Sismour et al. 2004). The **disoC:disoG** pair was later amplified in a PCR using a commercially available variant of the DNA polymerase I from *Thermus aquaticus* (Taq), and, while the reaction appeared to be reasonably efficient, the selectivity is too low for practical use (Johnson et al. 2004). Recently, Benner and coworkers replaced dT with the 2-thio analog of dT, which is less likely to mispair with the phenolic tautomer of **disoG**, in a PCR containing the **disoC:disoG** pair (Sismour and Benner 2005). With use of Taq polymerase, the most common DNA polymerase used in PCR, the approach was successful in improving the estimated fidelity of the PCR, from 93 to 98% fidelity per round. While all four natural nucleotides will be necessary for molecular biology applications, it is not an absolute requirement for nucleic acid biotechnology applications, such as DNAzymes and DNA aptamers, and the successful PCR amplification of this six-base system is likely to be of value for in vitro systems.

3.3 Bipolymerase Systems

As noted previously, large hydrophobic self-pairs are commonly limited by Kf-mediated extension of the unnatural terminus (see Sect. 2.3); however, other DNA polymerases may be more adept at extending these base pairs. For example, while Kf is unable to extend the **d7AI** self-pair (discussed in Sect. 2.3) (Ogawa et al. 2000b), mammalian polymerase β, which cannot synthesize the pair, can extend the **d7AI** pair as efficiently as it extends natural base pairs (Tae et al. 2001). Remarkably, when a Kf exo$^+$/polymerase β system was used, full-length DNA was synthesized with 400-fold selectivity relative to the most efficiently synthesized mispair formed with **d7AI**. Although the idea of a bipolymerase system may appear cumbersome, the idea is not without precedent. Most organisms, from bacteria to humans, use multiple polymerases to efficiently bypass DNA lesions with high efficiency and selectivity (Goodman 2002; Prakash et al. 2005).

3.4 Directed Evolution of DNA Polymerases

While natural polymerases may be able to utilize unnatural substrates with moderate efficiency and selectivity, practical use of unnatural base pairs may require optimization of the nucleobases as well as the polymerase. Directed evolution has the capacity to improve the ability of DNA polymerases to synthesize DNA containing unnatural base pairs. The directed evolution of DNA polymerases to study the determinants of natural DNA synthesis is well studied and has been reviewed recently (Loh and Loeb 2005); however, the directed evolution of DNA polymerases to utilize unnatural substrates is relatively new. Two effective selection systems have been developed and implemented, and both appear well suited to the expansion of the genetic alphabet (Ghadessy et al. 2001; Xia et al. 2002). Since these are both selections, rather than screens, they are able to efficiently sort through large libraries containing at least 10^7–10^8 polymerase mutants. The two selection systems, as well as their early successes, are discussed next.

3.5 In Vitro Compartmentalized Self-Replication

In the method of in vitro compartmentalized self-replication (CSR) (Fig. 9), cells overexpressing polymerase library members are added to an aqueous solution containing gene-specific primers, nucleoside triphosphates, metal cofactors, and RNase (Ghadessy et al. 2001). Since the reaction is in vitro, components are added exogenously, allowing selection to be based on a potentially wide variety of unnatural activities, including unnatural base pair synthesis. The aqueous mixture is added to a mineral oil based detergent solution to create segregated compartments, each of which contains one cell and, thus, one polymerase mutant. Upon heating, the cells lyse, releasing the encoded DNA polymerase mutant. Importantly, the compartments are stable even upon heating at 90 °C for prolonged time periods; thus, polymerases which are active under the variable conditions enrich themselves by synthesizing their own genes using the added primers, and do so in proportion to their activity. Following disruption of the droplets by a simple extraction procedure (Tawfik and Griffiths 1998), the pool of active genes can be collected from the aqueous layer.

As a proof of principle, Holliger and coworkers used in vitro CSR to identify Taq mutants which were either more thermostable or more resistant to the polymerase

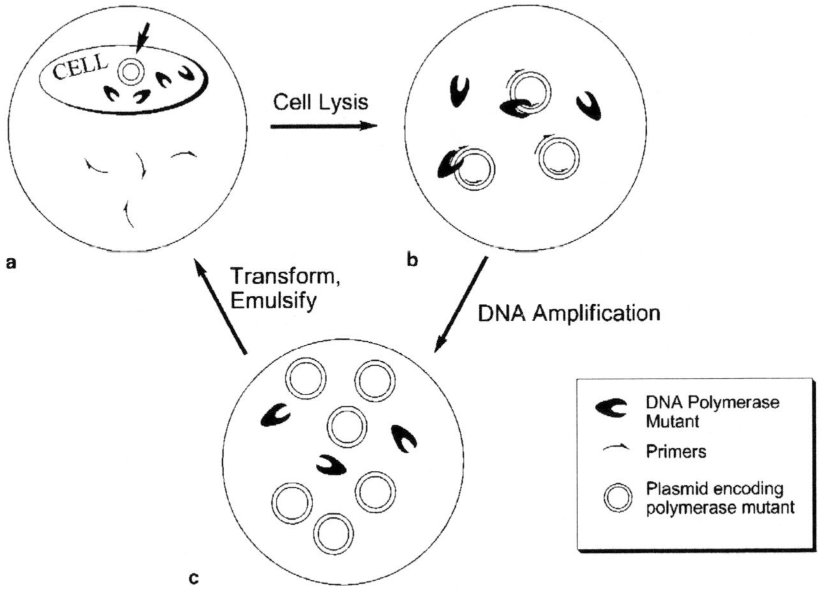

Fig. 9 In vitro compartmentalized self-replication. *a* cells are transformed with plasmids encoding a polymerase mutant (marked with an *arrow*) and are emulsified along with substrates for DNA synthesis. *b* cells are lysed without disrupting compartments, and polymerase mutants amplify their own genes using specific primers. *c* polymerase mutants that are active under given conditions will amplify their genes, which can be isolated and transformed into new cells

inhibitor heparin (Ghadessy et al. 2001). In a more recent study, in vitro CSR was used to increase the substrate specificities of Taq polymerase to include a wide range of modified natural triphosphates (Ghadessy et al. 2004). Using a library generated via error-prone PCR, Holliger and coworkers selected polymerases that are able to utilize primer termini which were mismatched at the 3′ end. Their goal was to relax the geometric requirements of extension to improve synthesis of DNA containing a broad spectrum of modified natural nucleobases. Polymerases which were isolated retained wild-type ability to synthesize natural DNA but gained the ability to extend some aberrant termini. In a steady-state single nucleotide incorporation assay, the most improved polymerase mutant, M1, is able to extend a dC:dC mispair more efficiently than wild-type Taq, by a factor of 427-fold. In PCR, M1 was able to incorporate a wide range of modified substrates, such as 5-nitroindol, fluorescently labeled dNTPs, biotin-labeled dNTPs, and phosphorothioates, none of which are efficiently synthesized by the wild-type polymerase.

3.5.1 Activity-Based Phage Display Selection

In the activity-based phage display selection system (Fig. 10), a mutant DNA polymerase and a DNA substrate are codisplayed on a single phage particle, and are selected on the basis of their ability to incorporate a biotinylated nucleoside

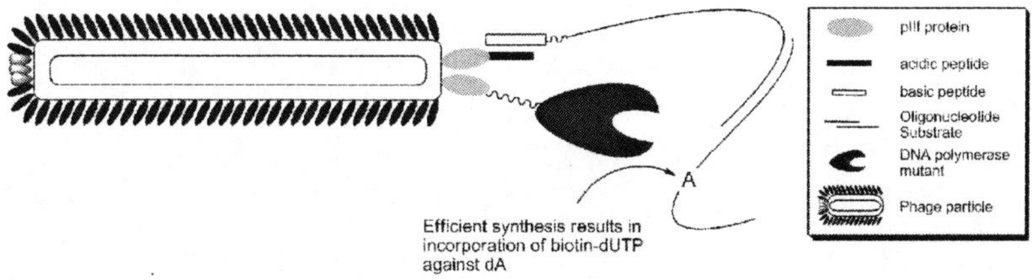

Fig. 10 Activity-based phage display selection system. A phage particle, containing a plasmid encoding a polymerase mutant, displays zero to one pIII-polymerase mutant fusion and four to five pIII-acidic peptide fusions, only one of which is shown. The acidic peptide is coupled to an oligonucleotide duplex through a basic peptide, which has been previously covalently linked to a primer template. The basic peptide forms a leucine zipper with the displayed acidic peptide and is further anchored with a disulfide bond. DNA-templated synthesis on the covalently attached primer requires incorporation of biotin–dUTP, labeling phage particles displaying active polymerases

triphosphate (Xia et al. 2002). *E. coli* is first transformed with a library of plasmids, each encoding a mutant polymerase as a fusion with the minor phage coat protein, pIII. The transformed *E. coli* cells are infected with X30 helper phage, which encodes a pIII-acidic peptide fusion. Each phage particle displays zero to one pIII-DNA polymerase fusions and four to five pIII-acidic peptide fusions, and also physically encapsulates the DNA polymerase encoding plasmid. An oligonucleotide, which will serve as the primer for the selection, is then linked to a basic peptide through a bismaleimide linker. After the basic peptide–primer conjugate has been annealed to a template oligonucleotide, the basic peptide–primer/template is selectively attached to the phage particles through a high-affinity leucine zipper formed between the acidic and basic peptides, which are then covalently linked through a disulfide bond. The phage particle, displaying both a DNA polymerase mutant as well as a DNA substrate, is resuspended in the reaction buffer, and challenged to synthesize DNA under user-controlled conditions. Successful DNA synthesis results in the incorporation of a biotin–dUTP into the covalently attached primer. Phage particles encoding the successful DNA polymerases can be enriched by recovery with streptavidin beads and released by DNase I cleavage.

As a proof of principle, the selection system was used to select polymerases that effectively synthesize RNA from a DNA template (Xia et al. 2002). The system has also been used to select DNA polymerases that synthesize DNA containing 2′OMe modified sugars (Fa et al. 2004). Recently, the selection system was used to select a variant of the Stoffel fragment from Taq (Sf) which better synthesizes DNA containing the d**PICS** self-pair (discussed in Sect. 2.3) (Leconte et al. 2005). While wild-type Sf is not able to extend the pair under even very forcing conditions, mutant P2 extends the d**PICS** self-pair at an improved, but still slow, rate. Although it was selected only for extension of the d**PICS** self-pair, P2 was also able to synthesize the d**PICS** self-pair with a second-order rate constant that is 320-fold higher than that of wild-type Sf, and approximately 200-fold more efficient than the rate at which P2 synthesizes the most efficient mispair, formed with dT. Thus, P2 synthesizes and extends the unnatural d**PICS** pair more efficiently than the wild-type enzyme, and, importantly, the improved activity does not come at the expense of the fidelity of natural synthesis. However, P2-mediated extension of the **dPICS** self-pair was not detectable under steady-state conditions, and thus, despite its significant improvement, is not yet viable for practical use.

4 Conclusion

The three approaches to base pair design, alternative hydrogen bonding, shape complementarity, and hydrophobic interactions, have already yielded promising third base pair candidates, which are both stable in duplex DNA and replicated with moderate efficiency and selectivity. Although these unnatural base pairs are currently not viable for either in vivo or in vitro use, they each provide promising scaffolds for further synthetic derivatization. The examination of alternative polymerases deserves further study, and may be enabled by the development of screening methods which have the potential to tap the plethora of different DNA polymerases available from nature. This technology could be especially valuable if interfaced with DNA polymerase evolution, which has proven to be capable of dramatically altering the activities of DNA polymerases. The advent of new strategies, such as development of screening and selection methods, and continued execution of multidisciplinary approaches, combining synthesis, enzymology, and directed evolution, should accelerate the search for an expanded genetic alphabet.

Acknowledgment This work was supported by the National Institutes of Health (GM060005).

References

Adelfinskaya O, Nashine VC, Bergstrom DE, Davisson VJ (2005) Efficient primer strand extension beyond oxadiazole carboxamide nucleobases. J Am Chem Soc 127:16000–16001

Astatke M, Ng K, Grindley NDF, Joyce CM (1998) A single side chain prevents Escherichia coli DNA polymerase I (Klenow fragment) from incorporating ribonucleotides. Proc Natl Acad Sci USA 95:3402–3407

Beese LS, Derbyshire V, Steitz TA (1993) Structure of DNA polymerase I Klenow fragment bound to duplex DNA. Science 260:352–355

Brotschi C, Haberli A, Leumann CJ (2001) A stable DNA duplex containing a non-hydrogen-bonding and non-shape-complementary base couple: interstrand stacking as the stability determining factor. Angew Chem Int Ed 40:3012–3014

Chiaramonte M, Moore CL, Kincaid K, Kuchta RD (2003) Facile polymerization of dNTPs bearing unnatural base analogues by DNA polymerase alpha and Klenow fragment (DNA polymerase I). Biochemistry 42:10472–10481

Clark SL, Remcho VT (2002) Aptamers as analytical reagents. Electrophoresis 23:1335–1340

Derbyshire V, Freemont PS, Sanderson MR, Beese L, Friedman JM, Joyce CM, Steitz TA (1988) Genetic and crystallographic studies of the 3′,5′-exonucleolytic site of DNA polymerase I. Science 240:199–201

Eger BT, Benkovic SJ (1992) Minimal kinetic mechanism for misincorporation by DNA polymerase I (Klenow fragment). Biochemistry 31:9227–9236

Elbeik T, Markowitz N, Nassos P, Kumar U, Beringer S, Haller B, Ng V (2004a) Simultaneous runs of the Bayer VERSANT HIV-1 version 3.0 and HCV bDNA version 3.0 quantitative assays on the system 340 platform provide reliable quantitation and improved work flow. J Clin Microbiol 42:3120–3127

Elbeik T, Surtihadi J, Destree M, Gorlin J, Holodniy M, Jortani SA, Kuramoto K, Ng V, Valdes R, Jr., Valsamakis A, Terrault NA (2004b) Multicenter evaluation of the performance characteristics of the Bayer VERSANT HCV RNA 3.0 assay (bDNA). J Clin Microbiol 42: 563–569

Fa M, Radeghieri A, Henry AA, Romesberg FE (2004) Expanding the substrate repertoire of a DNA polymerase by directed evolution. J Am Chem Soc 126:1748–1754

Fujiwara T, Kimoto M, Sugiyama H, Hirao I, Yokoyama S (2001) Synthesis of 6-(2-thienyl)purine nucleoside derivatives that form unnatural base pairs with pryidin-2-one nucleosides. Bioorg Med Chem Lett 11:2221–2223

Geyer CR, Battersby TR, Benner SA (2003) Nucleobase pairing in expanded Watson–Crick-like genetic information systems. Structure (Camb) 11:1485–1498

Ghadessy FJ, Ong JL, Holliger P (2001) Directed evolution of polymerase function by compartmentalized self-replication. Proc Natl Acad Sci USA 98:4552–4557

Ghadessy FJ, Ramsay N, Boudsocq F, Loakes D, Brown A, Iwai S, Vaisman A, Woodgate R, Holliger P (2004) Generic expansion of the substrate spectrum of a DNA polymerase by directed evolution. Nat Biotechnol 22:755–759

Goodman MF (2002) Error-prone repair DNA polymerases in prokaryotes and eukaryotes. Annu Rev Biochem 71:17–50

Henry AA, Olsen AG, Matsuda S, Yu C, Geierstanger BH, Romesberg FE (2004) Efforts to expand the genetic alphabet: identification of a replicable unnatural DNA self-pair. J Am Chem Soc 126:6923–6931

Hirao I, Fujiwara T, Kimoto M, Yokoyama S (2004a) Unnatural base pairs between 2- and 6-substituted purines and 2-oxo(1H)pyridine for expansion of the genetic alphabet. Bioorg Med Chem Lett 14: 4887–4890

Hirao I, Harada Y, Kimoto M, Mitsui T, Fujiwara T, Yokoyama S (2004b) A two-unnatural-base-pair system toward the expansion of the genetic code. J Am Chem Soc 126:13298–13305

Horlacher J, Hottiger M, Podust VN, Hübscher U, Benner SA (1995) Recognition by viral and cellular DNA polymerases of nucleosides bearing bases with nonstandard hydrogen bonding patterns. Proc Natl Acad Sci USA 92:6329–6333

Huang L, Turchi JJ, Wahl AF, Bambara RA (1993) Effects of the anticancer drug cis-diamminedichloroplatinum(II) on the activities of calf thymus DNA polymerase epsilon. Biochemistry 32:841–848

Hultschig C, Kreutzberger J, Seitz H, Konthur Z, Bussow K, Lehrach H (2006) Recent advances of protein microarrays. Curr Opin Chem Biol 10:4–10

Hwang GT, Romesberg FE (2006) Substituent effects on the pairing and polymerase recognition of simple unnatural base pairs. Nucleic Acids Res 34:2037–2045

Ishikawa M, Hirao I, Yokoyama S (2000) Synthesis of 3-(2-deoxy-b-D-ribofuanosyl)pyridin-2-one and 2-amino-6-(N,N-dimethylamino)-9-(2-deoxy-b-D-ribofuranosyl)purine derivatives for an unnatural base pair. Tetrahedron Lett 41:3931–3934

Jacobsen H, Klenow H, Overgaard-Hansen K (1974) The N-terminal amino-acid sequences of DNA polymerase I from *Escherichia coli* and of the large and the small fragments obtained by a limited proteolysis. Eur J Biochem 45:623–627

Johnson SJ, Beese LS (2004) Structures of mismatch replication errors observed in a DNA polymerase. Cell 116:803–816

Johnson SC, Sherrill CB, Marshall DJ, Moser MJ, Prudent JR (2004) A third base pair for the polymerase chain reaction: Inserting isoC and isoG. Nucleic Acids Res 32:1937–1941

Joyce GF (2004) Directed evolution of nucleic acid enzymes. Ann Rev Biochem 73:791–836

Kawai R, Kimoto M, Ikeda S, Mitsui T, Endo M, Yokoyama S, Hirao I (2005) Site-specific fluorescent labeling of RNA molecules by specific transcription using unnatural base pairs. J Am Chem Soc 127:17286–17295

Klussman S (ed) (2006) The aptamer handbook: functional oligonucleotides and their applications. Wiley-VCH, Weinheim

Kuchta RD, Mizrahi V, Benkovic PA, Johnson KA, Benkovic SJ (1987) Kinetic mechanism of DNA polymerase I (Klenow). Biochemistry 26:8410–8417

Kunkel TA, Schaaper RM, Loeb LA (1983) Depurination-induced infidelity of deoxyribonucleic acid synthesis with purified deoxyribonucleic acid replication proteins in vitro. Biochemistry 22:2378–2384

Leconte AM, Chen L, Romesberg FE (2005) Polymerase evolution: Efforts toward expansion of the genetic code. J Am Chem Soc 127:12470–12471

Leconte AM, Matsuda S, Hwang GT, Romesberg FE (2006) Efforts towards expansion of the genetic alphabet: pyridone and methyl pyridone nucleobases. Angew Chem Int Ed Engl 45:4326–4329

Lee JF, Stovall GM, Ellington AD (2006) Aptamer therapeutics advance. Curr Opin Chem Biol 10:282–289

Loh E, Loeb LA (2005) Mutability of DNA polymerase I: implications for the creation of mutant DNA polymerases. DNA Repair (Amst) 4:1390–1398

Lutz MJ, Held HA, Hottiger M, Hübscher U, Benner SA (1996) Differential discrimination of DNA polymerases for variants of the non-standard nucleobase pair between xanthosine and 2,4-diaminopyrimidine, two components of an expanded genetic alphabet. Nucleic Acids Res 24:1308–1313

Lutz MJ, Horlacher J, Benner SA (1998) Recognition of a non-standard base pair by thermostable DNA polymerases. Bioorg Med Chem Letts 8:1149–1152

Matsuda S, Romesberg FE (2004) Optimization of interstrand hydrophobic packing interactions within unnatural DNA base pairs. J Am Chem Soc 126:14419–14427

Matsuda T, Bebenek K, Masutani C, Hanaoka F, Kunkel TA (2000) Low fidelity DNA synthesis by human DNA polymerase-eta. Nature 404:1011–1013

Matsuda S, Henry AA, Romesberg FE (2006) Optimization of unnatural base pair packing for polymerase recognition. J Am Chem Soc 128:6369–6375

McMinn DL, Ogawa AK, Wu Y, Liu J, Schultz PG, Romesberg FE (1999) Efforts toward expansion of the genetic alphabet: Recognition of a highly stable, self-pairing hydrophobic base. J Am Chem Soc 121:11585–11586

Meyer AS, Blandino M, Spratt TE (2004) *Escherichia coli* DNA polymerase I. (Klenow fragment) uses a hydrogen-bonding fork from Arg668 to the primer terminus and incoming deoxynucleotide triphosphate to catalyze DNA replication. J Biol Chem 279:33043–33046

Mitsui T, Kimoto M, Sato A, Yokoyama S, Hirao I (2003a) An unnatural hydrophobic base, 4-propynylpyrrole-2-carbaldehyde, as an efficient pairing partner of 9-methylimidazo[(4,5)-b]pyridine. Bioorg Med Chem Lett 13:4515–4518

Mitsui T, Kitamura A, Kimoto M, To T, Sato A, Hirao I, Yokoyama S (2003b) An unnatural hydrophobic base pair with shape complementarity between pyrrole-2-carbaldehyde and 9-methylimidazo[(4,5)-b]pyridine. J Am Chem Soc 125:5298–5307

Mitsui T, Kimoto M, Harada Y, Yokoyama S, Hirao I (2005) An efficient unnatural base pair for a base-pair-expanded transcription system. J Am Chem Soc 127:8652–8658

Morales JC, Kool ET (1998) Efficient replication between non-hydrogen-bonded nucleoside shape analogs. Nat Struct Biol 5:950–954

Morales JC, Kool ET (1999) Minor groove interactions between polymerase and DNA: More essential to replication that Watson–Crick hydrogen bonds. J Am Chem Soc 121:2323–2324

Moran S, Ren RX-F, Rumney SI, Kool ET (1997) Difluorotoluene, a nonpolar isostere for thymine codes specifically and efficiently for adenine in DNA replication. J Am Chem Soc 119: 2056–2057

Moriyama K, Kimoto M, Mitsui T, Yokoyama S, Hirao I (2005) Site-specific biotinylation of RNA molecules by transcription using unnatural base pairs. Nucleic Acids Res 33:e129

Ng EW, Shima DT, Calias P, Cunningham ET, Jr., Guyer DR, Adamis AP (2006) Pegaptanib, a targeted anti-VEGF aptamer for ocular vascular disease. Nat Rev Drug Discov 5:123–132

Nimjee SM, Rusconi CP, Sullenger BA (2005) Aptamers: an emerging class of therapeutics. Annu Rev Med 56:555–583

Ogawa AK, Wu Y, Berger M, Schultz PG, Romesberg FE (2000a) Rational design of an unnatural base pair with increased kinetic selectivity. J Am Chem Soc 122:8803–8804

Ogawa AK, Wu Y, McMinn DL, Liu J, Schultz PG, Romesberg FE (2000b) Efforts toward the expansion of the genetic alphabet: Information storage and replication with unnatural hydrophobic base pairs. J Am Chem Soc 122:3274–3287

Pavlou AK, Reichert JM (2004) Recombinant protein therapeutics–success rates, market trends and values to 2010. Nat Biotechnol 22:1513–1519

Piccirilli JA, Krauch T, Moroney SE, Benner SA (1990) Enzymatic incorporation of a new base pair into DNA and RNA extends the genetic alphabet. Nature 343:33–37

Prakash S, Johnson RE, Prakash L (2005) Eukaryotic translesion synthesis DNA polymerases: specificity of structure and function. Annu Rev Biochem 74:317–353

Reha D, Hocek M, Hobza P (2006) Exceptional thermodynamic stability of DNA duplexes modified by nonpolar base analogues is due to increased stacking interactions and favorable solvation: correlated ab initio calculations and molecular dynamics simulations. Chemistry 12: 3587–3595

Sismour AM, Benner SA (2005) The use of thymidine analogs to improve the replication of an extra DNA base pair: a synthetic biological system. Nucleic Acids Res 33:5640–5646

Sismour AM, Lutz S, Park JH, Lutz MJ, Boyer PL, Hughes SH, Benner SA (2004) PCR amplification of DNA containing non-standard base pairs by variants of reverse transcriptase from Human Immunodeficiency Virus-1. Nucleic Acids Res 32:728–735

Spratt TE (2001) Identification of hydrogen bonds between *Escherichia coli* DNA polymerase I (Klenow fragment) and the minor groove of DNA by amino acid substitution of the polymerase and atomic substitution of the DNA. Biochemistry 40:2647–2652

Switzer C, Moroney SE, Benner SA (1989) Enzymatic incorporation of a new base pair into DNA and RNA. J Am Chem Soc 111:8322–8323

Switzer CY, Moroney SE, Benner SA (1993) Enzymatic recognition of the base pair between isocytidine and isoguanosine. Biochemistry 32:10489–10496

Tae EL, Wu YQ, Xia G, Schultz PG, Romesberg FE (2001) Efforts toward expansion of the genetic alphabet: replication of DNA with three base pairs. J Am Chem Soc 123:7439–7440

Tawfik DS, Griffiths AD (1998) Man-made cell-like compartments for molecular evolution. Nat Biotechnol 16:652–656

Tombelli S, Minunni M, Mascini M (2005) Analytical applications of aptamers. Biosens Bioelectron 20:2424–34

Watson JD, Crick FH (1953) Molecular structure of nucleic acids; a structure for deoxyribose nucleic acid. Nature 171:737–738

Wu Y, Ogawa AK, Berger M, McMinn DL, Schultz PG, Romesberg FE (2000) Efforts toward expansion of the genetic alphabet: optimization of interbase hydrophobic interactions. J Am Chem Soc 122:7621–7632

Xia G, Chen L, Sera T, Fa M, Schultz PG, Romesberg FE (2002) Directed evolution of novel polymerase activities: mutation of a DNA polymerase into an efficient RNA polymerase. Proc Natl Acad Sci USA 99:6597–6602

Yu C, Henry AA, Romesberg FE, Schultz PG (2002) Polymerase recognition of unnatural base pairs. Angew Chem Int Ed 41:3841–3844

Zhang X, Lee I, Berdis AJ (2005) The use of nonnatural nucleotides to probe the contributions of shape complementarity and pi-electron surface area during DNA polymerization. Biochemistry 44:13101–13110

Understanding Membrane Proteins. How to Design Inhibitors of Transmembrane Protein–Protein Interactions

J.S. Slusky, H. Yin, and W.F. DeGrado(✉)

Contents

Abstract Recent experiments demonstrated that buried membrane-protein hydrogen-bonding is less energetically favorable than the values that are discussed in section 3.3 of this chapter 1. Double mutant cycle analysis allows for the analysis of the

---

W.F. DeGrado
Department of Biochemistry and Biophysics, School of Medicine, University of Pennsylvania, Philadelphia, PA 19104, USA; Department of Chemistry, University of Pennsylvania, Philadelphia, PA 19104, USA, e-mail: wdegrado@mail.med.upenn.edu

C. Köhrer and U.L. RajBhandary (eds.) *Protein Engineering.*
Nucleic Acids and Molecular Biology 22,

interaction of two particular amino acids to the exclusion of their interaction with the rest of the protein. Such analysis of bacteriarhodopsin showed that amino acids in the correct orientation for hydrogen bonding contribute stabilization energies between −1.7 and +0.4 kcal mol^{-1} with an average over 8 pairs of −0.6 kcal mol^{-1}. This is approximately the same value as that of a hydrogen bond in a soluble protein, and is consistent with the notion that the core of membrane proteins and soluble proteins are biochemically similar. However, it should be noted that the burial of polar amino acids exerts a stablizing effect on membrane proteins beyond hydrogen bonding as well. Though hydrogen bonds may contribute less to the stability of a membrane protein than previously understood, this does not necessarily argue against the importance of the membrane protein hydrogen bond. The membrane protein's unfolded state is more similar-and therefore closer in energy-to its folded state than the unfolded soluble protein is to its folded state. Therefore, although the energy of the hydrogen bond is the same, the relative contribution of the hydrogen bond to the stability of the fold is greater in membrane proteins.

This same study also showed a disparity between hydrogen bonding at lipid-exposed positions and at positions in the core of the protein. Distances between atoms and the strength of the bond between those atoms are tightly correlated. Analysis of the distance of hydrogen bond donors and acceptors in membrane proteins and in soluble proteins revealed that hydrogen bonding in buried positions of membrane proteins have a similar distance distribution to that of solvent exposed positions on soluble proteins. However, lipid-exposed hydrogen bonds are shorter, and thus likely exert a greater stabilizing effect than buried hydrogen bonds in the membrane. This is consistent with the change in dielectric constant for the two environments.

Protein–protein interactions in the membrane are just beginning to be explored. Recently, significant advances have been made in disrupting protein–protein interactions in the membrane through protein design. These advances have allowed for the manipulation of biological processes in vivo, and have been shown to be useful probes for understanding the features that stabilize protein–protein interaction in the membrane. By bringing together information on how individual amino acids modulate transmembrane structure, what forces are responsible for oligomerization in the membrane, and how to computationally encode those concepts, a method has been established to create and disrupt protein–protein interactions in the membrane. This review aims to describe the necessity and utility of such probes, as well as provide a "how-to manual" for the design of such probes.

1 Introduction: The Significance of and Impediments to Designing Transmembrane Protein Inhibitors

The importance of creating probes to disrupt membrane protein interaction is due to the control of membrane proteins over a multitude of biological pathways, e.g., cell signaling, catalysis, and import/export. However, there are many difficulties inherent in studying membrane proteins. Nature's rule book for creating functional membrane protein structure is in many ways different from that of soluble protein structure.

High-resolution protein structure in the membrane has been historically more impenetrable owing to difficult expression and crystallization (von Heijne 1999).

While the driving force for the folding of soluble proteins is the hydrophobic effect, this force is absent for the regions of a protein embedded in the membrane. Additionally, the secondary structure is generally more rigid in a membrane environment because the lipids lack the ability to form hydrogen bonds with membrane protein backbones; thus, all hydrogen bonding must be satisfied internally. These factors result in differences in the distribution and function of amino acids utilized by proteins.

Although structural information is harder to come by, it is certainly the case that the great majority of transmembrane (TM) domains in cellular membranes adopt an α-helical conformation, including ion channels, transporters, and redox proteins. Understanding how these helices interact with each other and with the membrane is of critical importance. The differences between membrane proteins and soluble proteins are manifested in the conformations of these α-helices. The structural preferences of membrane helix–helix packing were first discussed by Bowie (1997) and were characterized as having less acute packing angles and to be more likely to be antiparallel than soluble helix pairs.

An individual TM helix can interact with another TM helix to initiate diverse cellular functions. For example, in phospholamban, a 52 amino acid integral membrane protein that regulates the Ca^{2+} pump in cardiac muscle cells, changes in TM α-helical packing are important in the process of transfer of Ca^{2+} ions across the membrane and affect the activity of the Ca^{2+}-ATPase (Lee 2002). Heterodimer forming helices of G-protein-coupled receptors (GPCRs) have been implicated in a number of pathways (Milligan 2004). The TM helix of spleen focus forming virus gp55-P envelope glycoprotein specifically binds the TM helix of the murine erythropoietin receptor (EpoR), triggering proliferation of erythroid progenitors and inducing erythroleukemia (Constantinescu et al. 1999). Interactions such as these are clearly important to probe, despite the technical difficulties of characterizing membrane proteins.

In this chapter we discuss methods of learning about such interhelical interactions by disrupting them. Many of the methods discussed involve the use of synthetic peptides. As membrane proteins are difficult to study owing to impediments in expression and crystallization, synthetic peptides can be a useful alternative to study membrane protein behavior. These peptides incorporate themselves into membrane mimetic environments such as nonpolar solvents, detergent micelles, and phospholipid bilayers and can readily adopt secondary structures (Zhang et al. 1992; Ren et al. 1997; Killian 1998). Peptides that correspond to TM regions of membrane proteins are much easier to synthesize and purify compared with their corresponding full-length proteins. Detection methods for this sort of peptide are also more straightforward than for the full-length protein. The ability to manipulate the peptide's sequence and add chemical labels allows for detection with fluorescence spectroscopy, electron spin resonance, and nuclear magnetic resonance (Zhang et al. 1992, 2001; Ren et al. 1997; Killian 1998; Lew et al. 2000). The relative ease of reconstituting synthetic peptides into membrane vesicles also allows for the investigation of lipid–protein interactions involving lipids that exist in very low concentrations in natural membranes or that form bilayers of a variety of hydrophobic thickness (Ren et al. 1997; Killian 1998).

Many aspects of membrane protein function have been elucidated through synthetic peptides. Synthetic inhibitors of TM interactions have been used to test the role of membrane protein interaction in signal transduction, assembly, and catalysis. A synthetic peptide of the TM domain integrin αIIb has been found to inhibit the interaction between integrin αIIb and integrin β3 in vitro and in vivo (Yin et al. 2006). Similarly, a nine-residue peptide representing the core sequence of T-cell receptor α (TCRα) interferes with the usual TCR–CD3 interaction both in vitro and in vivo (Manolios et al. 1997). A D-amino acid core sequence of TCRα was also found to interfere with TCR–CD3 binding (Quintana et al. 2007), as well as a part D-amino acid/part L-amino acid core sequence (Gerber et al. 2005). Additionally, a peptide corresponding to wild-type diacylglycerol kinase TM-2 inhibited the enzymatic activity of diacylglycerol kinase through formation of an inactive protein–peptide complex (Partridge et al. 2003).

More broadly, there are two ways to disrupt protein–protein interactions in the membrane. One is to add a natural binding partner to increase the likelihood of a peptide binding to that partner instead of to another membrane-located partner. This was illustrated in the previous paragraph. The second method of disrupting interaction is to design or select a nonnative inhibitor. Such inhibitors have the advantage of being able to have attributes that may not exist in natural partners, including differences in specificity and in affinity. This chapter will focus on the methods and considerations used to design such inhibitors.

Recently, it has been shown that peptides that bind with high specificity and affinity to TM regions of proteins can now be designed using de novo computational methods (Yin et al. 2007). This technique should have a wide array of applications both for diagnostics and for therapeutics. Although a variety of methods exist for the design or selection of antibodies and other proteins that recognize the water-soluble regions of proteins, companion methods for targeting TM regions are not as available. Thus, methods of membrane protein design may be of use for creating proteins with antibody-like utility.

2 Considerations for Peptide Design, Part I: How Amino Acids Modulate Transmembrane Structure

There are many factors that must be considered when designing a peptide that can inhibit protein–protein interactions in the TM: (1) the peptide must be capable of inserting itself into the membrane; (2) once it is inserted it must be thermodynamically stable into the membrane at a depth and orientation that would make it available for binding to its target; and (3) once it is at the proper depth and location, the designed peptide must prefer to bind to its target rather than to the lipid in which it is solvated. As will be discussed in the present section, the amino acids that comprise the peptide modulate all of these interactions. Each amino acid modulates structure and function by affecting the overall ability of the peptide to insert itself, the peptide's helical propensity, and the peptide's propensity to interact with different lipids or other peptides.

2.1 *Insertion*

Popot and Engelman (1990) have proposed a two-stage model to describe membrane protein folding – first insertion, then oligomerization. When designing a membrane peptide, one must consider its ability to insert itself from aqueous solutions into micelles, lipid bilayers, and preformed vesicles. Although natural proteins need to be able to insert themselves into the membrane through the translocon and need to be stable in the membrane, synthetic peptides have an additional requirement for function. In addition to being kinetically able to insert themselves into the membrane and to having thermodynamic stability in the membrane, synthetic peptides must also be water-soluble such that they do not aggregate prior to insertion (Fig. 1).

Appending solubility-enhancing groups to the C- and N-termini has been shown to facilitate this step (Harris et al. 2001; Gerber et al. 2004; Yan-Chun Tang 2004). The most commonly used solubility-enhancing groups include poly(ethylene glycol) and amino acid residues that carry charged side chains, such as Lys and Arg. The advantage of these modifications is threefold:

1. TM peptides usually have poor solubility in aqueous solution. With a solubility-enhancing auxiliary, it is possible to dissolve these peptides and study them in micelles or phospholipid systems.
2. The solubility-enhancing group at both ends of a TM peptide can also facilitate insertion.
3. Polar groups can prevent these hydrophobic peptides from aggregating.

Each amino acid has a different free energy of insertion into the membrane. There have been a number of methods of characterizing the contribution of each

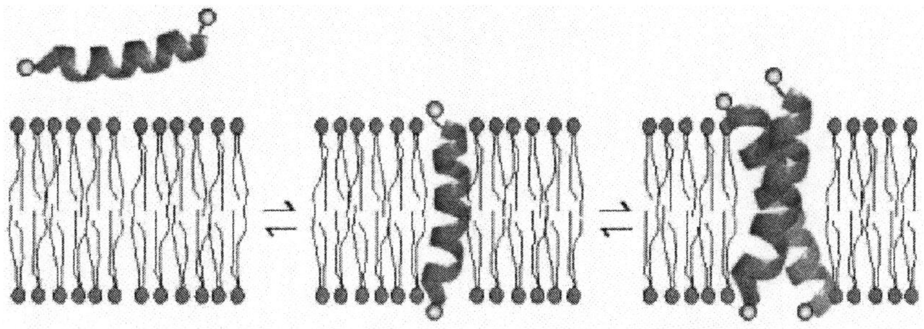

Fig. 1 A synthetic peptide must be soluble in water and capable of inserting itself in the membrane before it will be able to bind its target. A helical synthetic peptide with solubilizing groups affixed to its termini. *Left* free in solution, *center* inserted in the membrane, *right* oligomerizing

type of amino acid to the free energy of insertion of a protein. Calculating the free energy of partitioning from water to octanol has long been used as the best method for predicting the contribution towards the energy of insertion of an individual amino acid. This partitioning free energy was first calculated for the individual amino acids (Eisenberg and McLachlan 1986), and then for the amino acids when they were within the sequence AcWLXLL (where X is each of the 20 amino acids) (Wimley et al. 1996). The energies calculated by the latter method have recently been corrected slightly to lessen the originally assumed backbone contribution (Jayasinghe et al. 2001).

Recently, an in vivo method for understanding the free energy of insertion of amino acids was developed by the von Heijne group. In this method, peptides are biosynthetically incorporated into bilayers using the rough microsome membrane system, and the free-energy cost of inserting a TM helix into the phospholipid bilayer is calculated (Hessa et al. 2005). The design is as follows. The helix of interest is expressed between two glycosylation sites which can only be glycosylated if they are in the lumen of the endoplasmic reticulum. If the helix inserts itself into the lipid bilayer, one of the glycosylation sites will be in the cytoplasm and will therefore not be glycosylated. Measuring the fraction of singly glycosylated molecules to doubly glycosylated molecules allows for the calculation of the free energy of insertion. This method allowed for an assessment of the biological ΔG of insertion into the membrane for each amino acid.

Bioinformatics statistical potentials have also recently been used to predict the contribution of an amino acid to the protein's free energy of insertion. These methods were developed by inverting the Boltzmann equation to calculate a pseudo free energy from each amino acid's propensity to be located in the membrane as calculated from a database of high-resolution crystal structures (Ulmschneider et al. 2005; Senes et al. 2007). The three types of scales of characterizing the contribution of each amino acid to the free energy of insertion of a protein correlate fairly well.

2.2 *Insertion and Helical Propensity*

Li and Deber (1994) have studied the character of amino acids and have explored their helix propensity in the membrane as well as their ability to insert themselves into the membrane. In their study, they compared the propensity of different amino acids to form α-helices in aqueous solution, sodium dodecyl sulfate and lysophosphatidylglycerol micelles, and dimyristoylphosphatidylglycerol phospholipid vesicles. The peptides used in this study had a sequence of H_2N-SKSKAXAAXAWAXAKSKSKS-OH and the helical content was determined by circular dichroism measurements. The propensity of the X residue in promoting helicity followed the order Ile > Leu > Val > Met > Phe > Ala > Gln > Tyr > Thr > Ser > Asn > Gly > Pro. The potentially hydrophobic region (ten residues) of the peptide is too short to span the bilayer, so it is plausible that the "inserted" state measured in these studies corresponds to a helix that is embedded in the membrane, but with its axis parallel to the membrane surface rather than spanning the bilayer. Using such a short peptide shows the partitioning, and thus validates the expectation that a peptide that is inserted into an apolar environment will preferentially form a helix. This is important as it allows for the assumption that a peptide that inserts itself into the membrane will be α-helical, and leads to the conclusion that, when modeling such peptides, an unbound α-helix can be considered the unfolded state.

It is notable that β-branched residues (Val and Ile) that are known to promote β-sheet formation in water favor α-helices in the membrane. This observation is rationalized by the ability of these residues to destabilize helical conformation in water, retarding the premature folding of the TM domains in the aqueous compartment of the cell before insertion. However, when designing membrane helices it is important to avoid using multiple sequential β-branched amino acids, as such sequences have been shown to form amyloidogenic conformations (Lopez de la Paz and Serrano 2004).

2.3 Hydrophobic Mismatch: Characteristics That Control Tilt and Insertion Depth in the Membrane

A designed peptide should be thermodynamically stable at an appropriate depth and orientation within the membrane to facilitate binding to its target. The length and amino acid content of a peptide and the length and content of the lipids in the membrane with which the peptide is expected to interact can have significant effects on a peptide's preferred depth and tilt in a membrane. Caputo and London (2003a) have examined the interactions of amino acids with various types of lipid bilayers to determine features required for efficient insertion of synthetic peptides. Using lipid vesicles, they studied the effect of individual amino acids on peptide insertion into the membrane using a poly-Leu helix substituted with A, F, G, S, D, K, H, P, GG, SS, PG, PP, KK, or DD. A Trp was placed in the center of the peptide to allow measurement of insertion using fluorescence λ_{max} and dual quenching analysis (Caputo and London 2003b). All of the peptides with single substitutions efficiently inserted themselves into dioleoylphosphatidylcholine (DOPC) vesicles and adopted a TM state. In the thicker dierucoylphosphatidylcholine (DEuPC) vesicles, TM states were destabilized by a mismatch between helix length and bilayer thickness, especially for peptides with P, K, H, or D substitutions. In contrast to single substitutions, certain consecutive double substitutions strongly interfered with formation of TM states, regardless of hydrophobic mismatch. In both DOPC and DEuPC vesicles, DD and KK substitutions abolished the normal TM state, but GG and SS substitutions had little effect. A peptide with a PP substitution maintained the TM state in DOPC vesicles, but in DEuPC vesicles, the level of formation of the TM state was significantly reduced. Upon disruption of the normal TM insertion, peptides moved close to the bilayer surface, with the exception of the KK-substituted peptide in DOPC vesicles, which formed a truncated TM segment.

The effect of hydrophobic mismatch on the orientation of a peptide in the membrane has been explored with nuclear magnetic resonance. Peptides with hydrophobic stretches of 17 amino acids were shown to have an increase in tilt angle upon decreasing bilayer thickness. More tilt was seen for hydrophobic stretches flanked by Lys than for those flanked by Trp, likely owing to Trp's greater affinity for the interfacial region of the membrane and Lys's ability to snorkel (Ozdirekcan et al. 2005). In that study the maximum tilt observed was 12° off from perpendicular to the membrane. A much greater degree of tilt was seen by Park and Opella (2005). By using a TM helix of Vpu from HIV-1, with a hydrophobic stretch of 22 amino acids and sufficiently short lipid bilayers, they observed tilts of up to 51°.

2.4 Protein–Protein Versus Protein–Lipid Propensity

The design of peptides that interact with proteins hinges on the use of amino acids that will have a greater propensity to interact with other proteins than they will with the solvating lipid. It is widely known that the proportion of hydrophobic amino acids in membrane proteins is much higher than that of soluble proteins. This difference enables highly successful prediction of which parts of a protein will be in the TM region (Eisenberg et al. 1982; Rost et al. 1995, 1996; Jayasinghe et al. 2001) (Fig. 2). However, it is only in their surface residues that they differ most substantially – being very polar in water-soluble proteins and very apolar in membrane proteins. Despite this overall difference, membrane protein interiors have a similar distribution of polar and apolar amino acids to the interiors of soluble proteins (Rees et al. 1989). It should be noted that these interior positions are most often conserved (Liu et al. 2004), especially the hydrophilic residues (Freeman-Cook et al. 2004). The differences in amino acid composition between the lipid exposed and protein interacting faces of the TM helices also allow for the prediction of which residues participate in the protein–protein interface (Adamian and Liang 2001, 2005).

While the interiors of water-soluble and membrane proteins are similar, there are some essential differences. In membrane proteins, small amino acids (particularly Gly, Ala, and Ser) are most likely to participate in the binding face of a TM helix (Javadpour et al. 1999). This has been attributed to a number of factors. First, the smaller the residue, the greater the ability to create a cavity on the surface of the TM, which facilitates binding through shape complementarity, creating a larger van der Waals surface. In fact, amino acid volume correlates well with normalized occurrence at the helix interface (Fig. 3). Gly, as the smallest residue, is especially prevalent in mediating particular helix–helix interactions and may be involved in C_α-hydrogen bonding, discussed later in this chapter. Also, the docking of two helices results in an energetically unfavorable loss in entropy of the side chains. In water-soluble proteins, this energetic feature can be balanced by favorable hydrophobic interactions, which is not available in membrane proteins. Thus, helix–helix

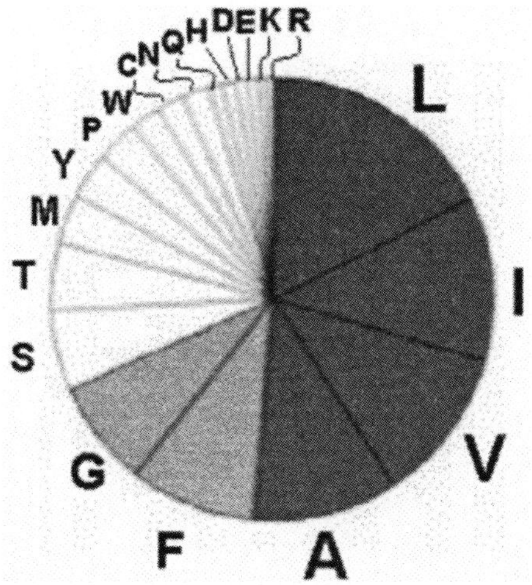

Fig. 2 Amino acid composition of membrane proteins as annotated in Swiss-Prot. *Shading* emphasizes that half of residues in the transmembrane are L, I, V, or A; two thirds are represented with the inclusion of F and G. (Adapted from Senes et al. 2000)

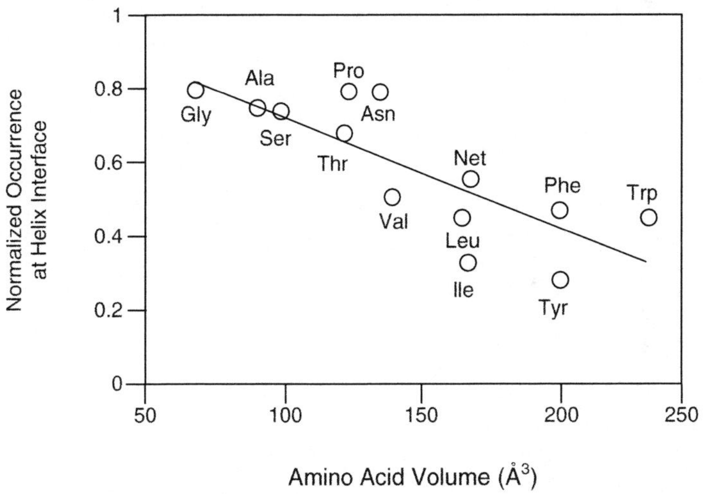

Fig. 3 Correlation between normalized occurrence of amino acids at the helix interface and amino acid volume – the smaller the volume of the amino acid, the more likely it is to occur at the helix interface in the membrane. (Adapted from Javadpour et al. 1999)

interfaces in membrane proteins are rich in side chains such as Gly and Ala or the β-branched amino acids – especially Val – which lack multiple side chain rotamers that need to be fixed when the helices coalesce. Small residues such as Ser and Thr might be preferred, both because of their size and because they can create hydrogen bonds with the neighboring helix.

2.5 *Sequence Motifs Predict Structure*

Small residues such as Gly have been found to interact with β-branched residues with an unusually high frequency (Senes et al. 2000), but there are more extreme examples of sequence conservation. Sequence motifs such as the GX3G motif (Russ and Engelman 2000) and GX3G-like motifs have been especially prevalent in the membrane and are known for modulating tight and specific interhelical interactions in the membrane. A more thorough review of these sequence motifs can be found elsewhere (Senes et al. 2004). The GX3G motif contains a Gly at the *i* and $i + 4$ positions within a helical peptide. It has been shown that similar patterns exist for smaller residues Ala and Ser replacing one or both of the Gly residues. It seems that the two small residues next to each other on the binding surface create a cavity, and that the use of this cavity for "knobs into holes" style packing is a strategy chosen by nature to provide a highly complementary binding site. GX3G motifs have been found in both homodimeric and heterodimeric helix bundles in membrane proteins. Examples of membrane protein TM regions that contain a GX3G-like motif at the helix-packing site include the epidermal growth factor receptor ErbB (Mendrola et al. 2002), the γ-secretase component APH-1 (Lee et al. 2004), F_0F_1-ATP synthase (Arselin et al. 2003), GPCRs (Overton et al. 2003), and integrins (Schneider and Engelman 2004).

Although it has long been understood that there are both particular sequence motifs and particular structure motifs in TM helices, connections between the two have only recently been pieced together. Walters and DeGrado (2006) showed that for all the high-resolution membrane protein structures, there are only a very limited number of structural motifs for helical dimers – with approximately three quarters of all dimer pairs falling into one of five structure motifs. More importantly, these structures have particular sequence motifs. For example, a GX3G motif generally corresponds to a helix pair that packs in a parallel, left-handed manner with an approximately 40° crossing angle. This connection between structure and sequence holds tremendous possibility for both structure prediction and protein design.

3 Considerations for Peptide Design, Part II: Forces Responsible for Oligomerization

In Sect. 2, we discussed the effects of sequence on creating structure and oligomerization. Although it is increasingly possible to predict structure from sequence, e.g., knowing that Gly prefers the interface between two TM helices, does not tell us explicitly about the physical forces that nature is using to create that interface. To fully understand how to design peptides that interact with membrane proteins, we need to know nature's energy function for oligomerization in the TM. Interhelix oligomerization in the membrane can only occur when a helix will have more stabilizing interactions with another membrane helix than it will with the membrane lipids. These interactions include contributions from enthalpic factors such as van der Waals forces, electrostatics, and hydrogen bonds; methods for characterizing these contributions will be described later. It has generally been more difficult to quantify the entropic factors involved in this process. In soluble proteins, the hydrophobic effect is dominated by entropic interactions, and in the design of soluble proteins much of the hydrophobic effect has been captured using knowledge-based contact potentials. Lipophobic effects can similarly be captured by membrane protein contact potentials, which will be discussed in Sect. 4.

3.1 van der Waals Packing

van der Waals contacts play an important role in oligomerization. Membrane proteins have been shown to have a higher packing value than soluble proteins, which corresponds to greater van der Waals packing (Eilers et al. 2000). This increase of protein interaction seems to confer a level of stability that compensates for the lack of hydrophobic effects. Mutational analysis of the helical homodimer glycophorin A TM has shown that although most Ala mutations on the interface of the dimer are destabilizing, many double Ala mutations stabilize the dimer (Doura and Fleming 2004). The existence of stabilizing double mutations that obliterate the GX3G motif demonstrates that changing multiple interactions allows for a shift in conformation such that a different van der Waals interaction drives oligomerization.

3.2 Electrostatics

Because there are so few charged residues in the membrane, there has been less biophysical characterization of membrane electrostatics than there has been of other interatomic forces in the membrane. Recent attempts to recoup native sequences when repacking membrane backbones have shown no significant increase in sequence recovery by adding an electrostatic term, though a 2% increase in sequence recovery was seen by adding a hydrogen-bonding term. This argues for the importance of modeling the orientational dependencies of such energies (Barth et al. 2007). Although charged residues are difficult to recover from backbones; when they do exist in the membrane they create salt bridges that can cause oligomerization. Notably, it has been shown that in the adaptive immune response, one NKG2D homodimer pairs with a single DAP10 dimer by formation of two Asp-Arg salt bridges in the TM, and that the mutation of the charged residues precludes oligomerization (Garrity et al. 2005).

3.3 Hydrogen Bonding

Hydrogen bonding is generally believed to be important in TM helix–helix interactions. How hydrogen bonds shape membrane protein structure has been discussed extensively elsewhere (White 2005). Studies have shown that the side chains of polar residues, such as Asn or Asp, provide a strong thermodynamic driving force for membrane helix–helix association – especially self-association (Choma et al. 2000; Gratkowski et al. 2001; Zhou et al. 2001). This propensity for polar amino acids to drive oligomerization has similarly been shown with molecular dynamics simulations (Ash et al. 2004).

The strength of hydrogen bonding varies at different depths within the membrane. This has been shown through both statistical and experimental studies. Hydrogen bonding of Asn residues was shown to stabilize a trimeric bundle by at least 2 kcal mol^{-1} per monomer when the polar residues were placed at the center of the membrane, but only by 0.1 kcal mol^{-1} when the polar residues were at the interfacial region (Lear et al. 2003). It has also been reported that replacing Ile for Asn in a dimer destabilizes the oligomerization by approximately 1.5 kcal mol^{-1} per monomer (Acharya et al. 2002). And most recently, an assay measuring the free energy of TM helix insertion though glycosylation showed that the formation of an Asn-Asn or Asp-Asp hydrogen bond stabilizes a peptide in the membrane by approximately 1 kcal mol^{-1} (Meindl-Beinker et al. 2006). These data are consistent with the understanding that as the polarity of the environment of the polar side chain increases the stability of the hydrogen bond decreases.

Hydrogen bonding of hydroxyl-containing amino acids has also been implicated in oligomer formation. In fact, by some calculations, Ser, Tyr, and Thr are, in that order, the most common residues participating in interhelical TM hydrogen bonds (Adamian and Liang 2002) (Fig. 4). These residues can bond with other polar side chains or oxygen from the backbone. Ser residues were used to hold together the earliest de novo designed TM helical bundles (Lear et al. 1988; Chung et al. 1992). Ser residues placed at the interface in every heptad of a coiled coil have been shown to be sufficient for stabilization of a TM dimer (North et al. 2006). And ToxCAT, which measures TM helix association in the *Escherichia coli* inner membrane, has revealed the prevalence of SXXSSXXT and SXXXSSXXT in oligomer formation through what seems like a cooperative network of interhelical hydrogen bonds (Dawson et al. 2002).

The C_α hydrogen of amino acids in general, and Gly in particular, frequently approaches acceptors with distances and geometries that are typical of hydrogen bonding. The weak hydrogen-donating capabilities of this carbon are strengthened by its neighboring amide groups that increase the acidity of the C_α atom. The existence of these bonds in membrane protein structures were originally proposed by the Smith laboratory (Javadpour et al. 1999), and since then many crystal structures have shown Gly residues in the correct orientation and distance for this bond to occur (Kleiger et al. 2001, 2002; Senes et al. 2001; Loll et al. 2003). The strength of this bond has recently been measured with IR spectroscopy and

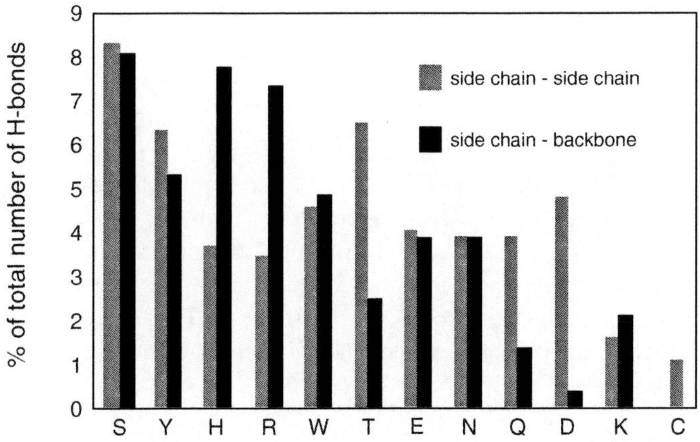

Fig. 4 Percentage of hydrogen bonds formed in the membrane by each type of side chain. *Gray bars* represent side chain–side chain hydrogen bonding, *black bars* represent side chain–backbone hydrogen bonding. (Adapted from Adamian and Liang 2002)

was found to have an energy of approximately 0.9 kcal mol^{-1} (Arbely and Arkin 2004). Molecular dynamics studies have estimated that that the energy of the C_α–H···O interaction varies widely depending on the angle of the C_α–H···O atoms, with some C_α-hydrogen bonds stabilizing structures by almost 1 kcal mol^{-1} and other interactions being dominated by the repulsion between the oxygen and a nearby backbone nitrogen and resulting in destabilizing energies (Mottamal and Lazaridis 2005). These calculations corroborate experimental results showing the stabilization of a protein via the removal of a possible C_α–H···O interaction (Yohannan et al. 2004).

Recent computational treatment of hydrogen bonding in the membrane has incorporated these data, modulating strength of the hydrogen bond based on membrane depth and atomic burial, and including a weaker hydrogen-bonding term for hydroxyl-containing amino acids and an even weaker term for C_α-hydrogen bonds (Barth et al. 2007).

4 Methods for Design: Computational Potential Energy Functions

To computationally design peptides, potential energy functions must be developed which reflect the forces proteins experience in the membrane. In light of all of the various different ways of viewing what causes oligomerization, a variety of methods have been used to model membrane proteins in silico. Computational modeling generally relies on the summation of various energy terms to calculate the stability of a given sequence in a particular conformation. Summations often take the form of an equation:

$$E_{\text{total}} = E_{\text{vdW}} + E_{\text{Hbond}} + E_{\text{electrostatic}} + E_{\text{solvation}} + E_{\text{self}} + E_{\text{contact}}.$$

The terms can include physical potentials and statistical potentials. Physical potentials such as van der Waals (E_{vdW}) interactions, hydrogen bonding (E_{Hbond}), and electrostatics ($E_{\text{electrostatic}}$) are identical in the membrane and in soluble solution. Therefore, equations for modeling these interactions can be found in papers on soluble protein modeling (Gordon et al. 1999) and will not be covered here. In the following sections we discuss potentials that describe interactions particular to membrane proteins ($E_{\text{solvation}}$, E_{self}, E_{contact}).

4.1 Solvation

The purpose of a *solvation potential* is to account for the effect of the solvent – in this case, the lipid – on the protein. The desolvation of water molecules from the protein during insertion into the TM is generally not calculated when calculating the stability of the protein, as the unfolded state is generally considered to be in the membrane. It has been shown experimentally that a protein's solvation by the membrane determines its depth within the membrane as well as its orientation with respect to the membrane (Ozdirekcan et al. 2005; Park and Opella 2005). A good solvation potential is one that is capable of predicting these effects.

The simplest method of calculating the effect of a lipid bilayer on the orientation of a protein is through statistical methods. Two similar residue-based statistical methods have been shown to be able to predict protein orientation in the bilayer (Ulmschneider et al. 2005, 2006; Senes et al. 2007). These methods use high-resolution crystal structures to calculate the propensity of each amino acid in any particular region of the membrane bilayer. The propensity is then converted to a pseudo energy using the Boltzmann equation. The two methods differ in that in the method of Ulmschneider et al. the cytoplasmic and extracellular orientations are treated independently, whereas in the method of Senes et al. the two orientations are treated symmetrically. Treating the membrane as asymmetric yields a truer fit for charged residues though at the cost of a reduced count size and therefore less statistical certainty. Another difference is the way in which the propensity is calculated. Nonetheless, the two methods arrive at similar scoring functions. Baker's group has used a similar more coarse-grained method, in which the membrane is divided into only four bins (representing inner hydrophobic, outer hydrophobic, interface, and polar membrane layers), in structure prediction in concert with a number of other potential energy functions (Yarov-Yarovoy et al. 2006). More recently, the same group has replaced this statistical function with a more physical-based function relying on implicit solvation energies, including implicit membrane model 1 (IMM1; described below) to model desolvation between buried and membrane-exposed protein positions and calculated transfer free energies of amino acid side chain analogs from vacuum to water and from vacuum to cyclohexane to model desolvation from water to the membrane. This change along with others allowed for more accurate structure prediction (Barth et al. 2007).

Implicit solvation is a much more computationally intensive method with which to calculate solvation effects on peptides. This method calculates generalized effects of the lipid on each atom of the protein. The two methods currently in use are the generalized Born method (Spassov et al. 2002; Im et al. 2003) and IMM1 (Lazaridis 2003). IMM1 is the less computationally intensive of the two, using a slightly less rigorous approximation for electrostatics, and a simpler *reference energy* (see Sect. 4.2). Both methods have been shown to successfully predict a peptide's orientation in the membrane (Im and Brooks 2005; Lazaridis 2005). Poisson–Boltzmann calculations are possible, and can yield information on position and orientation (Berneche et al. 1998; Kessel et al. 2000) but are also computationally expensive.

The most rigorous methods used to describe solvation within the membrane have explicit lipid and water molecules. These calculations can yield the most detailed information, but at a high computational cost. Simulations using molecular dynamics (Sansom et al. 1995; Shepherd et al. 2003) can only predict effects in the nanosecond time scale, and therefore using this method to determine conformation of a peptide within a bilayer will be considerably biased by the starting conditions of the calculation.

4.2 Self-Energy

A *self-energy*, also known as *reference energy*, is used to correct for the unfolded state. As discussed earlier, because of the stability of isolated helices populated by two-stage folding in membrane proteins, the unfolded state is usually approximated by a monomeric helix.

4.3 Contact Potentials

A *contact potential* is a statistical potential with energy based on the prevalence of the proximity of two bodies. The information contained in proximity includes hydrogen-bonding, electrostatics, hydrophobic, and solvophobic interactions. Such potentials have been developed for soluble proteins both for atom types (Mitchell et al. 1999; Lu and Skolnick 2001; McConkey et al. 2003; Summa et al. 2005) and for residue types (Miyazawa and Jernigan 1996; Zhang, Chen et al. 2006). Contact potentials have the benefit of being much less computationally intensive than physical potentials and can accurately predict soluble protein structure (Simons et al. 1997). So far, there has been very little use of contact potentials in the computational modeling of membrane proteins. A very coarse grained contact potential that gives contacting residues a score of −1, −1/2, −1/4, or 0, was shown to predict the native structure of the glycophorin A homodimer (Fleishman and Ben-Tal 2002). More recently, residue contact propensities were derived for membrane proteins on the basis of high-resolution crystal structures (Adamian and Liang 2001). A similar contact potential was recently used for structure prediction (Yarov-Yarovoy et al. 2006; Barth et al. 2007), but these potentials have yet to be used for protein design.

5 Putting It All Together: Membrane Structure Prediction and Membrane Protein Design

Using (1) the understandings of how amino acids modulate membrane protein interaction, (2) the knowledge of what forces are involved in membrane protein oligomerization, and (3) the computational tools that predict some of the more complex forces, one can carry out meaningful structure prediction and protein design. Protein structure prediction and protein design can be seen as two sides of the same coin. In structure prediction one has the sequence and desires to know what the lowest-energy conformation will be; with design one has the shape and desires a sequence that will adopt that shape with low energy. Both goals generally require conformational searches of both the side chains and the backbone. Often the backbone is designed or selected first and then – in the case of design – the side chain identity and conformations are determined. However, sometimes, the design or structure prediction is iterative, with backbone structure and side chain structure informing each other (Rohl et al. 2004). Because of these parallels, similar strategies and potential energy functions are utilized for both.

5.1 Structure Prediction

Many methods have been used to predict membrane protein structure. The TM dimer of glycophorin A can be predicted with an elegant energy function that simply sums two terms: a negative term for the contact of two residues known to promote helix packing and a positive term for the burial of large residues (Fleishman and Ben-Tal 2002). Combining a van der Waals energy function with a scoring function from mutational data has been shown to predict helical dimer structure (Metcalf et al. 2007). More complicated methods, including those that combine statistical energy functions and physical energy functions, have been used to predict more complicated membrane protein structure. A modification of the Rosetta algorithm (Rohl et al. 2004), which builds the conformation of a proteins using a fragment assembly method, was used to predict the structure of some portions of a wide range of proteins (Yarov-Yarovoy et al. 2006). The energy function used differs from the standard Rosetta energy function in the addition of a statistical environmental energy. This term is based on the propensity of a given amino acid being within a particular burial state and depth within the membrane. The energy function also includes a membrane-specific contact potential.

More recently, this function was modified by replacing the statistical solvation/environmental potential with an implicit solvation and environmental potential, and a more detailed hydrogen-bonding treatment. This led to structure prediction of near atomic accuracy for a number of helical proteins (Barth et al. 2007). A modified form of TASSER (Zhang and Skolnick 2004), a program that combines homology modeling with statistical and physical energy functions, has been used to model GPCRs (Zhang et al. 2006). These functions have yet to be used in design.

5.2 *Membrane Peptide Design*

Recently, TM helical peptides that can bind to targets with high specificity and affinity have been designed de novo (Yin et al. 2007). The method consists of four steps (Fig. 5): (1) selection of a helical-pair structural motif based on sequence; (2) selection of a nativelike helical-pair backbone within the chosen structural motif; (3) threading the sequence of the targeted TM helix onto one of the two helices of the selected pair; and (4) selection of the amino acid sequence of the designed peptide helix with a side chain repacking algorithm.

Because the natural backbones that were used were positioned such that they had nativelike interhelical hydrogen bonding, hydrogen bonding did not need to be encoded in the potential energy function. The potential that was used was very simple and consisted of only two terms: a van der Waals term and a statistical depth-dependent potential (Senes et al. 2007) in lieu of a solvation term. The ability

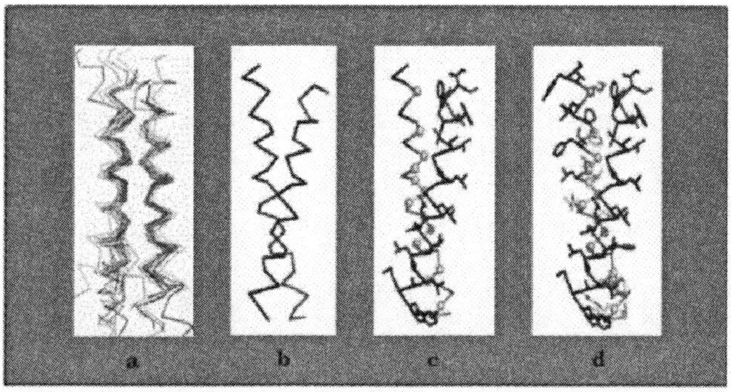

Fig. 5 Design process for peptides that bind to transmembrane proteins. **a** Superimposed helical-pair backbones of one structural motif. **b** Selected backbone. **c** Target structure is threaded onto one helix (front helix), positions for repacking are selected on the other helix. Repacked positions are highlighted in *light gray* with *spheres* on C_α. **d** The designed helix (farther back) is repacked to have good van der Waals contacts with the target helix. Repacked positions shown in *light gray* with *spheres* on C_α

of the designed peptides to discriminate between similar targets testifies to the importance of the shape complementarity in membrane protein oligomerization.

To illustrate the method, peptides were designed that specifically recognize the TM helices of two closely related integrins (αIIbβ3 and αvβ3). The affinity and specificity of this design was verified using a large number of experimental methods. Fluorescence resonance energy transfer assays and analytical ultracentrifugation in micelles demonstrated the affinities of the designed peptides for their targets. A dominant negative ToxCAT assay in *E. coli* membranes established that the designed peptides acted specifically on their targets to the exclusion of other homologous proteins. Platelet assays and force spectroscopy demonstrated that the designed peptides activated their targets in vivo. More work will be needed to verify that this method can be used for a wider variety of protein targets, but it seems likely that this method has many diagnostic and therapeutic applications.

6 Conclusion

The interactions between helices in the lipid bilayer are critical for many areas of cellular function. The ability to manipulate these interactions can only come through understanding how individual amino acids modulate TM structure, what forces are responsible for oligomerization in the membrane, and how to computationally encode those concepts. Recent design of TM peptides that bind to TM proteins with high specificity and affinity tests and extends our understanding of the principles of membrane protein structure and function. Furthermore, it suggests that sequence-specific recognition of helices in TM proteins can be achieved through optimization of the geometric complementarity of the target–host complex. Not only does this provide insight into protein–protein interactions in the membrane, it also provides a method to design antibody-like reagents that target a variety of membrane proteins. We believe that this method for targeting TM helices should have applications as widespread as antibodies themselves.

Acknowledgment We thank the National Institutes of Health (GM60610 and training grant 5-T32-GM08275) for support of this work.

References

Acharya A, Ruvinov SB, Gal J, Moll JR, Vinson C (2002) A heterodimerizing leucine zipper coiled coil system for examining the specificity of a position interactions: amino acids I, V, L, N, A, and K. Biochemistry 41:14122–14131

Adamian L, Liang J (2001) Helix–helix packing and interfacial pairwise interactions of residues in membrane proteins. J Mol Biol 311:891–907

Adamian L, Liang J (2002) Interhelical hydrogen bonds and spatial motifs in membrane proteins: polar clamps and serine zippers. Proteins Struct Funct Genet 47:209–218

Adamian L, Nanda V, DeGrado WF, Liang J (2005) Empirical lipid propensities of amino acid residues in multispan alpha helical membrane proteins. Proteins Struct Funct Bioinformat 59:496–509

Arbely E, Arkin IT (2004) Experimental measurement of the strength of a C_a–H···O bond in a lipid bilayer. J Am Chem Soc 126:5362–5363

Arselin G, Giraud MF, Dautant A, Vaillier J, Brethes D, Coulary-Salin B, Schaeffer J et al (2003) The GxxxG motif of the transmembrane domain of subunit e is involved in the dimerization/oligomerization of the yeast ATP synthase complex in the mitochondrial membrane. Eur J Biochem 270:1875–1884

Ash WL, Stockner T, MacCallum JL, Tieleman DP (2004) Computer modeling of polyleucine-based coiled coil dimers in a realistic membrane environment: insight into helix–helix interactions in membrane proteins. Biochemistry 43:9050–9060

Barth P, Schonbrun J, Baker D (2007) Toward high-resolution prediction and design of transmembrane helical protein structures. Proc Natl Acad Sci USA 104:15682–15687

Berneche S, Nina M, Roux B (1998) Molecular dynamics simulation of melittin in a dimyristoylphosphatidylcholine bilayer membrane. Biophys J 75:1603–1618

Bowie JU (1997) Helix packing in membrane proteins. J Mol Biol 272:780–789

Caputo GA, London E (2003a) Cumulative effects of amino acid substitutions and hydrophobic mismatch upon the transmembrane stability and conformation of hydrophobic alpha-helices. Biochemistry 42:3275–3285

Caputo GA, London E (2003b) Using a novel dual fluorescence quenching assay for measurement of tryptophan depth within lipid bilayers to determine hydrophobic alpha-helix locations within membranes. Biochemistry 42:3265–3274

Choma C, Gratkowski H, Lear JD, DeGrado WF (2000) Asparagine-mediated self-association of a model transmembrane helix. Nat Struct Biol 7:161–166

Chung LA, Lear JD, Degrado WF (1992) Fluorescence studies of the secondary structure and orientation of a model ion channel peptide in phospholipid-vesicles. Biochemistry 31: 6608–6616

Constantinescu SN, Liu XD, Beyer W, Fallon A, Shekar S, Henis YI, Smith SO et al (1999) Activation of the erythropoietin receptor by the gp55-P viral envelope protein is determined by a single amino acid in its transmembrane domain. EMBO J 18:3334–3347

Dawson JP, Weinger JS, Engelman DM (2002) Motifs of serine and threonine can drive association of transmembrane helices. J Mol Biol 316:799–805

Doura AK, Fleming KG (2004) Complex interactions at the helix–helix interface stabilize the glycophorin A transmembrane dimer. J Mol Biol 343:1487–1497

Eilers M, Shekar SC, Shieh T, Smith SO, Fleming PJ (2000) Internal packing of helical membrane proteins. Proc Natl Acad Sci USA 97:5796–5801

Eisenberg D, Weiss RM, Terwilliger TC (1982) The helical hydrophobic moment - A measure of the amphiphilicity of a helix. Nature 299(5881):371–374

Eisenberg D, McLachlan AD (1986) Solvation energy in protein folding and binding. Nature 319:199–203

Fleishman SJ, Ben-Tal N (2002) A novel scoring function for predicting the conformations of tightly packed pairs of transmembrane alpha-helices. J Mol Biol 321:363–378

Freeman-Cook LL, Dixon AM, Frank JB, Xia Y, Ely L, Gerstein M, Engelman DM et al (2004) Selection and characterization of small random transmembrane proteins that bind and activate the platelet-derived growth factor beta receptor. J Mol Biol 338:907–920

Garrity D, Call ME, Feng J, Wucherpfennig KW (2005) The activating NKG2D receptor assembles in the membrane with two signaling dimers into a hexameric structure. Proc Natl Acad Sci USA 102:7641–7646

Gerber D, Sal-Man N, Shai Y (2004) Two motifs within a transmembrane domain, one for homodimerization and the other for heterodimerization. J Biol Chem 279(20): 21177–21182

Gerber D, Quintana FJ, Bloch I, Cohen IR, Shai Y (2005) D-enantiomer peptide of the TCRα; transmembrane domain inhibits T-cell activation in vitro and in vivo. FASEB J 19:1190–1192

Gordon DB, Marshall SA, Mayo SL (1999) Energy functions for protein design. Curr Opin Struct Biol 9:509–513

Gratkowski H, Lear JD, DeGrado WF (2001) Polar side chains drive the association of model transmembrane peptides. Proc Natl Acad Sci USA 98:880–885

Harris JM, Martin NE, Modi M (2001) Pegylation - A novel process for modifying pharmacokinetics. Clinical Pharmacokinetics 40(7):539–551

Hessa T, Kim H, Bihlmaier K, Lundin C, Boekel J, Andersson H, Nilsson I et al (2005) Recognition of transmembrane helices by the endoplasmic reticulum translocon. Nature 433: 377–381

Im W, Brooks CL (2005) Interfacial folding and membrane insertion of designed peptides studied by molecular dynamics simulations. Proc Natl Acad Sci USA 102:6771–6776

Im W, Feig M, Brooks CL (2003) An implicit membrane generalized born theory for the study of structure, stability, and interactions of membrane proteins. Biophys J 85:2900–2918

Javadpour MM, Eilers M, Groesbeek M, Smith SO (1999) Helix packing in polytopic membrane proteins: role of glycine in transmembrane helix association. Biophys J 77:1609–1618

Jayasinghe S, Hristova K, White SH (2001) Energetics, stability, and prediction of transmembrane helices. J Mol Biol 312:927–934

Joh NH et al (2002) Modest stabilization by most hydrogen-bonded side-chain interactions in membrane proteins. Nature 453(7199):1266–1270

Kessel A, Cafiso DS, Ben-Tal N (2000) Continuum solvent model calculations of alamethicin-membrane interactions: thermodynamic aspects. Biophys J 78:571–583

Kleiger G, Perry J, Eisenberg D (2001) 3D structure and significance of the G Phi XXG helix packing motif in tetramers of the E1 beta subunit of pyruvate dehydrogenase from the archeon Pyrobaculum aerophilum. Biochemistry 40(48):14484–14492

Kleiger G et al (2002) GXXXG and AXXXA: Common alpha-helical interaction motifs in proteins, particularly in extremophiles. Biochemistry 41(19):5990–5997

Killian JA (1998) Hydrophobic mismatch between proteins and lipids in membranes. Biochim Biophys Acta 1376:401–415

Lazaridis T (2003) Effective energy function for proteins in lipid membranes. Proteins Struct Funct Genet 52:176–192

Lazaridis T (2005) Implicit solvent simulations of peptide interactions with anionic lipid membranes. Proteins Struct Funct Bioinformat 58:518–527

Lear JD, Wasserman ZR, Degrado WF (1988) Synthetic amphiphilic peptide models for protein ion channels. Science 240:1177–1181

Lear JD, Gratkowski H, Adamian L, Liang J, DeGrado WF (2003) Position-dependence of stabilizing polar interactions of asparagine in transmembrane helical bundles. Biochemistry 42: 6400–6407

Lee AG (2002) Ca2+-ATPase structure in the E1 and E2 conformations: mechanism, helix–helix and helix-lipid interactions. Biochim Biophys Acta-Biomembr 1565:246–266

Lee SF, Shah S, Yu C, Wigley WC, Li H, Lim M, Pedersen K et al (2004) A conserved GXXXG motif in APH-1 is critical for assembly and activity of the γ-secretase complex. J Biol Chem 279:4144–4152

Lew S, Ren J, London E (2000) The effects of polar and/or ionizable residues in the core and flanking regions of hydrophobic helices on transmembrane conformation and oligomerization. Biochemistry 39:9632–9640

Li SC, Deber CM (1994) A measure of helical propensity for amino acids in membrane environments. Nat Struct Biol 1:368–373

Liu W, Eilers M, Patel AB, Smith SO (2004) Helix packing moments reveal diversity and conservation in membrane protein structure. J Mol Biol 337:713–729

Loll B et al (2003) Functional role of C-alpha-H center dot center dot center dot O hydrogen bonds between transmembrane alpha-helices in photosystem I. J Mol Biol 328(3):737–747

Lopez de la Paz M, Serrano L (2004) Sequence determinants of amyloid fibril formation. Proc Natl Acad Sci USA 101:87–92

Lu H, Skolnick J (2001) A distance-dependent atomic knowledge-based potential for improved protein structure selection. Proteins-Structure Function and Genetics 44(3):223–232

McConkey BJ, Sobolev V, Edelman M (2003) Discrimination of native protein structures using atom-atom contact scoring. Proceedings of the National Academy of Sciences of the United States of America 100(6):3215–3220

Manolios N, Collier S, Taylor J, Pollard J, Harrison LC, Bender V (1997) T-cell antigen receptor transmembrane peptides modulate T-cell function and T-cell-mediated disease. Nature Med 3:84–88

Meindl-Beinker NM, Lundin C, Nilsson I, White SH, von Heijne G (2006) Asn- and Asp-mediated interactions between transmembrane helices during translocon-mediated membrane protein assembly. EMBO Rep 7:1111–1116

Mendrola JM, Berger MB, King MC, Lemmon MA (2002) The single transmembrane domains of ErbB receptors self-associate in cell membranes. J Biol Chem 277:4704–4712

Metcalf DG, Law PB, DeGrado WF (2007) Mutagenesis data in the automated prediction of transmembrane helix dimers. Proteins Struct Funct Bioinformat 67:375–384

Mitchell JBO, et al (1999) BLEEP- potential of mean force describing protein-ligand interactions: I. Generating potential. J Computational Chem 20(11):1165–1176

Milligan G (2004) G protein-coupled receptor dimerization: function and ligand pharmacology. Mol Pharmacol 66:1–7

Miyazawa S, Jernigan RL (1996) Residue-residue potentials with a favorable contact pair term and an unfavorable high packing density term, for simulation and threading. J Mol Biol 256(3):623–644

Mottamal M, Lazaridis T (2005) The contribution of C–H···O hydrogen bonds to membrane protein stability depends on the position of the amide. Biochemistry 44:1607–1613

North B, Cristian L, Fu Stowell X, Lear JD, Saven JG, DeGrado WF (2006) Characterization of a membrane protein folding motif, the Ser zipper, using designed peptides. J Mol Biol 359:930

Overton MC, Chinault SL, Blumer KJ (2003) Oligomerization, biogenesis, and signaling is promoted by a glycophorin A-like dimerization motif in transmembrane domain 1 of a yeast G protein-coupled receptor. J Biol Chem 278:49369–49377

Ozdirekcan S, Rijkers DTS, Liskamp RMJ, Killian JA (2005) Influence of flanking residues on tilt and rotation angles of transmembrane peptides in lipid bilayers. A solid-state ^{2}H NMR study. Biochemistry 44:1004–1012

Park SH, Opella SJ (2005) Tilt angle of a trans-membrane helix is determined by hydrophobic mismatch. J Mol Biol 350:310

Partridge AW, Melnyk RA, Yang D, Bowie JU, Deber CM (2003) A transmembrane segment mimic derived from *Escherichia coli* diacylglycerol kinase inhibits protein activity. J Biol Chem 278:22056–22060

Popot JL, Engelman DM (1990) Membrane-protein folding and oligomerization – the 2-stage model. Biochemistry 29:4031–4037

Quintana FJ, Gerber D, Bloch I, Cohen IR, Shai Y (2007) A structurally altered D,L-amino acid TCRα transmembrane peptide interacts with the TCRα and inhibits T-cell activation in vitro and in an animal model. Biochemistry 46:2317–2325

Rees DC, Deantonio L, Eisenberg D (1989) Hydrophobic organization of membrane-proteins. Science 245:510–513

Ren J, Lew S, Wang Z, London E (1997) Transmembrane orientation of hydrophobic alpha-helices is regulated both by the relationship of helix length to bilayer thickness and by the cholesterol concentration. Biochemistry 36:10213–10220

Rohl CA, Strauss CEM, Misura KMS, Baker D (2004) Protein structure prediction using Rosetta numerical computer methods, Pt D, vol. 383. Methods Enzymol 383:66–93

Rost B, et al (1995) Transmembrane Helices Predicted at 95-Percent Accuracy. Protein Science 4(3):521–533

Rost B, Fariselli P, Casadio R (1996) Topology prediction for helical transmembrane proteins at 86% accuracy. Protein Science 5(8):1704–1718
Russ WP, Engelman DM (2000) The GxxxG motif: A framework for transmembrane helix–helix association. J Mol Biol 296:911–919
Sansom MSP, Son HS, Sankararamakrishnan R, Kerr ID, Breed J (1995) 7-Helix bundles – molecular modeling via restrained molecular-dynamics. Biophys J 68:1295–1310
Schneider D, Engelman DM (2004) Involvement of transmembrane domain interactions in signal transduction by alpha/beta integrins. J Biol Chem 279:9840–9846
Senes A, Gerstein M, Engelman DM (2000) Statistical analysis of amino acid patterns in transmembrane helices: the GxxxG motif occurs frequently and in association with beta-branched residues at neighboring positions. J Mol Biol 296:921–936
Senes A, Ubarretxena-Belandia I, Engelman DM (2001) The C alpha-H center dot center dot center dot O hydrogen bond: A determinant of stability and specificity in transmembrane helix interactions. Proceedings of the National Academy of Sciences of the United States of America 98(16):9056–9061
Senes A, Engel DE, DeGrado WF (2004) Folding of helical membrane proteins: the role of polar, GxxxG-like and proline motifs. Curr Opin Struct Biol 14:465–479
Senes A, Chadi DC, Law PB, Walters RFS, Nanda V, DeGrado WF (2007) E-z, a depth-dependent potential for assessing the energies of insertion of amino acid side-chains into membranes: derivation and applications to determining the orientation of transmembrane and interfacial helices. J Mol Biol 366:436–448
Shepherd CM, Vogel HJ, Tieleman DP (2003) Interactions of the designed antimicrobial peptide MB21 and truncated dermaseptin S3 with lipid bilayers: molecular-dynamics simulations. Biochem J 370:233–243
Simons KT, Kooperberg C, Huang E, Baker D (1997) Assembly of protein tertiary structures from fragments with similar local sequences using simulated annealing and Bayesian scoring functions. J Mol Biol 268:209–225
Spassov VZ, Yan L, Szalma S (2002) Introducing an implicit membrane in generalized Born/solvent accessibility continuum solvent models. J Phys Chem B 106:8726–8738
Summa CM, Levitt M, DeGrado WF (2005) An Atomic Environment Potential for use in Protein Structure Prediction. J Mol Biol 352(4):986
Ulmschneider MB, Sansom MSP, Di Nola A (2005) Properties of integral membrane protein structures: Derivation of an implicit membrane potential. Proteins-Structure Function And Bioinformatics 59(2):252–265
Ulmschneider, M.B., M.S.P. Sansom, and A. Di Nola (2006) Evaluating tilt angles of membrane-associated helices: Comparison of computational and NMR techniques. Biophy J. 90(5): 1650–1660
von Heijne G (1999) A day in the life of Dr K. or how I learned to stop worrying and love lysozyme: a tragedy in six acts. J Mol Biol 293:367–379
Walters RF, DeGrado WF (2006) Helix-packing motifs in membrane proteins. Proc Natl Acad Sci USA 103:13658–13663
White SH (2005) How hydrogen bonds shape membrane protein structure. Adv Protein Chem 72:157–172
Wimley WC, Creamer TP, White SH (1996) Solvation energies of amino acid side chains and backbone in a family of host–guest pentapeptides. Biochemistry 35:5109–5124
Yan-Chun Tang CMD (2004) Aqueous solubility and membrane interactions of hydrophobic peptides with peptoid tags. Peptide Science 76(2):110–118
Yarov-Yarovoy V, Schonbrun J, Baker D (2006) Multipass membrane protein structure prediction using Rosetta. Proteins Struct Funct Bioinformat 62:1010–1025
Yin H, Litvinov RI, Vilaire G, Zhu H, Li W, Caputo GA, Moore DT et al (2006) Activation of platelet αIIbβ3 by an exogenous peptide corresponding to the transmembrane domain of αIIb. J Biol Chem 281:36732–36741
Yin H, Slusky JS, Berger BW, Walters RS, Vilaire G, Litvinov RI, Lear JD et al (2007) Computational design of peptides that target transmembrane helices. Science 315:1817–1822

Yohannan S, Faham S, Yang D, Grosfeld D, Chamberlain AK, Bowie JU (2004) A C_a–H...O hydrogen bond in a membrane protein is not stabilizing. J Am Chem Soc 126:2284–2285
Zhang J, Chen R, Liang J (2006) Empirical potential function for simplified protein models: Combining contact and local sequence-structure descriptors. Proteins: Structure, Function, and Bioinformatics 63(4):949–960
Zhang Y, Skolnick J (2004) Automated structure prediction of weakly homologous proteins on a genomic scale. Proc Natl Acad Sci USA 101:7594–7599
Zhang YP, Lewis RN, Hodges RS, McElhaney RN (1992) FTIR spectroscopic studies of the conformation and amide hydrogen exchange of a peptide model of the hydrophobic transmembrane alpha-helices of membrane proteins. Biochemistry 31:11572–11578
Zhang YP, Lewis RN, Hodges RS, McElhaney RN (2001) Peptide models of the helical hydrophobic transmembrane segments of membrane proteins: interactions of acetyl-K2-(LA)12-K2-amide with phosphatidylethanolamine bilayer membranes. Biochemistry 40:474–482
Zhang Y, DeVries ME, Skolnick J (2006) Structure modeling of all identified G protein-coupled receptors in the human genome. Plos Computat Biol 2:88–99
Zhou FX, Merianos HJ, Brunger AT, Engelman DM (2001) Polar residues drive association of polyleucine transmembrane helices. Proc Natl Acad Sci USA 98:2250–2255

Index

R